AF566802

A BIOCHEMICAL PHYLOGENY OF THE PROTISTS

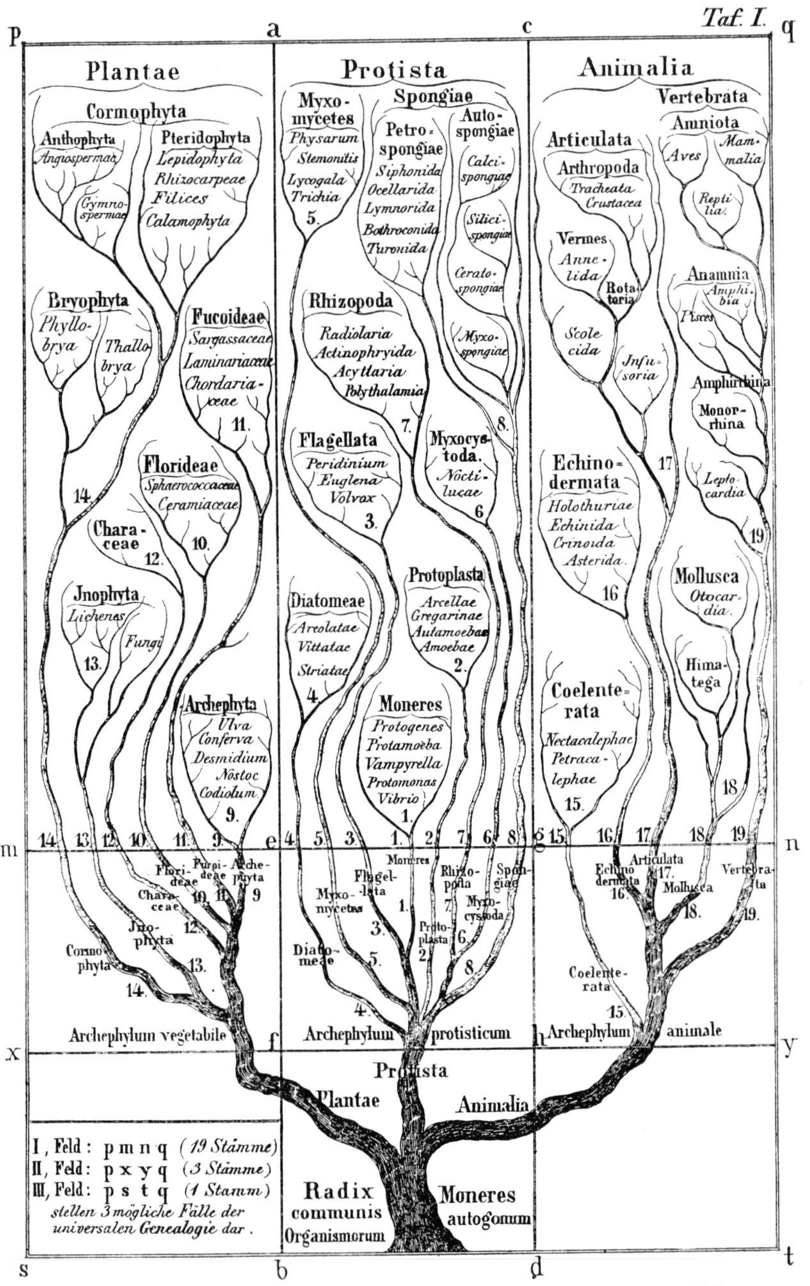

Ernst Haeckel's phylogenetic tree, rendered more than 100 years ago, should be compared with its modern counterpart on page 221. Reproduced from Plate I of Haeckel (1866) by permission of Walter de Gruyter & Co.

A BIOCHEMICAL PHYLOGENY OF THE PROTISTS

MARK A. RAGAN
Department of Biology
Dalhousie University
Nova Scotia, Canada

DAVID J. CHAPMAN
Department of Biology
University of California
Los Angeles, California

ACADEMIC PRESS New York San Francisco London 1978
A Subsidiary of Harcourt Brace Jovanovich, Publishers

ACADEMIC PRESS, INC.
111 Fifth Avenue, New York, New York 10003

United Kingdom Edition published by
ACADEMIC PRESS, INC. (LONDON) LTD.
24/28 Oval Road, London NW1

Library of Congress Cataloging in Publication Data

Ragan, Mark A
A biochemical phylogeny of the protists.

Bibliography: p.
Includes indexes.
1. Unicellular organisms–Classification.
2. Chemotaxonomy. I. Chapman, D. J., joint author.
II. Title. [DNLM: 1. Biochemistry. 2. Evolution.
3. Microbiology. QW4 R141b]
QR12.R34 576 76-52739
ISBN 0–12–575550–3

PRINTED IN THE UNITED STATES OF AMERICA

Contents

Preface

Biochemical characters have been used for more than a century in the reconstruction of phylogenies. It is only in the last two decades, however, that the underlying concepts of biochemistry and molecular biology have begun to exert a significant influence upon phylogenetics. Perhaps the most spectacular example of this progress has been the realization that amino acid sequences of most proteins are a direct reflection of the nucleic acid-based genome, and hence, of the phylogenetic history of organisms.

The proliferation of biochemical data is especially welcome in phylogenetic examination of the protists, where there is a chronic shortage of objective phylogenetic characters. Yet biochemical characters have been utilized by different authors to support quite different and sometimes mutually exclusive phylogenies. Moreover, there is considerable difficulty in approaching much of the relevant primary literature, not only because it is often extremely specialized (if not arcane), but also because it is well dispersed throughout numerous books and scientific journals (we have drawn data from 180 journals and 118 monographs, symposia, and theses). These problems loom especially large for the student and the nonspecialist approaching biochemical phylogenetics for the first time. In this book we seek to examine a broad spectrum of biochemical characters; to point out which ones have been useful in phylogenetics, and the underlying bases of such usefulness; and to illustrate methods of deducing phylogenies from biochemical data. Our efforts will have been justified to the extent that this book serves not as an arbiter of phylogenetic questions, but as a stimulus and guide to further thought and research.

A preliminary draft of some of this work had its origin at the University of Chicago in February 1972; helpful discussions were held at that time with Drs. J. H. Law and T. H. Steck. Dr. G. S. Getz kindly provided then-unpublished data. During preparation of the final draft, valuable advice has been received from Drs. N. J. Antia, T. Christensen, J. S. Craigie, W. F. Doolittle, L. J. Goad, T. W. Goodwin, D. O. Hall, M. V. Laycock, H. Matsubara, K. K. Rao, and J. W. Schopf. Drs. J. S. Craigie, M. V. Laycock, and P. J. McLaughlin have supplied unpublished data. Many colleagues have provided us with access to manuscripts in advance of publication. To all of the above we express our thanks. We also thank the scientists and publishers who have generously permitted us to use copyrighted materials. These have been acknowledged throughout the text. However, we must assume all responsibility for the interpretation of these data. Our special thanks go to Marjorie McDonald. Her typing skills brought the final manuscript to the light of day.

M. A. R. wishes to thank the Isaac Walton Killam Trust and the National Research Council of Canada for financial support while the final draft was being completed. D. J. C. expresses his thanks to the National Science Foundation and the Regents of the University of California for their support of his research over the years.

MARK A. RAGAN
DAVID J. CHAPMAN

A BIOCHEMICAL PHYLOGENY OF THE PROTISTS

1

Introduction

1.1 WHAT ARE THE PROTISTS?

The word "protist" was coined by Haeckel (1866) to describe the morphologically simple forms of life, including bacteria, fungi, many algae, protozoa, and sponges. Seventy years later Chatton (1937) emphasized the two basic types of cellular organization, which he designated as "procaryotic" and "eucaryotic." This basic division is now a well-established tenet of taxonomic and phylogenetic thinking. Copeland (1938) and Stanier and van Niel (1941) subsequently reclassified the prokaryotic protists (bacteria and blue-green algae) as "monera," and retained the term "protista" for eukaryotic forms. Later Dougherty and Allen (1953) recognized the lower protists (prokaryotes), "mesoprotists" (red algae), and "metaprotists" (remaining eukaryotic algae, fungi, and protozoa). In more recent years the Dinophyceae (dinoflagellates) have also been considered "mesoprotists," particularly in the Russian literature and by Dodge (1965), but in a different context from that used by Dougherty and Allen (1953). These latter workers considered the Rhodophyceae (red algae) to be "intermediate" between the prokaryotic Cyanophyceae (blue-green algae) and other eukaryotic algae. The distinction was made primarily on the basis of biochemistry (especially pigments) and the lack of flagellate structures in the Rhodophyceae. Dodge (1965), on the other hand, established his "mesoprotists" or "mesocaryotes" by the sole criteria of nuclear structure and nuclear behavior. We do not intend to argue the merits of the distinction into "mesoprotists" or "mesocaryotes," except to mention that the terms have not received the universal acceptance accorded the prokaryote and eukaryote. The word "protist" will be used here to encompass the bacteria, blue-green algae, actinomycetes, eukaryotic algae (including red algae and dinoflagellates), fungi, water molds, euglenoids, and protozoa. This approach retains the original sense of the term and avoids the tendency to create new and unnecessary terminology. It is, in effect, a convenience term, and does not imply or suggest a taxonomic or systematic entity. Other simple forms of life, including viruses and pleuropneumonia-like organisms, will not be discussed in this treatment. Viruses have been considered by Dougherty (1955), Evans (1960), and Joklik (1974).

We have chosen as convenient reference points, three taxonomic schemes for the eukaryotic organisms under discussion (Appendix). Original taxonomic designations have been retained, in preference to more recent name changes or combinations. This approach may not represent adherence to systematic protocol or rules of nomenclature. However, we believe that less confusion will result if the old name, under which the biochemical or chemical investigations were carried out, is retained. For example, we will retain *Anacystis nidulans* (not *Lauterbornia*) and *Porphyridium cruentum* (not *P. purpureum*).

Many protists have been very incompletely studied by biochemists, e.g., *Kakabekia*-like organisms (Siegel and Giumarro, 1966; Siegel *et al.*, 1967), the Chloromonadophyceae, and cyanellae symbionts. The emphasis on certain algal groups, photosynthetic prokaryotes, and fungi, is symptomatic of our state of knowledge and the emphasis placed upon these organisms as experimental material in biochemical studies. Wherever possible, however, the lesser known groups have been considered, and projections, ideas, or suggestions with regard to these organisms have been put forward.

1.2 WHY ARE THE PROTISTS INTERESTING?

If one considers organisms around us that are readily visible, it becomes clear that for the most part they fall into two major categories: vascular plants and higher animals. Although a closer investigation would probably reveal mosses, ferns, earthworms, and insects, many morphological similarities within groups are nonetheless apparent among all these organisms. The biochemical processes characteristic of these organisms are even more uniform: all animals use very similar respiratory cytochromes, and all higher plants utilize identical chlorophylls in photosynthesis, and possess a very similar energy conversion or photosynthetic apparatus. It is in the protists that these relatively narrow ranges of body form, physiology, and especially biochemistry are found to vary most widely. The observed variations in morphology, physiology, and biochemistry often provide insights into the involvements of each in life processes.

Protists are very intimately involved in almost every aspect of most important ecological processes including oxygen production, disease, decay of organic matter, and nutrient cycling. The increasing incursion of industrialized man into previously undisturbed and perhaps finely balanced communities is likely to bring about significant changes in the lives of protists and consequently in human life.

1.3 SYSTEMATICS, TAXONOMY, AND PHYLOGENY

Since the days of Aristotle, scientists have engaged in the classification of organisms into hierarchical systems. Such activity not only helps to organize our knowledge of different forms of life, but should also allow people to make certain deductions about these organisms.

Taxonomy is the study of the bases, principles, procedures, and rules of classification (Heywood, 1973), or is the classificatory process itself (systematics). Apart from the nomenclatural rules, much of taxonomy is a matter of opinion. The desire for a "perfect" taxonomy must be tempered by pragmatism and by the need for convenience and ready applicability.

Phylogeny can be considered as a taxonomy in which the resulting system is thought to be representative of the historical evolution of the organisms considered. Descent from common ancestors with evolutionary modifications is studied in its various manifestations: morphological, biochemical, or otherwise. Consequently a phylogeny, adequately constructed, is a more powerful conceptual framework than is a taxonomy alone. The problem comes, however, in the modification "adequately constructed." What is an adequate construction, given that the fossil record is, and probably forever will be, incomplete or unintelligible concerning the details of protistan phylogeny? Is it intellectually valid to utilize biochemical characters in the reconstruction of phylogeny? What is to be done with data that do not appear to be in agreement with most other data? Can some criteria be considered more significant than others, and if so, on what basis? These questions will be considered in the following pages.

1.4 WHY A BIOCHEMICAL PHYLOGENY?

Why is it useful to construct a phylogeny of the protists using biochemical data? There are several answers to this question, none of which is complete in itself.

1. Biochemical data are *genetic,* being directly coded in the DNA ("primary semantides" or "primary semantophores" of Zuckerkandl and

Pauling, 1965a,b). Ribonucleic acids ("secondary semantides") and proteins ("tertiary semantides") are produced sequentially from the primary semantides, and consequently provide insight, although less directly, into the primary genetic makeup of the organism. There is of course environmental input into biochemical and physiological processes of living organisms, but where desired this can be disregarded or minimized by examination at the proper biochemical level.

2. In recent years there has been an enormous increase in the number of biochemical data available from protists. Many of these data have come from biochemists who utilize certain protists as favorable experimental systems, while other data have come from scientists interested in protists themselves. Construction of a biochemical phylogeny might help to organize some of these data, and suggest fruitful areas of further research.

3. Biochemical data have already been used to support a wide range of mutually conflicting and mutually exclusive phylogenies. It is important to examine these data to determine if they are internally consistent, and if so, how to use them properly in constructing phylogenies.

4. Biochemical methods may in certain circumstances be easier to apply than are more traditional examinations of morphology, ultrastructure, or life history.

Needless to say, there are also difficulties inherent in biochemical techniques and in their application to phylogenetics. It is not always easy to collect the necessary biochemical data, whether reaction pathways, molecular structures, or chemical compositions. There may be problems of the absence of a character (is it due to the repression of a gene, or to a critical mutational step that has occurred recently, or to insensitivity of the analytical method?). The culture conditions or an abnormal environment for the protist may cause some subtle, "unnatural" change in its biochemistry. Finally, only a percent or two of all known protists have even been studied at all by biochemists.

This raises the problem of the representative taxon: What species is representative of the genus, what genus of the family, and so on up the taxonomic ladder? Indeed, the question "Is there a representative taxon?" is rarely asked, and it is very easy to end up answering the question with circular reasoning. The concept of the "type" (e.g., type species of the genus; type genus of the family) in taxonomy is well understood. This, however, almost invariably applies only to the morphological realm. In view of the increasing use of chemical and biochemical data in taxonomy, it is appropriate to raise the question (Chapman and Ragan, 1977) of whether or not there should be a "chemical" type taxon, and if so, should this be the same as the morphological type. We believe these are basic questions, and are even more important when it is realized that taxonomy and phylogeny are closely interwoven, and that biochemistry and morphology receive different emphasis in the two disciplines. Although conceptually convenient, typological systematics is poorly suited to deal with evolutionary events, as the confusion in primate paleontology has been vividly demonstrating in recent years (Paleontology Correspondent, 1974).

Evolution is a historical process that can profitably be examined from a number of viewpoints. The aim of this biochemical phylogeny is to introduce one possible way of viewing evolution. Other approaches may be more powerful in describing Darwinian selection and evolution at the organismal and population levels. It is not intended that our phylogeny, based upon one approach, should be used to the exclusion of other phylogenies or methods of construction. They should not be mutually exclusive, but rather they should act as a check and balance upon each other, since the greatest rewards lie in the final synthesis of all possible approaches to evolution and construction of a phylogeny.

2

Biochemistry and Evolution

2.1 THE NATURE OF BIOCHEMICAL DATA AND THE CENTRAL DOGMA

Data are statements of information derived from observation and consideration of characters.* Biochemical data are not necessarily true, due to the possibility of experimental error in the observation of organisms and in the execution of experiments. Some biochemical data incorporate other data, as will be seen, and hierarchies of biochemical data exist. In these regards such data do not differ from others except in their subject matter.

Data are rarely used alone; they are interpreted into facts, and are then used in hypothesis building and hypothesis testing. Are biochemical facts in any sense different from nonbiochemical ones? The answer is a qualified affirmative. In phylogenetics, for example, it can be argued that there is less subjectivity in interpreting biochemical data than in the interpretation of morphological or other types of data. This may be the "statistical objectivity" of Turner (1967). This reduced degree of subjectivity, the argument goes, increases the chances that the resulting facts will be of lasting value, and will place biochemical phylogenies on more solid grounds than is the case with other phylogenies. The opposing view is that this reduced

* The term "character" is often, but incorrectly, used interchangeably with the word "datum." Hennig (1966) has defined the former term in the context of the "character-bearing semaphoront," the unit of biological systematics. Semaphoronts are considered to be "individuals in given short periods of their lifetime" (Hennig, 1965).

degree of subjectivity is in reality a reduced level of interpretation, and as a consequence the resulting biochemical phylogeny is based upon more limited facts than are other phylogenies.

If there is indeed a difference between biochemical data and other types of data, it is that a *natural framework of interpretation* presents itself with biochemical data. The natural framework is the *biosynthetic history* of the observed molecule.

Retracing the biosynthesis of a metabolite (back through the appropriate biosynthetic enzymes, through the mRNA, to the DNA sequence(s) responsible for its biosynthesis) solves several problems endemic to phylogenetics: First, it demonstrates homology, if any, between two molecules. Second, differences in biosynthetic pathways may readily be observed. Third, problems arising from gene repression are, at least in theory, avoided. Finally, this approach provides natural groupings, aiding in the logistics of phylogeny building and data processing. This process is a natural one in that it follows, in retrograde, the actual history of the biochemical compound through real time.* The validity and usefulness of this method are suggested here without proof, but evidence for its applicability will be examined in later pages.

The transfer of information from DNA through RNA to the cellular proteins, some of which act as enzymes in the biosynthesis of various metabolites, is certainly one of the most fundamental life processes. In appreciation of the basic importance of these biological molecules and of this information transfer, the above process is often termed the "Central Dogma" of molecular biology. Although apparent evolutionary modifications of this process (reverse transcriptases, proviruses) have been discovered more recently, the importance of the Central Dogma processes becomes increasingly obvious with further biophysical and biochemical research.

2.2 DATA WEIGHTING AND THE CENTRAL DOGMA

Experience has shown—and the following chapters will document—that certain characters are of considerable phylogenetic usefulness, and that others are relatively uninteresting phylogenetically. From this observation, by no means limited to biochemical data (Mayr, 1969), there arises the perennial question: Is one type of datum *intrinsically more likely* to be of use in phylogenies than is a second type? Is it possible to predict phylogenetic usefulness *a priori*? If so, it would be routinely possible to

* Other meanings of the term "natural" have been discussed by Sneath and Sokal (1973).

weight heavily the inherently more useful data and to ignore the lesser, especially if the latter appeared to contradict the more useful ones.

Unfortunately no generalized classification of characters has been discovered. Suggestions that character X or character Y is intrinsically likely to be of great value in phylogenetics usually acquire a host of modifying conditions. This is frequently seen whenever an overly zealous researcher maps out the phylogeny of all organisms from the distribution of a single character; in essence this constitutes an extreme form of data weighting, in which all other data are assigned zero weight. It has been recognized since the work of Adanson (1763) that all possible useful data should be included in systematics.

At another extreme are the numerical or phenetic taxonomists who claim that all data must be lumped together without weighting into a computer, which then prints out the best available scheme. Many in this school do not claim that the resulting systems are even phylogenies, but can justly point out that their approach is relatively (although not completely!) free from personal biases in the interpretational stages. Indeed some data, such as isoenzyme patterns, are suited to this treatment. But it has been pointed out that numerical taxonomy is "getting the least out of the most" (Turner, 1967).

Between the two extremes lies current opinion. "To be sure," the current wisdom goes, "some characters have proved to be more useful than others in the construction of phylogenies. To be sure, there is some data weighting, if only subjectively, in most phylogenies, and this is a valid if necessarily *a posteriori* phenomenon. But it would be foolhardy to attempt to *predict* which characters will be of greater use, and which will be relatively useless." Experience has borne out this point of view quite well over the years, although the question has not been adequately discussed for biochemical characters. It is possible that justifications for data weighting could be found in biochemistry and molecular biology even if none was forthcoming for the traditional morphological characters.

There have been suggestions in recent years that phylogenetically interesting biochemical characters possess certain attributes (Erdtman, 1968):

1. Widely distributed identical characters are of little phylogenetic interest.

2. Characters unique to individual species are of little use in phylogenies, due to the impossibility of relating them to similar features in other organisms, and their very limited distribution.

3. Structurally complex molecules can often be weighted more heavily than can structurally simple molecules.

There is a unifying relationship underlying these observations: the concept of biosynthesis (Birch, 1973a,b). Compounds may be perceived as the products of biosynthetic pathways. It may then be seen, in parallel with the above statements, that

1. If a compound is found in two organisms, but is biosynthesized by different pathways, its biosynthesis takes on considerable phylogenetic significance.
2. If a unique compound found in a given organism can be related biosynthetically (as a further elaboration, as a precursor, or as arising by the action of related enzymes) to a compound in another organism, there is the possibility that the two organisms are related phylogenetically. This approach is particularly useful when the biochemical pathway involves a very significant and unusual chemical modification of a molecule, such that one may assume with some justification that a very specific and discrete enzyme is involved in the establishment of the pathway. To a certain extent one is using an enzyme, albeit hypothesized, as the character. These tertiary semantides are nearer the genome than the episemantic molecules or metabolites. This "biosynthetic approach" has an advantage in that it may reduce the problems posed by the limitations of analytical techniques (the presence-or-absence question) and the inevitable question of possible environmental control or determination of the presence of a given compound. One is no longer using a single compound as the character (with the inherent problems), but rather a character represented by a series of compounds. Nevertheless this approach does not eliminate the possibility of gene repression.
3. Structurally complex compounds tend to be more significant because they presumably require more biosynthetic steps, hence more (specialized) enzymes (but cf. Herout, 1973). An extension of the concept of biosynthesis to include all related events from the level of DNA to metabolites could provide an even more powerful method for examining the relative usefulness of biochemical characters.

It might be argued that this approach does not come to grips with the problem of weighting individual biochemical characters, but instead merely increases the number of biochemical data by retracing the biosynthetic history of the compound [in the terminology of Sneath and Sokal (1973), amassing "logically correlated character complexes"]. Instead, by considering *in toto* a biosynthetic pathway or sequence containing the distinctive feature, rather than individual molecules, one is in fact reducing the number of characters and thus the number of biochemical data. Moreover, DNA is more than just another macromolecule; it is the primary genetic substance of the organism. All information necessary for cell development and activity is contained in the DNA, and if phylogeneticists could "read" DNA as the living cell does, they would be in a position to predict the structures of RNA molecules, enzymes, and (possibly) further metabolites. Data reflecting details of inherent informational potential in the DNA are likely to be of relative significance for phylogenetics, and this significance will stem from the nature of the character, not merely from the structure of the "character complex." In the absence of being able to "read" the DNA it is still highly desirable, however, to get as near to the genome as possible for phylogenetic—and taxonomic—purposes.

It is important to observe that this scheme may be valid only for weighting a metabolite and *its* DNA, mRNA, and biosynthetic enzymes, and that only "good" biochemical data (primary structures, etc.) are being considered. There is as yet no indication that this (or any) process can be used to weight biochemical characters pertaining to compounds of entirely different biosynthetic origins, nor to characters of less robust biochemical nature. Although it might be possible to trace the biosynthesis of all compounds back to glycolytic pathway intermediates or to acetyl coenzyme A, the phylogenetic usefulness of such an exercise would be severely limited by the great complexity of biosynthetic relationships in living organisms.

Similarly, a single biochemical compound can be "described" in numerous ways. Some descriptions are of greater phylogenetic interest than others because they could, at least in theory, predict other descriptions. For example, given the amino acid sequence of an enzyme and unlimited computer time, it should be possible to predict its tertiary structure and to state what its natural substrate is likely to be, just as it is currently possible to determine the amino acid sequence and maximum molecular weight of a polypeptide given its codon DNA base sequence. This could not be done on the basis of immunological properties, electrophoretic behavior, or amino acid composition data, even in the best of situations.

The experimental ease of investigating the more phylogenetically significant biochemical data decreases in two directions: one as the level of the DNA is approached, the second as molecules become more complex, irrespective of the level of analysis. It is possibly not a coincidence that these two directions correspond to the two areas of greatest inherent usefulness in the construction of phylogenies from biosynthetically related molecules.

2.3 PREBIOTIC PROCESSES AND "BIOCHEMICAL PREDESTINATION"

The correlation (or lack thereof) with paleontological records represents one of the few contacts phylogenetic speculation has with independent external data. Despite their importance to phylogenetics (Klein and

Cronquist, 1967; Margulis, 1969, 1970; Heller, 1973), very early fossils are difficult to detect and interpret (e.g., Knoll and Barghoorn, 1975). Even a more complete collection of early fossils would still pose difficulties in the study of *their* origin(s).

Another method is clearly needed to study the dawn of life, and one has been supplied by Miller (1953, 1955). In a series of experiments, he synthesized several biological amino acids by the introduction of electrical discharges (i.e., energy) into an anoxic mixture of simple gases generally considered to have been components of the primitive earth's atmosphere. Later workers have produced an amazing array of amino acids, polypeptides, nitrogenous nucleotide bases, sugars (including deoxypentoses), organic acids, many other poorly defined organic compounds, and pyrophosphates (Fox, 1965; Kenyon and Steinman, 1969) by starting with other simple organic chemicals and other sources of energy.

These two lines of attack—paleontological and reductionist, on one hand, constructionist, on the other—have not yet overlapped each other (Kenyon and Steinman, 1969). The remaining gap may be reduced by continuing application of present methods, or may require some novel experimental approach. In the absence of experimental data, biological problems represented by the gap between abiogenctically produced organic polymers and functional evolutionary systems will have to be considered from a more theoretical viewpoint.

Perhaps the most interesting of the "gap" problems is the linkage of nonrandom nucleic acid and nonrandom protein to produce the first system for transfer and storage of (pre-)biological information (Section 4.1). Another interesting problem involves the emergence of order, or nonrandomness, in a manner we can call biological or at least prebiological (as opposed to inorganic crystallization). Attempts have been made to consider these and other problems independently, but despite considerable ingenuity these attempts have met with limited success. In this context the reductionist approach, characteristic of many biologists (including molecular biologists), may prove to be counterproductive. Since the problem at hand concerns organization, artificial subdivision of the problem may be a fatal flaw to such an approach.

The theoretical approach of Eigen (1971, 1973) is of particular interest. Drawing heavily upon the nonequilibrium thermodynamics developed mainly by Prigogine (1947, 1969), Eigen has shown that material, "self-instructive catalytic hypercycles" can arise spontaneously. These self-organized and self-organizing systems are capable of many processes strongly analogous to life processes, including growth, competition, selection, adaptation, individualization, and self-control. Self-organization and evolution are not only consistent with the laws of thermodynamics, but given certain chemical properties of matter (themselves potentially explicable by quantum mechanics) and certain physical conditions likely to have existed on the primitive earth, they are the direct consequences of these laws.

It appears that both experimentation and theorization point to one common result: the appearance of self-organizing systems was not only possible but indeed likely under conditions likely to have been found on the primeval earth. Kenyon and Steinman (1969) label this conclusion "biochemical predestination," a catchy if somewhat overstated name for a very important concept whose usefulness will, it is hoped, lie in the further work it inspires and guides.

2.4 DARWINIAN AND NON-DARWINIAN EVOLUTION

Darwinian evolution—the spontaneous appearance of natural variation in a population and the survival of the fittest of these organisms as parents of succeeding generations—has been a pervasive and useful conceptual framework for biologists for many years. Almost all modern biologists agree that Darwinian processes are responsible for much evolutionary change at the organismal level. Kimura and Ohta (1973a), for instance, state that Darwinian evolution appears to explain changes in "morphological and physiological traits or other measures that are related to survival and fertility." This would occur especially when natural populations are near their saturation levels. It does not follow, however, that evolution at the molecular level is necessarily Darwinian, nor that prebiotic processes were examples of Darwinian evolution. In modern as well as prebiotic systems, the whole is always more than the sum of its constituent parts. Darwinian evolutionary processes may be operative only on the resulting organized system, or may operate on the individual parts as well; but it is important to note that the latter does not necessarily follow from the former.

The original theory of Darwin and Wallace (1858) has been reexamined recently by Williams (1970) and Papentin (1973a,b,c). Purged of Lamarckian overtones, this theory is remarkably consistent with modern molecular biology and biochemistry. Biochemistry has, however, pointed out phenomena that some biologists feel should be explained in terms of extra-Darwinian processes.

The discovery that several different deoxyribonucleotide triplets may code for the same amino acid immediately suggests that mutations in the DNA will not always bring about any changes in protein structure (and consequently in protein function) (Watson, 1970). Such mutations could be of significance (in a Darwinian sense) only if there are selective forces operating directly on the DNA or the RNA, whether through molecular sequences, conformations, codon abundances, precursor pools, or whatever. Farquhar and McCarthy (1973) have found that DNA coding for histones has a smaller proportion of third-position mutations than would be expected from truly neutral mutation, and have suggested that this could be the result of selection at the level of DNA or RNA. Only a few other examples have been reported, although this may be in large part due to the difficulty of performing the requisite experiments.

A number of other observations have been offered as evidence favoring neutral (non-Darwinian) mutation and fixation processes, particularly in proteins. These include the relative abundance of the various amino acids found in proteins, the observed rates of fixation of amino acid substitutions in enzymes, the approximate constancy of the rate of substitution in a given protein, and the multiplicity of isoenzymes in many organisms. These will be briefly examined in turn.

2.4.1 Relative Abundances of Amino Acids in Proteins

If Darwinian processes were important at the protein level, it would be expected that certain amino acids, types of amino acids, or combinations of amino acids would be strongly selected for at critical sites in certain proteins. It might therefore be possible for the resulting amino acid composition data to reflect these nonrandom requirements. A non-Darwinian prediction would correlate the observed amino acid ratios with the ratios in codon numbers for the different amino acids [e.g., because serine has six codons and methionine only one, serine might be expected to be about six times more common in proteins than methionine (King and Jukes, 1969)]. Although the correlation is not perfect (Elton, 1973; Jukes *et al.*, 1975), there does seem to be a rough relationship between amino acid molar abundance in proteins and codon frequency. Hypermutable and immutable sites likewise are approximately predicted from a Poisson distribution (King and Jukes, 1969). The difficulty with this approach, however, is that even a perfect correlation would not necessarily require that protein evolution has been non-Darwinian (Clarke, 1970).

2.4.2 Rates of Fixation of Mutations

The rates of amino acid substitution into certain proteins appear to be extremely high. Because the argument of Haldane (1957) assigns a "cost" to allele substitution (due to selective deaths through reduced fertility), and since there is necessarily an upper limit to the number of selective deaths a population can bear, it has been possible to compute the genetic load required for the rapid rates of DNA change responsible for this substitution. Kimura (1968), Maynard Smith (1968), and others have shown that an inordinately high genetic load would be required.

There are several possible explanations for this difficulty, termed "Haldane's Dilemma." One is that unjustified assumptions were made during the calculations, and that consequently there is no dilemma at all (Maynard Smith, 1968, pp. 795–796). A second alternative is that the substitutions at the DNA level are nearly neutral, are fixed by random processes, and therefore cause no genetic load. Grant and Flake (1974) have reviewed other (ecological) explanations, viz., the molecular level may be the wrong level of analysis.

2.4.3 Constancy in Mutation Fixation Rates

A very important argument in favor of the neutrality hypothesis would be the approximate constancy in the rate of amino acid fixation into individual proteins over long time periods. Although not all mutation and fixation rates are absolutely constant (Jukes and Holmquist, 1972; Goodman and Moore, 1974; van den Berg and Beitema, 1975), there does appear to be an approximate constancy for some proteins in organisms of about the same degree of morphological complexity. The neutrality hypothesis is in accordance with this observation (Kimura and Ohta, 1971; Van Valen, 1973, 1974), although it might be less easily explained by traditional Darwinian processes. For further discussion the reader is referred to Section 5.2.

2.4.4 Isoenzyme Multiplicity

Sokal (1974b) has noted that some population biologists consider non-Darwinian evolution to be responsible for the variety of isoenzymes observed in many organisms. From the viewpoint of cellular biochemistry, however, different isoenzymes may be seen to function in different intracellular compartments (e.g., chloroplastic vs. cytosoluble isoenzymes), or in different metabolic processes (e.g., in glycolysis vs. gluconeogenesis). Although it cannot be asserted that all isoenzymes are currently under Darwinian selective pressure, it is likely that further research will provide more examples of the direct involvement of different isoenzymes in morphogenesis and in metabolism (Masters and Holmes, 1972).

The Darwinian vs. non-Darwinian debate is far from resolved (Wills, 1973). What has emerged is that evolution at the molecular level is indeed amenable to study, and that an entire spectrum of selective pressures may exist. It would be of great interest to examine possible correlations between the degree of selective neutrality and the biochemical level of the expression of a mutation. The overall relationship is likely to be complex, but interesting. The evolutionary changes described in the following pages are valid, whether brought about by Darwinian or extra-Darwinian processes. In many cases they have been interpreted (either here or in the original literature) as Darwinian in nature; this is very often done from force of habit, as adequate data for such assertions are infrequently encountered.

2.5 BIOCHEMISTRY AND MORPHOLOGY

Although much of the following discussion will deal with molecular levels of analysis, it is important that one does not lose sight of the morphological characters associated with evolution and phylogeny. It is to some degree artificial to speak of morphological phylogenetics and biochemical phylogenetics as if they were independent undertakings, although the current understanding of the relationships between biochemistry and morphology is slight in comparison with the amount of knowledge concerning either subject individually. In the past few decades, however, considerable progress has been made in the elucidation of this interrelationship. The one gene–one polypeptide hypothesis of Beadle and Tatum (1941) and the elucidation of the genetic code by Crick and Watson in the early 1950's were fundamental to this progress. More recently, studies of the biochemical foundations of pleiotropy and biological regulation have begun to illustrate the complexity of morphogenetic expression.

In the contrast between morphological and biochemical phylogenetics, it is important to distinguish between the morphology of the whole organism ("macromorphology") and subcellular "micromorphology" (e.g., ultrastructural considerations). The characters involved with the latter are much more conservative, evolutionarily speaking. One finds, certainly among the algae, that the phylogeny (and taxonomy) of the lower taxa (species, genera, and families) is based heavily upon the macromorphological characters, while the conservative micromorphological characters are used more in the erection of the phylogeny of the higher taxa (orders, classes, and divisions). However, a biochemical phylogeny and a micromorphological (e.g., ultrastructural) phylogeny should of course resemble each other. In view of the relationship between biochemical processes and composition on one hand and ultrastructure of cellular components on the other, it could hardly be otherwise. Our understanding of these interrelationships is such that separate phylogenies are usually proposed. Eventually both approaches will merge to give a single phylogeny. In the current state of affairs, however, one can profitably be used as a check on the other.

2.6 ONTOGENY AND PHYLOGENY

The development of morphological characters, or ontogeny, has been a classic field of research and speculation for several hundred years. The first suggestions that ontogenic development could reflect evolutionary descent were made by Tiedemann (1810's) and Serres (1824), well before the work of Darwin was published. This theory, often referred to as "ontogenic recapitulation of phylogeny," or simply recapitulation, received support from the work of von Baer (1828), who interpreted the theory to relate developmental stages of an animal to embryonic stages of its ancestral, less highly evolved relatives. These early ideas have been discussed in more detail by de Beer (1958).

The recapitulation theory was greatly popularized by Haeckel (1866, 1898, 1904), and was noted by Darwin (1872), who stated that "embryology will often reveal to us the structure, in some degree obscured, of the prototypes of each great class." Müller (1864) was probably the first to extend the theory to recapitulation of the adult stages of ancestral forms, and with this misinterpretation the theory began to fall out of favor in many circles. Difficulties with the recapitulation theory were pointed out by His (1874), Schimper (1885), and others. Although the theory was never disproved, the degree of subjectivity required in the interpretation of these embryonic-ancestral stages convinced many that more fruitful lines of research should be sought elsewhere.

Interestingly, vestiges of the recapitulation theory have emerged occasionally in the biochemical literature, for instance in papers on sugar phosphorylation (Horecker, 1963), the urea cycle (Cohen and Brown, 1960; Bennett and Frieden, 1962), pleuropneumonia-like organisms (Dougherty, 1955), cell wall chemistry in the Zygomycetes (Bartnicki-Garcia, 1970), and culture forms of Trypanosomidae (Danforth, 1967). If any relationship does exist between ontogeny and phylogeny, it doubtless proceeds through the intermediacy of the DNA. Phylogenetic development could presumably be recorded in the DNA, and the pattern of gene derepression during ontogeny might for some reason bring about the observed recapitulation. In general, there is far too little information available on changes in biochemical features during development to allow any definite conclusions to be drawn about this old theory. The important point is that through biochemistry a less subjective approach to the recapitulation theory can be made.

2.7 A NOTE ON THE BIOCHEMICAL METHOD

The biosynthetic approach underlying the subsequent discussion has been mentioned already, and will be further examined in upcoming pages. A large number of assumptions and specialized terms necessarily permeate a detailed examination of phylogeny; a few of these will be stated here in the hope of providing a presentation of maximum clarity.

Homology is the fundamental relational concept upon which all comparative biology is based. Homology requires common genetic ancestry and can consequently be defined in terms of phylogeny, hence in terms of evolution (Bock, 1969, 1974). Although two characters (e.g., two molecules) may be homologous, it may be difficult to detect or to establish this homology. The different types of criteria that have been applied to biomolecules will be described in the following pages. If there is an indication that resemblance between characters may not have arisen from a common genetic ancestry, the supporting details will be spelled out in full, and terms such as "similarity" or "resemblance" will be used in place of "homology" (Margoliash, 1970).

A "mutation" will be understood to be a change in the primary structure of DNA. In keeping with the genetic theme, the phylogeny to be presented may be regarded as having some cladistic attributes.

The "minimum evolutionary length" assumption underlies most treatments of nucleotide and amino acid sequence data, including our own, and to a lesser degree may pervade the entire argument. It must be borne in mind that the logical extreme of this assumption is a sort of molecular orthogenesis which certainly does not exist in nature.

Finally, it is important to realize that any phylogeny, systematic scheme, or evolutionary theory is but a postulate based on the interpretation of data. Since these schemes themselves cannot be put to the test nor experimentally verified, one must avoid labeling them as correct or incorrect, or necessarily assuming that the viewpoint with the most numerous or most eloquent advocates is the one most representative of natural history.

3

Phylogenetics

3.1 PHYLOGENETIC TREES

> Phylogenetic trees, having no roots, are easily blown over (Klein and Cronquist, 1967).

Since the first phylogenetic tree was drawn by Haeckel (1866, see Frontispiece), these trees have been regarded as curious outgrowths of the equally curious field of speculative phylogeny, a field often characterized as consisting of innumerable "facts," hypotheses, and theories, but of few hard data. Phylogenetic trees have been called "the most noxious of all weeds" by some detractors (Ingold, 1959). The advantages of phylogenetic trees are obvious enough: they "facilitate information retrieval" (Klein and Cronquist, 1967) and, by allowing predictions of similarities and dissimilarities among organisms, present a useful and rapid framework for the planning of new experiments. Various difficulties tend to complicate these benefits, however, and distract one from the potential utility of the trees:

1. Only very carefully constructed and interpreted phylogenetic trees can be considered to show a time axis and temporal change. Trees based on numbers of amino acid or nucleotide differences will underestimate numbers of mutational events and hence will provide unreliable time estimates (Williams, 1974a,b).
2. Branches of the trees are often of arbitrary length, but may subconsciously be interpreted as depicting some measure of evolutionary development or time since the divergence of that line.
3. The trees are often interpreted as showing a rigorously gradual

evolution of organisms from simpler forms (i.e., of being unduly Lamarckian). Although evolution has indeed been a continuous process, both the fossil record and the currently observable organisms suggest that there are interesting discontinuities in organismal structure and in adaptive strategies of these organisms, e.g., symbioses, allopolyploids, hybrids, episome-transmitted characters, and "radical mutants" if any. An interesting discussion of this point is given by Rosen (1973).

4. There is considerable theoretical and methodological difficulty inherent in the choice and delineation of the groups of organisms pictured in the trees (Heywood, 1973). In this regard, an important difficulty has been pointed out by Hennig (1966), who has shown that characters observed in single species cannot legitimately be used to describe phylogenetic relationships among higher taxa.

5. Confusion may arise from unstated thoeretical and methodological assumptions underlying the phylogenetic tree, for instance whether it was designed by a classical (genetic-evolutionary) taxonomist, or by a pheneticist, or by a cladist or phylogeneticist of Hennig's school.

3.2 CHOICE OF EXPERIMENTAL ORGANISMS

In considering the phylogeny of taxa above the species level, one encounters the problem of choice of organisms. This is a critical matter. Ideally one would examine every species and genus in considering, for example, the phylogeny of families. Very obviously this is impractical, and one is left with the question of representative taxa (Section 1.4).

One of the useful features of phylogenetic trees is that they can help suggest organisms for biochemical (or other) investigation. Certain taxa could be interesting on account of their phylogenetic position, their "primitive" or "advanced" state with regard to certain characters, their divergence from the mainstream of evolutionary development, or for other reasons. An examination of phylogenetic trees can also prevent unjustified conclusions from being drawn using characters of specialized or aberrant organisms, e.g., considering *Euglena gracilis* to be a "representative alga."

It must be reiterated that phylogenies are highly dependent upon the assumption that the examined organisms are representative of the higher taxa under consideration. While intellectually rejecting typological taxonomy, biochemical taxonomists are often compelled by lack of extensive data to make this a working assumption. Organisms available in pure culture have provided biochemists with large quantities of information otherwise not easily available. Domesticated protists, however, often have different nutritional requirements than do related organisms in natural populations, due, for instance, to (artificial) selection during their isolation. Even for widely studied characters, such as pigmentation or cell wall composition, only a few percent of all known protists have been adequately examined (specialists would term this an optimistic overestimate). There is certainly a place for surveys of individual biochemical characters in a wide range of protists, in spite of the reduced depth to which characters can be examined in such studies.

Finally, it should be emphasized that phylogenies represent evolutionary history, or at least attempt this. By their very nature, however, they rely very heavily (especially biochemically) upon extant organisms or taxa. If, for example, one hypothesizes that one extant taxon, A, is the ancestral stock of existing taxon B, one is stating only that the ancestral organism, if it existed today, would be classified within taxon A. There is not the suggestion that present organisms, fossil evidence to the contrary, did necessarily exist in earlier times. Hence the suffix "-type" is used as a reminder of this distinction.

3.3 THE ENDOSYMBIOTIC THEORY OF ORGANELLE EVOLUTION

> It is puzzling that the ability to liberate oxygen in photosynthesis is shared by groups so unlike the phytoflagellates as the blue-green algae and red algae (Hutner and Provasoli, 1951).

Without a doubt the greatest discontinuity observed among living organisms is the dichotomy in cellular structure separating the prokaryotes and the eukaryotes.* This dichotomy was pointed out by Ferdinand Cohn (1867), who stated:

> In der Klasse der Algen sind zwei verschiedene Haupttypen vereinigt, die von homologen niedersten Formen beginnend in ihren höheren Entwicklungsstufen weiter auseinander treten, und sich am leichtesten durch das Vorhandensein, resp. das Fehlen von Schwärmzellen, die durch Geisseln oder Flimmercilien bewegt werden, charakterisiren lassen. . . .

Further (Cohn, 1875):

> Vielleicht möchte sich die Bezeichnung *Schizophytae* für diese erste und einfachste Abtheilung lebender Wesen empfehlen, die mir, den höheren Pflanzengruppen gegenüber, natürlich abgegrenzt erscheint, wenn auch die Merkmale, durch welche sie charakterisirt ist, mehr negativer als positiver Art sind.

* Prokaryotes have no membrane-bound internal organelles, no 9 + 2 flagella, and no nuclear membrane. Eukaryotes have all these characters, except the Rhodophyceae, which lack 9 + 2 structures.

Although this dichotomy has been formulated primarily on morphological grounds (Stanier and van Niel 1941, 1962), many biochemical characters also reflect this difference. Because there are no organisms which are obviously "intermediate forms" (direct descendants of the transitional organisms between prokaryotes and eukaryotes), it has been suggested that this transition was a relatively sudden event.* The mechanisms suggested to explain a sudden transition are contained in the "endosymbiotic theory" of organelle origin. The gradual transition theory has been called the "direct filiation theory" (Stanier, 1970), and will be referred to here as the gradualist theory.

Although other explanations could probably be contrived to explain the dichotomy, these two can marshal the most supporting data and have received the widest acceptance. Furthermore, it is not clear that these two theories need be as mutually exclusive as they are often thought to be: there is no *a priori* reason to refuse to consider both an endosymbiotic origin of chloroplasts and a gradual development and differentiation of mitochondria. It is also worth pointing out that proof that some chloroplasts had arisen from symbionts would not require that mitochondria, or even that all other chloroplasts, had had similar origins.†

As pointed out exhaustively by Margulis (Sagan, 1967; Margulis, 1968, 1969, 1970) and others, one of the foremost values of the symbiotic theory is that it generates testable hypotheses and highlights the interaction between nuclear and organelle-based nucleic acids. The history and foundations of the endosymbiotic theory will be discussed, and the application of this theory to phylogenetics will then be briefly analyzed.

3.3.1 Historical Development of the Endosymbiotic Theory

Probably the first suggestions that chloroplasts were once free-living organisms that entered the protoeukaryotic cell by symbiotic association were made by Schimper (1883), who stated:

> Sollte es sich definitiv bestätigen, dass die Plastiden in den Eizellen nicht neu gebildet werden, so würde ihre Beziehung zu dem sie enthaltenden Organismus

---

* Uzzell and Spolsky (1974) have criticized this reasoning as constituting a revival of special creation, in the sense that an argument is being made on the apparent absence of intermediates. This criticism neither provides a place for the cyanellae, nor provides any rationale why typically eukaryotic biochemical features "took over" the nuclear genetic system but left the organellar genetic systems much less modified. Moreover, the endosymbiotic theory is not universally quantified, as is special creation; it could be rendered very unlikely by the failure of a large number of subsidiary hypotheses (see Lewontin, 1972).

† It is likely, in light of their morphological and biochemical similarities, that all chloroplasts had a similar origin. The origination of chloroplasts from mitochondria has been discussed by Guilliermond (1914, 1921) and Raven (1970; cf. Crouse *et al.*, 1974).

einigermassen an eine Symbiose erinnern. Möglicherweise verdanken die grünen Pflanzen wirklich einer Vereinigung eines farblosen Organismus mit einem von Chlorophyll gleichmässig tingierten ihren Ursprung.

In 1885, Schimper again wrote:

Die Chromatophoren treten vielmehr von Anfang als vollständig selbständige plasmatische Körper auf; sie verhalten sich sowohl, was ihre Reproduktion wie auch ihren Chemismus betrifft, viel eher wie einige Organismen, denn als Theile des Plasmakörpers; sie zeigen keine Beziehung zu den Cytoplasma oder dem Zellkern, und behalten ihre wichtigsten Eigenthümlichkeiten trotz vielerlei Metamorphosen durch die ganze Pflanzenwelt hindurch.

His conclusions were based primarily on the ability of the chloroplasts to divide and to distribute themselves to each new cell produced. These properties had first been investigated by Nägeli (1846), and later by Schmitz (1883) and Meyer and Reinke in the 1880's. Schimper himself extended these observations, particularly into the higher plants. The "chemical autonomy" to which Schimper alluded is the ability of the chloroplasts to produce starch or oil under the influence of light, without visible participation of the nucleus.

In the remaining years of the last century, Schimper's suggestions were rejected by many researchers, but were accepted by a few others (Altmann, 1890). Still others carried the idea to another extreme by postulating the "independence" of numerous other subcellular bodies such as the "physode" (Crato, 1892, 1896).

The next major paper supporting the endosymbiotic theory came from Mereschkowsky (1905), who noted several resemblances between chloroplasts and blue-green algae. He suggested that chloroplasts arose from the latter by a process of invasion and symbiosis. His five points of resemblance were similarities in size, color, shape, and general simplicity of construction; similarities in the distribution of chlorophyll within each; lack of definite nuclei, but the presence of "protonuclei" in chloroplasts (pyrenoids) and in blue-green algae (*Nukleinkörner*); respiratory assimilation of carbon dioxide in the light by each; and reproduction by simple fission. In support of the symbiotic theory Mereschkowsky furthermore pointed out the genetic continuity of chloroplasts, their apparent ability to function somewhat autonomously from the nucleus, and the presence of symbiotic green and blue-green algae in colorless hosts.

Famintzin (1907) pointed out the lichens and zoochlorellae as examples of evolutionary symbioses, and compared these phycobionts to chloroplasts. Other early work has been excellently reviewed by Schürhoff (1924) and Wallin (1927). More recent reviews of chloroplast origin have been provided by Echlin (1966), Goksøyr (1967), Klein and Cronquist (1967), Sagan

(1967), Allsopp (1969), Margulis (1968, 1969, 1970), Carr and Craig (1970), Cohen (1970, 1973), Raven (1970), D. L. Taylor (1970), Jeffrey (1971), Luft (1971), Raudaskoski (1971), Schnepf and Brown (1971), Stubbe (1971), Whitton *et al.* (1971), Flavell (1972), Lee (1972), Chadefaud (1972, 1974), Schiff (1973), Metzner (1973), F. J. R. Taylor (1974, 1976a), Uzzell and Spolsky (1974), Bogorad (1975), and Buetow (1976).

3.3.2 Biochemical Characters and the Endosymbiotic Theory

Mereschkowsky's arguments in support of the endosymbiotic theory of chloroplast origin were sufficient to keep the idea from falling into utter oblivion during the fifty years following the publication of his research, but were insufficient to convince most biologists of that time. Equally significant in keeping the theory alive were the discoveries of extranuclear inheritance (Baur, 1909; Correns, 1909) and of a relatively wide range of "blue-green alga-like" inclusions in a variety of hosts. The most notable of these were assigned to the genera *Glaucocystis, Paulinella, Cyanophora,* and *Glaucosphaera* (see Pringsheim, 1958; Geitler, 1959). These "syncyanosen," or cyanellae, were at first considered to be curiosities, and little biochemical research was conducted on them until the early 1960's, when laboratory cultures became available. Their rediscovery was started by Hall and Claus (1963, 1967), followed by a number of laboratories simultaneously: Chapman (1966), Echlin (1967), Schnepf (1965), Schnepf and Koch (1966), Schnepf and Deichgraber (1966). Since then most of the earlier known cyanellae have been the subject of ultrastructural and, in a few cases, biochemical investigation.

It was not until the discovery of chloroplast DNA* that interest in the endosymbiotic theory was revived on a large scale. Mitochondrial DNA was demonstrated during the early 1960's, and a corresponding symbiotic origin theory developed for mitochondria. The theories and supporting data concerning mitochondrial origin have recently been reviewed and discussed (Gibor and Granick, 1964; Sagan, 1967; Nass, 1969; Boardman *et al.*, 1970; Margulis, 1970; Raven, 1970; Raudaskoski, 1971; Schnepf and Brown, 1971; Tewari, 1971; Flavell, 1972; Getz, 1972; Raff and Mahler, 1972, 1973; Cohen, 1973; Mahler, 1973; Meyer, 1973; Uzzell and Spolsky, 1973, 1974; Fridovich, 1974a; Spencer and Cross, 1975; Tuan and Chang, 1975; Reijnder, 1975; Mahler and Raff, 1975; F. J. R. Taylor, 1976a), and will not be considered in any detail here. The emphasis has always been on the

* Usually credited to Ris and Plaut (1962), but foreshadowed by the publications of Ris (1961) and of Sagan and Scher (1961), as well as by a long record of equivocal genetic and light-microscopic investigation.

origin of the chloroplast and/or the mitochondrion. Although Sagan (1967) and Margulis (1970) did consider the origin of the eukaryote system, some recent publications have focused attention on the nature of the host in the origin of eukaryote cells (Hall, 1973a) and the origin of the nucleus (Cavalier-Smith, 1975).

The presence of chloroplast DNA opens the possibility that these organelles might be genetically autonomous from the nuclear DNA. To demonstrate such a hypothesis, it would be necessary to demonstrate not only that DNA is present, but that it functions to control its own replication (Boardman *et al.*, 1970; Givan and Leach, 1971; Tewari, 1971; Kirk, 1972; Bogorad, 1975). Complete biosynthetic autonomy, as opposed to genetic autonomy, would involve the synthesis of biological molecules by enzymes encoded in chloroplast DNA and translated on chloroplast ribosomes. If the chloroplast were biosynthetically autonomous, it would be able to synthesize all or most of its own ribosomes, RNA, nucleotide polymerases, chlorophylls, carotenoids, membrane lipids, permease systems, metabolic enzymes, oxidases, cytochromes, etc. Biosynthesis of all these compounds would be inhibited by antibiotics interfering exclusively with chloroplast DNA, RNA, or ribosomes, and these biosynthetic capabilities would be inherited in a non-Mendelian fashion.*

This criterion for "autonomy" would represent an extreme case, and loses sight of the supposed evolutionary advantage of the symbiosis in the first place. Duplication of nuclear-controlled biosynthetic processes would negate one of the main sources of evolutionary advantage of the symbiosis (from the chloroplast's viewpoint). In the case of long-standing symbioses, it would be expected that certain functions of both the chloroplast and the nucleus would have been lost to the other genome because mutations deleting duplicated functions would no longer be so selectively disadvantageous. As the two genomes lost duplications in function, their interdependence, and the possibility of synchronous interaction, could be greatly increased (Hungate, 1955).

It is at about this point that partisans of the two theories of chloroplast origin begin to have disagreements. How interdependent can the chloroplast and the nucleus be without weakening the proposed symbiotic origin? Most scientists would agree that endosymbiotic blue-green alga-like organisms could have lost certain "peripheral" capabilities, such as perhaps the elaboration of structurally specialized xanthophylls or the ability to produce self-sufficiency levels of ATP obtainable from the cytoplasm. But when

* The chloroplast DNA has been shown to code for much of its own rRNA and for the heavy subunit of ribulose-1,5-diphosphate carboxylase, plus a few as yet unidentified membrane proteins (Blair and Ellis, 1973).

faced with reports of nuclear location of genes for some chloroplast ribosomal proteins (Kloppstech and Schweiger, 1973a,b), NADP-linked triose-phosphate dehydrogenase (Bovarnick *et al.*, 1974), ferredoxin (Section 5.10), photosynthetic electron transport intermediates and photosynthetic pigments, biosynthetic enzymes (Kirk, 1972), the small subunit of Fraction I protein (Section 6.6.1), and Photosystem II protein (Brown *et al.*, 1974), many would find their tentative faith in the endosymbiotic theory stretched beyond the breaking point. Some of these phenomena seem explicable only by transfer of genes from chloroplast to nuclear DNA, which appears distressingly *ad hoc* in the absence of confirmatory evidence.

A further difficulty with the endosymbiotic theory has been posed by Stanier (1970): viz., the original protoeukaryote, still without chloroplasts and mitochondria, presumably depended on glycolysis for its energy supply. This cell line has obviously been responsible for production of specialized morphological types although it was metabolically "primitive." It therefore seems paradoxical that another line of prokaryotes, the protomitochondria, evolved an advanced metabolic system including the TCA cycle and oxidative electron transport enzymes, but itself underwent only minimal morphological specialization in the process.

It is interesting that these problems do not apply directly to chloroplasts. There are, however, several ways out of the dilemma:

1. The mitochondria might have become fixed in the protoeukaryote at a very early stage in evolution (Sagan, 1967). Thus the morphological diversification might have been carried out by chloroplast-free but mitochondria-containing cells. This argument would be consistent with the occurrence today of many organisms that possess mitochondria but no chloroplasts, and is, furthermore, consistent with the greater autonomy of the chloroplast than of the mitochondrion.

2. The ability to perform endocytosis might have developed in the protoeukaryote line (Stanier, 1970). This would presumably have removed the evolutionary pressure toward development of advanced types of metabolism by increasing the available supply of organic carbon. Instead, evolution of Golgi and microtubular systems might have been strongly selected for, as these could be involved in endocytosis. Endocytosis would also provide a means of establishing the endosymbiosis at a later date. This hypothesis is supported by the ability of many simple protists to perform endocytosis.

3. Aerobic metabolic systems could have been developed by both the protoeukaryote and the protomitochondrion. The evolutionary advantage selecting for endosymbiosis would then arise from the increased ability to perform oxidative phosphorylation by the Mitchell (1969) chemiosmotic

mechanism, which requires compartmentalization by biomembranes. This hypothesis, and its consequences, have been discussed by Hall (1973a).

Although the theoretical difficulties involved in mitochondrial origin by endosymbiosis may have been largely overcome, there has been minimal factual verification as yet (Raff and Mahler, 1972, 1973). However, certain evidence favoring the symbiotic origin of mitochondria has been reported by Nass (1969), Margulis (1970), Küntzel and Schäfer (1971), Steinman and Hill (1973), and Fridovich (1974a).

3.3.3 The Endosymbiotic Theory in Phylogenetics

All chloroplasts that have been examined possess numerous similarities to each other, making it reasonable that chloroplasts, or protochloroplasts, arose at most only a few times during evolution. These similarities are often quite fundamental to chloroplast function, and encompass both biochemical and morphological characters. Phylogenies, whether based on the symbiotic theory or not, usually involve only a single origin of chloroplasts. The gradualist theory suggests that chloroplast evolution has paralleled evolution of the rest of the cell ever since the appearance of the former in the cell. The endosymbiotic theory, however, is not bound by this restriction.

Momentarily ignoring the origin of organized nuclei and of mitochondria, let us assume that protoeukaryotic hosts A and B acquire protochloroplastic symbionts a and b, respectively, giving rise to photosynthetic eukaryotes Aa and Bb. If a and b are relatively similar, it might be very difficult experimentally to discern this case from that in which protochloroplast c invades host C, giving rise to Cc, which then diverges evolutionarily into two lines, A′a′ and B′b′. The converse could equally well be true: Protochloroplast d invades host D, which eventually diverges to two very dissimilar lines Ed and Fd. But if protochloroplasts a and b are dissimilar, then all descendants of Aa will differ considerably from all descendants of Bb, barring occasional convergences. In the latter case A and B could even be very similar.

Organelles, including chloroplasts, can obviously be lost during the course of evolution, although it may be difficult to imagine the selective desirability of such mutation in some cases. Protozoologists have often seen no difficulty in theorizing the evolution of many colorless protists from photosynthetic ancestors, although many phycologists tend to disagree. Lwoff (1951) has stated that "such irreversible evolution has taken place very often."

Many colorless algae retain some structures similar to those of the chloroplast although they are incapable of chlorophyll biosynthesis, but in some instances they are capable of carotenoid biosynthesis. These include *Polytoma uvella, Polytomella agilis, Prototheca zopfii, Prototheca ciferri,* and *Astasia longa* (Dodge and Crawford, 1971). The colorless cryptomonad *Chilomonas paramecium* likewise possesses proplastid-like structures (Sepsenwol, 1973). However, the dinoflagellates *Oxyrrhis marina, Crypthecodinium cohnii,* and *Katodinium glandulum,* as well as the chrysomonad *Paraphysomonas vestita,* appear to have no plastidic structures at all (Dodge and Crawford, 1971).

4

Nucleic Acids

4.1 EVOLUTION OF THE GENETIC CODE

Studies on biochemical evolution must at one point or another treat the original appearance of life on this planet. Since the celebrated experiments of Miller (Section 2.3), researchers have had considerable success in the abiogenetic synthesis of many important biomonomers. Considerable research has centered on the role of nucleic acids in these early evolutionary stages, because both DNA and RNA play such an important role in current life processes. Likewise, other researchers have focused on the role of proteins. As pointed out by Eigen (1971, 1973) and others, however, neither "information" (DNA) nor "function" (protein) can be considered independently of the other. Although it is true that neither DNA nor protein was necessarily involved in the origin of life, these two molecules must have become involved at a very early stage in the evolutionary process. It is only through the interaction of nucleotide polymer and protein that potential information becomes actual information.

The origin of nucleotide–protein interaction has been conceptualized in three somewhat different manners: as the result of natural physical-chemical interactions between the two macromolecules (Lacey and Pruitt, 1969; Hoffmann, 1974), as a "frozen accident" (Crick, 1968; Raszka and Mandel, 1972; Papentin, 1973b), and as the result of the coevolution of nucleic acids and emerging amino acid biosynthetic pathways (Wong, 1975). The physical chemical approach hypothesizes that the present-day genetic code has been selected because it is better suited than are the other arrangements, whereas the accident hypothesis suggests that the current code happened to be present in the earliest forms of life, and was preserved because a change to any other code would have been so drastic that the organism could not have survived.

The current genetic code is doubtless the result of some evolutionary modification. There have been a number of approaches to the study of this modification, among the more interesting of which are those of Eigen (1971, 1973), Dillon (1973), and Woese (1973). At another level of argument, Goldberg and Wittes (1966) and Mackay (1967) have pointed out the ability of the genetic code to minimize the effect of randomizing influences on the structure of proteins. Similar studies have been described by Crick (1968), Orgel (1968), and Woese (1967, 1969).

What is clear is that the present-day genetic code is very much nonrandom and is involved in a nontrivial way with the molecular details of transcription and translation. Although it has been argued that the presence of the same code in all organisms does not necessarily require a monophyletic derivation of life (Kenyon and Steinman, 1969), it is evident that variations in molecular detail on this level are likely to be important markers in the phylogeny of the protists.

4.2 DNA: STRUCTURE AND COMPOSITION

Deoxyribonucleic acid is the source and repository of all mutations that occur in the cell. For this reason it is likely that molecular details of DNA structure will be of fundamental importance in biochemical phylogenies. Theoretically it may be possible to discover distant relationships among all protists and higher forms by the proper examination of information-bearing DNA.

The deoxyribonucleotide primary sequence of DNA would be expected to provide the most interesting information on evolution; unfortunately, it is exceedingly difficult to sequence DNA. Only in the past few years has it become possible to examine sequences even for small portions of some viral DNA's (Sanger *et al.*, 1973). It will be many years before there are sufficient DNA sequences known to be of use in phylogeny. The secondary structure of DNA appears to be quite constant, at least at the resolution at which it is normally examined. Phylogenetic trends in secondary structure have never been reported and may not exist. With these considerations in mind, researchers have turned to other methods of characterizing the structure of DNA for phylogenetic comparisons. These include the following approaches, and will be treated in turn: partial sequence techniques, guanidine–cytosine (GC) ratios, hybridization experiments, and other parameters.

4.2.1 Partial DNA Sequences

Large nucleic acid molecules can be degraded into smaller, more manageable fragments by nucleases or by partial hydrolysis, and most fragments can be separated by chromatography and/or electrophoresis. After identification of as many as possible of the resulting fragments, an "oligonucleotide catalog" (or "fingerprint") can be developed for that nucleic acid. Subsequently, different nucleic acids' catalogs can be compared by statistical or numerical methods. Fragments containing four to fourteen nucleotides have been studied, both from DNA (Sanger *et al.*, 1973) and from several different RNA's (Section 4.3.2).

Fragments containing only two nucleotides ("doublets") are easier to analyze, but contain significantly less information than do larger fragments. Patterns of overall doublet frequency delineated by numerical techniques agree moderately well with modern concepts of bacterial phylogeny, and doublet analysis may prove to be of some interest in viral phylogeny (Subak-Sharpe *et al.*, 1974). Nonetheless, the low information content of the doublets severely limits the potential of this technique.

4.2.2 Guanidine–Cytosine (GC) Ratios

Once Chargaff's rules had been experimentally determined, it became possible to describe DNA molecules by a few simple ratios among their deoxyribonucleotide bases. One of these ratios of bases (G + C/G + C + T + A), that of guanidine plus cytosine to the total (as mole percentage), has become accepted as an interesting parameter in the description of different DNA's. DNA GC ratios are known for more than a thousand different protists, with the fungi and bacteria especially well represented (Table 1).

DNA GC ratios have sometimes proved to be useful in reinforcing other data in taxonomies or occasionally in phylogenies. Within the genus *Chlorella*, for instance, Hellmann and Kessler (1974a) found the DNA GC ratio to be a useful taxonomic marker in distinguishing various species of these green algae. Storck and co-workers (see below) have found interesting patterns in the DNA GC ratios of a wide range of true fungi. In contrast, Mandel *et al.* (1965) found that the divisions within the genus *Chlorobium* (based on DNA GC ratios) did not correspond well with the two species then recognized in the genus. On a larger scale, Ashmarin *et al.* (1970) found certain correlations between nutritional modes of bacteria and their DNA GC ratio.

In an interesting series of papers, Storck and co-workers have examined correlations between DNA GC ratios and the taxonomic status of fungi.

TABLE 1

DNA GC Ratios

Organism	GC ratio (%)	Reference
Bacteria[a]		
Flexibacter elegans	48	Edelman *et al.*, 1967
F. rubrum	37	Edelman *et al.*, 1967
Flexothrix sp.	38	Edelman *et al.*, 1967
Leucothrix mucor (11 str.)	49	Edelman *et al.*, 1967
Saprospira grandis	47	Edelman *et al.*, 1967
S. thermalis	37	Edelman *et al.*, 1967
Vitreoscilla spp. (2 spp.)	44–45	Edelman *et al.*, 1967
Chlorobium spp. (17 str.)	50–58	Mandel *et al.*, 1965
Chromatium sp.	64	Belozersky, 1963
Rhodospirillum rubrum	61	Taylor and Storck, 1964
Cyanophyceae		
Agmenellum quadruplicatum	49	Schiff, 1973
Anabaena spp. (2 spp.)	44–46	Edelman *et al.*, 1967
Anabaenopsis sp.	42	Edelman *et al.*, 1967
Anacystis spp. (3 spp.)	55–60	Edelman *et al.*, 1967; Schiff, 1973
Calothrix parietina	43	Edelman *et al.*, 1967
Chlorogloea fritschii	43	Schiff, 1973
Coccochloris spp. (3 spp.)	50–71	Edelman *et al.*, 1967; Schiff, 1973
Dermocarpa violacea	44	Edelman *et al.*, 1967
Fremyella diplosiphon	42	Edelman *et al.*, 1967
Gleocapsa alpicola (3 str.)	35–48	Edelman *et al.*, 1967; Schiff, 1973
Gleobacter violaceus	64	Rippka *et al.*, 1974
Lyngbya spp. (2 spp.)	47–51	Edelman *et al.*, 1967
Microcoleus spp. (2 spp.)	45–48	Edelman *et al.*, 1967
Microcystis aeruginosa (2 str.)	45–66	Schiff, 1973
Nodularia sphaerocarpa	39	Edelman *et al.*, 1967
Nostoc spp. (6 spp.)	43–44	Edelman *et al.*, 1967
Phormidium luridum var. *olivacea*	47	Edelman *et al.*, 1967
Plectonema spp. (4 spp.)	43–48	Edelman *et al.*, 1967
Tolypothrix sp.	43	Edelman *et al.*, 1967
Synechococcus spp. (5 str.)	52–71	Schiff, 1973
Chloroplasts		
of *Acetabularia mediterranea*	36–58	Green, 1974
of *Chlamydomonas reinhardi*	33–39	Green, 1974
of *Chlorella ellipsoidea*	33–36	Green, 1974

TABLE 1 *(Continued)*

Organism	GC ratio (%)	Reference
of *Chlorella protothecoides*	29	Green, 1974
of *Chlorella pyrenoidosa*	26–35	Green, 1974
of *Euglena gracilis* strain Z	21–26	Schiff, 1973
of *E. gracilis* var. *bacillaris*	24–30	Schiff, 1973
of *Ochromonas danica*	31	Schiff and Epstein, 1965
of *Porphyra tenera*	37	Ishida *et al.*, 1969
of higher plants (11 spp.)	36–41	Tewari, 1971
FUNGI AND WATER MOLDS		
Oomycetes		
Pythiaceae (15 spp.)	49–58	Storck and Alexopoulos, 1970
Saprolegniaceae (12 spp.)	40–62	Storck and Alexopoulos, 1970
except *Sapromyces* sp.	27	Storck and Alexopoulos, 1970
Chytridiomycetes		
Blastocladiella emersonii	66	Mandel, 1968
Rhizophlyctis rosea	44	Mandel, 1968
Rhizophydium sp.	50	Mandel, 1968
Zygomycetes		
Mucorales		
Choanephoraceae (3 spp.)	39–40	Storck and Alexopoulos, 1970
Cunninghamellaceae		
Cunninghamella spp. (14 spp.)	28–34	Storck and Alexopoulos, 1970
Mycotypha spp. (8 spp.)	42–48	Storck and Alexopoulos, 1970
Kickxellaceae (8 spp.)	30–56	Storck and Alexopoulos, 1970
Mortierellaceae (5 spp.)	49–52	Storck and Alexopoulos, 1970
Mucoraceae		
Absidia spp. (7 spp.)	39–59	Storck and Alexopoulos, 1970
Actinomucor elegans (12 str.)	40–43	Storck and Alexopoulos, 1970
Circinella spp. (6 spp.)	36–54	Storck and Alexopoulos, 1970
Mucor spp. (27 spp.)	30–49	Storck and Alexopoulos, 1970

TABLE 1 (*Continued*)

Organism	GC ratio (%)	Reference
Phycomyces blakesleeanus (6 str.)	38–44	Storck and Alexopoulos, 1970
Rhizopus spp. (19 str.)	38–49	Storck and Alexopoulos, 1970
Zygorhynchus moelleri	39	Storck and Alexopoulos, 1970
Pilobolaceae (1 sp.)	46	Storck and Alexopoulos, 1970
Syncephalastraceae (14 str.)	47–52	Storck and Alexopoulos, 1970
Thamnidiaceae		
Chaetocladium brefeldii	41	Storck and Alexopoulos, 1970
Cokeromyces spp. (7 str.)	32–38	Storck and Alexopoulos, 1970
Helicostylium piriforme (2 str.)	50–54	Storck and Alexopoulos, 1970
Radiomyces spp. (10 str.)	44–50	Storck and Alexopoulos, 1970
Thamnidium spp. (20 str.)	37–61	Storck and Alexopoulos, 1970
Entomophthorales		
Basidiobolus ranarum	38	Storck and Alexopoulos, 1970
Ascomycetes (Hemiascomycetidae)		
Ascoideaceae (2 str.)	33–43	Storck and Alexopoulos, 1970
Endomycetaceae (5 str.)	34–54	Storck and Alexopoulos, 1970
Saccharomycetaceae		
Debaryomyces spp. (4 str.)	34–45	Storck and Alexopoulos, 1970
Hansenula spp. (45 str.)	29–45	Storck and Alexopoulos, 1970
Pichia spp. (5 str.)	26–46	Storck and Alexopoulos, 1970
Saccharomyces spp. (130 str.)	30–48	Storck and Alexopoulos, 1970; Nakase and Komagata, 1971a,b; Yarrow and Nakase, 1975

TABLE 1 (*Continued*)

Organism	GC ratio (%)	Reference
Schizosaccharomyces spp. (2 str.)	40–42	Storck and Alexopoulos, 1970
Ascomycetes (Euascomycetidae)		
Discomycetes		
Helotiales (1 sp.)	46	Storck and Alexopoulos, 1970
Pezizales (1 sp.)	50	Storck and Alexopoulos, 1970
Plectomycetes		
Eurotiales		
Eurotiaceae (40 str.)	48–57	Storck and Alexopoulos, 1970
Gymnoascaceae (21 str.)	50–55	Storck and Alexopoulos, 1970
Phaeotrichaceae (1 sp.)	54	Storck and Alexopoulos, 1970
Microascales (2 spp.)	53–56	Storck and Alexopoulos, 1970
Pyrenomycetes		
Chaetomiales (18 str.)	48–60	Storck and Alexopoulos, 1970
Clavicipitales (1 sp.)	53	Storck and Alexopoulos, 1970
Hypocreales (1 sp.)	52	Storck and Alexopoulos, 1970
Sphaeriales (21 str.)	50–55	Storck and Alexopoulos, 1970
Loculoascomycetes (3 spp.)	51–56	Storck and Alexopoulos, 1970
Imperfect fungi		
Melanconiales (2 spp.)	50–56	Storck and Alexopoulos, 1970
Moniliales		
Cryptococcaceae		
Candida spp. (38 str.)	30–60	Storck and Alexopoulos, 1970
Cryptococcus spp. (14 str.)[b]	46–66	Storck and Alexopoulos, 1970
Rhodotorula spp. (11 str.)[b]	48–70	Storck and Alexopoulos, 1970
Torulopsis spp. (7 str.)	34–52	Storck and Alexopoulos, 1970

TABLE 1 (*Continued*)

Organism	GC ratio (%)	Reference
Trichosporon spp. (4 spp.)	32–59	Storck and Alexopoulos, 1970
Moniliaceae		
Aspergillus spp. (60 str.)	46–61	Storck and Alexopoulos, 1970
Paecilomyces spp. (3 str.)	50–52	Storck and Alexopoulos, 1970
Penicillium spp. (92 str.)	47–61	Storck and Alexopoulos, 1970
Tuberculariaceae (13 str.)	50–53	Storck and Alexopoulos, 1970
Sphaeropsidales (3 spp.)	49–56	Storck and Alexopoulos, 1970
Basidiomycetes (Heterobasidiomycetidae)		
Heterobasidiomycetes (20 spp.)	50–65	Storck and Alexopoulos, 1970
Basidiomycetes (Homobasidiomycetidae)		
Agaricaceae (6 str.)	44–58	Storck and Alexopoulos, 1970
Hymenochaetaceae (4 str.)	50–51	Storck and Alexopoulos, 1970; Storck *et al.*, 1971
Lycoperdaceae (1 sp.)	51	Storck and Alexopoulos, 1970
Polyporaceae (48 str.)	51–60	Storck and Alexopoulos, 1970
Schizophyllaceae (3 spp.)	57–61	Storck and Alexopoulos, 1970
PROTOZOA		
Mastigophora		
Blastocrithidia culcis (2 str.)	56–58	Mandel, 1967; Tuan and Chang, 1975
Crithidia spp. (3 str.)	54–58	Mandel, 1967
Leishmania spp. (2 str.)	54–57	Mandel, 1967
Schizotrypanum cruzi	49	Shapiro, 1968
Trichomonas spp. (2 spp.)	29–34	Mandel, 1967
Trypanosoma spp. (4 spp.)	45–59	Mandel, 1968
Sarcodina		
Acanthamoeba castellani	56–58	Mandel, 1967; Marzzoco and Colli, 1974
Amoeba proteus	66	Mandel, 1967
Entamoeba invadens	23	Mandel, 1967
Naegleria gruberi	33	Mandel, 1967

TABLE 1 (*Continued*)

Organism	GC ratio (%)	Reference
Ciliatea		
Actinosphaerium nucleofilum	43	Gibson, 1966
Blepharisma americanum	37	Gibson, 1966
Colpidium spp. (3 str.)	32–35	Mandel, 1967
Didinium nasutum	36	Gibson, 1966
Dileptus anser	32	Gibson, 1966
Glaucoma chattoni	34	Schildkraut *et al.*, 1962
Paramecium spp. (8 str.)	29–39	Schildkraut *et al.*, 1962; Gibson, 1966; Villa and Storck, 1968
Stentor polymorphus	45	Gibson, 1966
Tetrahymena spp. (34 str.)	19–33	Mandel, 1968; Villa and Storck, 1968; Hill, 1972
Telosporea		
Plasmodium berghei	41	Whitfeld, 1953
SLIME MOLDS		
Plasmodiophoromycetes		
Labyrinthula sp.	58	Mandel, 1968
Acrasis rosea	36	Mandel, 1967
Acytostelium leptosomum	36	Mandel, 1967
Dictyostelium spp. (3 str.)	22–25	Schildkraut *et al.*, 1962: Mandel, 1968
Polysphondylium spp. (2 str.)	20	Mandel, 1967
Protostelium irregularis	34	Mandel, 1967
Fuligo varians	34	Belozersky, 1963
Lycogala sp.	42	Belozersky, 1963
Physarum polycephalum (3 str.)	38–42	Mandel, 1968; Villa and Storck, 1968
ANIMALS		
Invertebrates (12 spp.)	34–44	Sueoka, 1961
Vertebrates (16 spp.)	40–44	Sueoka, 1961
ALGAE		
Euglenophyceae		
Astasia longa	56	Mandel, 1967
Euglena gracilis	46	Gibson, 1966
E. gracilis strain Z (1)	48–53	Schiff, 1973
E. gracilis var. *bacillaris* (3)	48–55	Schiff, 1973
Rhodophyceae		
Ahnfeltia plicata	53	Pakhomova *et al.*, 1968
Astrephomene gubernaculifera	61–62	Brooks and Nasatir, 1966

TABLE 1 (*Continued*)

Organism	GC Ratio (%)	Reference
Griffithsia globulifera	42	Nasatir and Brooks, 1966
Phyllophora nervosa	57	Pakhomova *et al.*, 1968
Polyides rotundus	62	Pakhomova *et al.*, 1968
Rhodymenia palmata	49	Mandel, 1968
Bacillariophyceae		
Chaetoceros decipiens	39	Sueoka, 1961
Cyclotella cryptica	41	Mandel, 1968
Cylindriotheca fusiformis	45	Mandel, 1968
Melosira italica	46	Pakhomova *et al.*, 1968
Navicula spp. (2 str.)	50–58	Mandel, 1968
Nitzschia angularis	47	Mandel, 1968
Rhabdonema adriaticum	37	Mandel, 1968
Thalassiosira nordenscheldii	40	Sueoka, 1961
Phaeophyceae		
Cystoseira barbata	59	Sueoka, 1961
Dictyota fasciola	59	Pakhomova *et al.*, 1968
Chrysophyceae		
Cricosphaera spp. (2 spp.)	59–60	Mandel, 1968
Ochromonas danica	48	Mandel, 1967
Haptophyceae		
Coccolithus huxleyi	65	Mandel, 1968
Isochrysis galbana	61	Schildkraut *et al.*, 1962
Prymnesium parvum	58	Schildkraut *et al.*, 1962
Xanthophyceae		
Monodus subterraneus	52	Mandel, 1968
Cryptophyceae		
Chroomonas sp.	61	Rae, 1976
Rhodomonas lens	59	Rae, 1976
Dinophyceae		
Exuviaella cassubica	37	Rae, 1976
Amphidinium carterae	40	Rae, 1976
Crypthecodinium cohnii	41	Rae, 1976
Peridinium triquetrum	53	Rae, 1976
Symbiodinium microadriaticum	46	Rae, 1976
Chloromonadophyceae		
Gonyostomum semen	35	Rae, 1976
Vacuolaria virescens	34	Rae, 1976
Charophyceae		
Chara sp.	50	Mandel, 1968
Nitella sp.	49	Mandel, 1968
Chlorophyceae		
Acetabularia mediterranea (5 str.)	37–53	Green, 1974; Bonotto *et al.*, 1975

TABLE 1 ***(Continued)***

Organism	GC Ratio (%)	Reference
Ankistrodesmus spp. (21 str.)	59–70	Sueoka, 1961; Hellmann and Kessler, 1974b
Asteromonas gracilaris	50	Pakhomova *et al.*, 1968
Chlamydomonas spp. (6 str.)	60–68	Sueoka, 1961; Mandel, 1968
Chlorella spp. (88 str.)	43–79	Hellmann and Kessler, 1974a
Chlorogonium elongatum	54	Mandel, 1968
Cladophora sp.	47	Pakhomova, 1974
Dunaliella spp. (2 spp.)	50	Pakhomova *et al.*, 1968
Hydrodictyon reticulatum	54–61	Sueoka, 1961; Pakhomova *et al.*, 1968
Lagerheimia ciliata	60	Pakhomova *et al.*, 1968
Polytoma spp. (3 str.)	42–65	Mandel, 1967, 1968
Polytomella papillata	41	Jones and Thompson, 1963
Rhizoclonium japonicum	54	Mandel, 1968
Scenedesmus spp. (31 str.)	52–64	Sueoka, 1961; Pakhomova *et al.*, 1968; Hellmann and Kessler, 1974b
Spirogyra sp.	39	Mandel, 1968
Stigeoclonium tenue	63	Pakhomova *et al.*, 1968
Ulothrix fimbriata	46	Mandel, 1968
Volvox carteri	50	Kochert and Sansing, 1971
Higher plants (93 spp.)	30–49	Sueoka, 1961; Belozersky, 1963; Edelman *et al.*, 1967; Biswas and Sarkar, 1970

[a] Space does not allow even a representative sample of bacteria and actinomycetes to be included in this table. Reported DNA GC ratios for prokaryotes other than those listed here range from 23% for some Mycoplasmatales (Freundt, 1974) to 74% for some *Streptomyces* spp. (Bradley and Bond, 1974). For further details consult Belozersky and Spirin (1958), Sebald and Véron (1963), Taylor and Storck (1964), Hill (1966), and Buchanan and Gibbons (1974).

[b] The taxonomic position of these organisms is not certain (see text).

Zygomycetes tend to have the lowest ratios, and the zygomycete range barely overlaps that of the Euascomycetidae. Interestingly, *Dipodascus uninucleatus*, which on morphological grounds has been suggested as a link between Zygomycetes and Ascomycetes, has a DNA GC ratio (43%) well into the zygomycete range (29% to 48%) and below the euascomycetidean range (46% to 60%) (Storck, 1966; Storck *et al.*, 1971). The "overlap" technique is intuitively expected to show possible phylogenetic relationships. Although in itself it is not a powerful test for homology, it seems to be useful as a corroboratory test within some protistan taxa.

The dichotomy in DNA GC ratios between ascomycetous yeasts (Hemiascomycetidae) and mycelial fungi (Euascomycetidae) is interesting. Although there is some overlap (Table 1), in reality the majority of yeasts have DNA GC ratios below about 45%, and most Euascomycetidae have GC ratios above 50%. The lack of overlap can be regarded as an indication of a relatively distant evolutionary divergence compared to the divergence among members of each group. More overlap is seen between the yeasts and the Zygomycetes than between mycelial Ascomycetes and the Zygomycetes, contrary to morphological and some other biochemical indications.

The highest DNA GC ratios among the fungi are found in the Basidiomycetes, usually considered by morphological criteria as the most advanced fungi. Within the Homobasidiomycetidae good correlations have been found between DNA GC ratios and taxonomic positions for some thirty-two fungi examined (Storck *et al.*, 1971). The range for the poroid Agaricaceae coincides with that for the Euascomycetidae, indicating that the range of DNA GC ratios is not a unique description of a taxonomic group. The yeastlike organisms of the genera *Cryptococcus*, *Rhodotorula*, and *Sporobolomyces*, which have been assigned to various higher taxa over the years, possess DNA GC ratios more characteristic of the Heterobasidiomycetidae than of the Hemiascomycetidae, and possible heterogeneity within each of these genera has been noted (Storck *et al.*, 1969). Other yeastlike organisms have recently been reassigned to the Heterobasidiomycetidae on morphological grounds (McCully and Robinow, 1972a,b).

The examination of these base ratios may provide a useful means of relating derived organisms to a parental group. The example suggested is the classification of the imperfect fungi or reassignment of the taxa to other fungal groups, but other applications may be found in the colorless algae, Cyanophyceae, and bacteria such as *Hyphomicrobium* spp. Similarities between the DNA GC ratios of flagellated protozoa and unicellular algae are suggestive of phylogenetic relationships but are far from conclusive.

Increase in the DNA GC content has been correlated with an increase in the frequency of occurrence of the "unusual" deoxyribonucleotide base 5-methylcytosine (Pakhomova *et al.*, 1968). So-called unusual bases (Table 2) are found in the DNA of most investigated organisms, and there has been a suggestion that the increase in frequency of occurrence of these bases is positively correlated with an overall increase in evolutionary complexity

TABLE 2

Selected Bases Other than G,C,T,A in DNA of Protists

Base	Reference
In bacteria	
5-Methylcytosine	Rae, 1973
6-Methylaminopurine	Mazin and Sulimova, 1974
In Cyanophyceae	
5-Methylcytosine	Whitton *et al.*, 1971
6-Methylaminopurine	Whitton *et al.*, 1971
N^6,N^6-Dimethylaminopurine	Pakhomova, 1974
In Rhodophyceae, Bacillariophyceae, and Phaeophyceae	
5-Methylcytosine	Pakhomova *et al.*, 1968
6-Methylaminopurine	Pakhomova *et al.*, 1968
In Dinophyceae	
5-Hydroxymethyluracil	Rae, 1973, 1976
In Chlorophyceae	
5-Methylcytosine	Pakhomova *et al.*, 1968
6-Methylaminopurine	Pakhomova *et al.*, 1968
N^6,N^6-Dimethylaminopurine	Pakhomova, 1974
In Euglenophyceae	
N^6,N^6-Dimethylaminopurine	Pakhomova, 1974
5-Methylcytosine	Shapiro, 1968
In Protozoa	
5-Methylcytosine	Sueoka, 1961; Belozersky, 1963
3-Methylcytosine	Pakhomova, 1974
N^6-Methyladenine	Cummings *et al.*, 1974
1-Methylguanine	Pakhomova, 1974
N^2-Methylguanine	Pakhomova, 1974
N^2,N^2-Dimethylguanine	Pakhomova, 1974
7-Methylguanine	Pakhomova, 1974
In Zygomycetes	
6-Methylaminopurine	Il'in *et al.*, 1972

(Mazin and Sulimova, 1974). Interestingly enough, 5-methylcytosine is present in the Cyanophyceae, but is considered to be absent from the DNA of the chloroplasts of the Chlorophyceae and higher plants (Ellis *et al.*, 1973). Discrepancies in the GC determinations of dinoflagellates have now been explained by the presence of hydroxymethyluracil (Rae, 1976). This unusual base, which appears to be absent from other algae, was found in all dinoflagellates examined and appears to be a distinguishing feature of the division.

4.2.3 Hybridization Experiments

Because DNA is double-stranded in most organisms and because its bases are paired in complementary fashion, it is possible to form hybrid double-stranded DNA structures using individual strands from different sources. Much attention has focused on DNA–RNA hybridization (Section 4.3.3), but some work on DNA–DNA hybridization has also appeared, and is subject to many of the same difficulties as are the DNA–RNA hybridization experiments.

Much of the uncertainty in interpreting hybridization experiments comes from the question: How similar must sequences be before hybridization is possible? No two DNA molecules from different organisms have identical nucleotide sequences, but limited hybridization is often observed. Some of this hybridization may be due to the presence of strongly conserved DNA sequences, such as those coding for rRNA (Groot *et al.*, 1975). In general, at least 75% base pairing is required for DNA–DNA hybridization at ambient temperatures of $T_m = 25°C$ (De Ley and Kersters, 1975). Particularly tenuous hybrids can be identified by thermal renaturation studies on the heteroduplex. There is also some correlation between overall DNA GC content and the ease of hybridization. Specific examples of difficulties arising from hybridization experiments and their interpretation have been given by Basden *et al.* (1968), Pace (1973), Schiff (1973), and Bradley and Bond (1974).

4.2.4 Other Parameters

The replication of DNA molecules has been examined in many systems, both prokaryotic and eukaryotic. It is usually stated that the replication of all DNA is semiconservative; indeed, all prokaryotic and all eukaryotic nuclear DNA appear to be so replicated. On one hand, chloroplastic DNA is thought to be replicated semiconservatively in *Chlamydomonas reinhardi* and in *Euglena gracilis,* although not all workers in this field are convinced. On the other hand, the DNA of *Euglena gracilis* mitochondria appears to be replicated in dispersive fashion (nonsemiconservatively) (Richards and Ryan, 1974). These data, although consistent with the endosymbiotic and gradualistic theories of chloroplast origin, may suggest a unique origin of euglenoid mitochondria; no simple prokaryote has yet been found to possess dispersive DNA replication.

The molecular weights of many protistan DNA's are known, but no phylogenetic trends are evident in these data. A better indicator of both genetic complexity and evolutionary history may be DNA "molecular complexity." Although the DNA's of chloroplasts and prokaryotes have approximately the same molecular weight, chloroplastic DNA has a considerably lower molecular complexity.

4.3 RNA: STRUCTURES AND COMPOSITION

Ribonucleic acid is found in all organisms except the DNA viruses. It appears to exist in three different forms in the protists: messenger RNA (mRNA), transfer or soluble RNA (tRNA), and ribosomal RNA (rRNA). The existence of the three classes, differing as they do in gross structure and in function, is probably the result of evolutionary modification for the control of the expression of genetic information.

The cellular RNA is biosynthesized by condensation of ribonucleotides along the DNA, which acts as a template. Not all of the DNA is so transcribed; therefore even the most thorough studies of RNA will be limited to uncovering the information in only part of the DNA. As ultimately frustrating as this may prove to be, this situation can be exploited for the experimental purification of biosynthetic products of individual genes. This situation also suggests that the composition of the RNA obtained from organisms at different times, or grown under differing culture conditions, may vary as different genes are being expressed.

The topological relationships of the genes for various molecules of RNA have been reviewed by Pace (1973). Although the overall organization appears to be similar in prokaryotes and in eukaryotic nuclei, the 5 S gene in the former is included in the rRNA polycistronic transcriptional unit, whereas the 5.8 S gene in the eukaryotes may be included instead (Ginsburg and Steitz, 1975). Such a dichotomy, if true of other organisms as well, can only provide the generality that prokaryotes are more like each other than they are like eukaryotes in this regard. Although it has been demonstrated that chloroplast DNA codes for some chloroplast RNA (Rawson and Haselkorn, 1973), the transcriptional organization of chloroplast rRNA genes has not been fully elucidated. In chloroplasts of *Euglena gracilis* Z and of pea, the two major rRNA components are coded by separate but clustered cistrons (Scott, 1973; Thomas and Tewari, 1974).

4.3.1 RNA Base Sequences

As with DNA, the most phylogenetically amenable information about RNA would be the primary sequence. Due to the size of tRNA's, several of these molecules have been sequenced fully (Cedergren *et al.*, 1972). Results with isoaccepting species for serine and valine are particularly interesting:

yeast seryl-tRNA appears to be more like the seryl-tRNA from rat liver than like the one from *Escherichia coli.* Valyl-tRNA's from yeast and *Torulopsis utilis* are more similar to each other than either is to any of the three valyl-tRNA's from *E. coli.* Yeast leucyl-tRNA does not appear to be strongly homologous with the leucyl-tRNA's from *E. coli,* but the tRNA species for valine, methionine, phenylalanine, tryptophan, tyrosine, alanine, and isoleucine are more similar to each other than could be expected by chance. In *E. coli,* sequences for threonyl-, glycyl-, isoleucyl-, and two valyl-tRNA's exhibit more than 66% homology among themselves (Cedergren *et al.*, 1972; Dayhoff, 1972a; Clarke and Carbon, 1974).

Low molecular weight rRNA's sedimenting at 5 S and 5.8 S have been sequenced as well. Kimura and Ohta (1973b) and Jordan *et al.* (1974) have concluded from these data that 5 S rRNA from *Chlorella pyrenoidosa,* yeast, and human are more similar to each other than any is to bacterial 5 S rRNA (see also Hori, 1976). Sequences from *Anacystis nidulans* and *Oscillatoria tenuis* demonstrate not only that blue-green algal 5 S rRNA's are more closely allied to bacterial 5 S rRNA's than to eukaryotic ones, but also that considerable sequence divergence exists both among the prokaryotes and within the Cyanophyceae (Corry *et al.*, 1974). The possibility of a common evolutionary origin for tRNA and 5 S rRNA has been the subject of some dispute (Mullins *et al.*, 1973; Holmquist *et al.*, 1973; Sogin *et al.*, 1973).

4.3.2 Partial Sequence and Base Ratio Techniques

As in the case of DNA, the difficulty in obtaining nucleotide sequences of larger RNA molecules has led to increased interest in the phylogenetic implications of simpler but unfortunately less easily interpretable parameters. Oligonucleotide catalogs or fingerprints (Section 4.2.1) of prokaryotic and chloroplastic 16 S rRNA, and of eukaryotic 18 S rRNA, have revealed similarities among 16 S molecules from *Euglena gracilis* Z chloroplasts, *Porphyridium cruentum* chloroplasts, *Bacillus subtilis, Anacystis nidulans, Rhodopseudomonas spheroides,* and four gram-negative enteric bacteria. By this criterion, the two chloroplastic molecules were not significantly more similar to each other than either was to *A. nidulans* or *B. subtilis* 16 S rRNA. Fingerprints from *P. cruentum* and yeast 18 S rRNA species did not show marked similarity (Zablen *et al.*, 1975; Bonen and Doolittle, 1975). The 16 S rRNA fingerprint of *R. spheroides* bears a somewhat greater resemblance to those of enteric bacteria than to those from *B. subtilis* or *A. nidulans* (Zablen and Woese, 1975).

Ribonucleic acid GC ratios (Table 3) have been used, often with ambiguous results, although some interesting relationships are seen among

TABLE 3

RNA GC Ratios[a]

Organism	GC (%)	Reference
Bacteria[b]		
Chlorobium thiosulfatophilum	54	Belozersky, 1963
Chromatium sp.	55	Belozersky, 1963
Cyanophyceae		
Anacystis nidulans	56–59	Biswas and Myers, 1960
Anacystis sp. (+)	52	Lava-Sanchez *et al.*, 1972
Mastigocladus laminosus	57	Evreinova *et al.*, 1961
Dinophyceae		
Gyrodinium cohnii (+)	46	Rae, 1970
Phaeophyceae		
Fucus vesiculosus (+)	54	Lava-Sanchez *et al.*, 1972
Chlorophyceae		
Spirogyra sp. (+)	50	Lava-Sanchez *et al.*, 1972
Ulva lactuca (+)	50	Lava-Sanchez *et al.*, 1972
Volvox carteri	50	Kochert and Sansing, 1971
Euglenophyceae		
Euglena gracilis	55–58	Mandel, 1967
E. gracilis (+)	57	Brown and Haselkorn, 1971
Eumycota		
Zygomycetes		
3 spp.	46–51	Belozersky, 1963
2 spp. (+)	48–50	Lava-Sanchez *et al.*, 1972
Ascomycetes		
Hemiascomycetidae (7 spp.)	47–54	Storck, 1965
Euascomycetidae		
3 spp.	50–52	Storck, 1965
2 spp. (+)	50–54	Lava-Sanchez *et al.*, 1972
Imperfect fungi		
13 spp.	48–54	Belozersky, 1963; Storck, 1965
1 sp. (+)	50	Lava-Sanchez *et al.*, 1972
Basidiomycetes		
12 spp.	50–60	Belozersky, 1963; Storck, 1965
4 spp. (+)	48–54	Lava-Sanchez *et al.*, 1972
Protozoa		
Acanthamoeba castellani (+)	56	Lava-Sanchez *et al.*, 1972
Amoeba spp. (3 spp.)	50–58	Mandel, 1967
Lycogala sp.	56	Belozersky, 1963
Paramecium caudatum (+)	49	Lava-Sanchez *et al.*, 1972
Plasmodium berghei	44	Whitfeld, 1953

TABLE 3 (*Continued*)

Organism	GC (%)	Reference
Tetrahymena pyriformis (6 str.)	39–46	Belozersky, 1963; Hill, 1972
T. pyriformis (+)	36	Kumar, 1969
Trichomonas vaginalis	32	Mandel, 1967
Chloroplasts		
Volvox carteri (+)	50	Kochert and Sansing, 1971
Higher animals (60 spp.) (++)	43–65	Belozersky, 1963; Lava-Sanchez *et al.*, 1972
Higher plants (51 spp.) (++)	51–60	Belozersky, 1963; Lava-Sanchez *et al.*, 1972

[a] Total RNA, except: (+) rRNA only, or average of reported rRNA species; (++) includes both total RNA (for some species) and rRNA (for other species).

[b] Space limitations do not allow a representative sample of bacteria to be listed. Reported values range from 48% to 59%. For further details consult Belozersky and Spirin (1958), Woese (1961), Midgley (1962), Belozersky (1963), Pace and Campbell (1967), and Tamura (1967).

RNA GC ratios of Zygomycetes and Hemiascomycetidae (Storck, 1965). Lava-Sanchez *et al.* (1972) have demonstrated a correlation between rRNA base composition and trends toward morphological complexity. There appears to be only a weak positive correlation between RNA GC ratios and DNA GC ratios, at least in the bacteria (Belozersky and Spirin, 1958). So-called uncommon bases have been reported from many organisms (Table 4), but no phylogenetic trends are obvious. All known RNA's contain uracil in the place of the thymidine found in DNA, but the evolutionary implications of this are unclear (Lesk, 1969).

4.3.3 Hybridization Experiments

DNA–RNA hybridization techniques, alluded to earlier, show only a 20% to 30% relationship between the bacteria *Escherichia coli* (gram-negative) and *Bacillus subtilis* (gram-positive). Even within the genus *Bacillus* itself, the apparent similarity ranges from 100% to 50% (Pace, 1973).

Also using DNA–RNA hybridization, Bendich and McCarthy (1970) reported partial hybridization between nucleic acids from the toad *Xenopus* and from some higher plants. These authors, interestingly, suggested that on the basis of hybridization studies, the higher plants may be more closely related to the fungi than either group is to *Euglena*. Other results using hybridization techniques imply that rRNA from pea is more similar to cucumber DNA than to pea DNA; such unexpected results suggest that the

TABLE 4

Selected Uncommon Bases in Protistan RNA's

Base	Reference
In Bacteria	
2-Methyladenosine	Pace, 1973
N^6-Methyladenosine	Pace, 1973
N^6,N^6-Dimethyladenosine	Pace, 1973
N^4,O^2-Dimethyladenosine	Pace, 1973
2-Thiomethyl-N^6-(Δ^2-isopentenyl)adenosine	Dillon, 1973
Threonylcarbamoyladenosine	Cedergren and Cordeau, 1973
3-Methylcytidine	Cedergren and Cordeau, 1973
5-Methylcytidine	Pace, 1973
1-Methylguanosine	Pace, 1973
N^2-Methylguanosine	Pace, 1973
7-Methylguanosine	Pace, 1973
5-Methyluridine	Pace, 1973
4-Thiouridine	Cedergren and Cordeau, 1973
2′-*O*-Acetyluridine	Cedergren and Cordeau, 1973
2′-*O*-Methyluridine	Cedergren and Cordeau, 1973
5-Methyl-2-thiouridine	Cedergren and Cordeau, 1973
5-(β-D-Ribofuranosyl)uracil	Pace, 1973
"Y^+ base"	Dillon, 1973
In Cyanophyceae	
2-*O*-Methylribocytidine	Biswas and Myers, 1960
In Ascomycetes (Yeast)	
2-Thiomethyl-N^6-(Δ^2-isopentenyl)adenosine	Dillon, 1973
5-Methylcytidine	Cedergren and Cordeau, 1973
N-Acetylcytidine	Cedergren and Cordeau, 1973
2′-*O*-Methylcytidine	Cedergren and Cordeau, 1973
1-Methylinosine	Cedergren and Cordeau, 1973
1-Methylguanosine	Cedergren and Cordeau, 1973
2′-*O*-Methyl-5-carboxymethyluridine	Gray, 1975

interpretation of hybridization data may be less straightforward than was originally assumed.

Hybridization experiments have also been conducted with DNA from chloroplasts. Pigott and Carr (1972a,b) reported hybridization between *Euglena gracilis* plastid DNA and the rRNA from certain Cyanophyceae and, less strongly, with other cyanophycean and with bacterial rRNA's. On the basis of a 100% hybridization value for the interaction of euglenoid chloroplast DNA with euglenoid chloroplast rRNA, similarity with blue-green algae ranged from 47% (with *Gloeocapsa alpicola*) to 11% (with *Anabaena cylindrica*). Bacteria gave values between 6% and 1.5%, but only 1% hybridization was observed with cytoplasmic (nonchloroplastic) rRNA from *Euglena* itself. These and other experiments have been critically reviewed by Tewari (1971) and Schiff (1973).

4.3.4 Aminoacyl-tRNA Synthetases

The recognition of RNA molecules by aminoacylating enzymes may be of interest in examining the autonomy and phylogeny of chloroplasts. Aliev and Filippovich (1968) reported that aminoacyl-tRNA synthetases (EC 6.1.1) from higher plant chloroplastic and cytoplasmic fractions were able to recognize the corresponding tRNA's, but showed much lessened aminoacylation activities with some of the tRNA's from the opposite fraction. More frequently, synthetases from chloroplasts (but not from mitochondria) of tobacco have been discovered to catalyze preferentially the aminoacylation of chloroplast tRNA's (Guderian *et al.*, 1972); similar results have been noted for bean chloroplasts, where cytoplasmic leucyl-tRNA is recognized only by the cytoplasmic leucyl-tRNA synthetase. But in bean, the situation is more complex; some chloroplast leucyl-tRNA is charged exclusively with the cytoplasmic synthetase and is hybridized only to nuclear DNA, but the major part of the chloroplastic leucyl-tRNA is recognized only by the chloroplastic enzyme (but hybridizes both to the chloroplastic and, less easily, to the nuclear DNA). The chloroplastic tRNA for leucine is easily charged by the leucyl-tRNA synthetase from *Escherichia coli*, indicating "a certain resemblance" with the bacterial leucyl-tRNA (Guillemaut *et al.*, 1973).

Similarly, aminoacyl-tRNA synthetases from *Euglena gracilis* Z and from *Anacystis nidulans* have been shown to be interchangeable in the aminoacylation of tRNA's from either the blue-green alga or *E. gracilis* chloroplasts. However, this was not observed with components from *A. nidulans* and from the extrachloroplastic cytoplasm of *E. gracilis* (Krauspe and Parthier, 1973, 1974; Parthier and Krauspe, 1974). Furthermore, Beauchemin *et al.* (1973) have described the virtual identity of components from *A. nidulans* and *E. coli.*

There are indications that some chloroplastic aminoacyl-tRNA synthetases may be encoded in the chloroplast genome of *Euglena gracilis* strain B (Reger *et al.*, 1970) and *E. gracilis* strain Z (Parthier *et al.*, 1972; Krauspe and Parthier, 1973), but not in *E. gracilis* var. *bacillaris* (Hecker *et al.*, 1974).

Another property of tRNA's that may be of use in phylogenetic studies is the interconversion between two chemically distinct but as yet uncharacterized forms, as observed for tryptophanyl-tRNA (Preddie *et al.*, 1973). In bacteria, tryptophanyl-tRNA is converted between forms I and II. Green algae contain two distinct isoaccepting tryptophanyl-tRNA's, one of which appears to be localized in the chloroplast and is capable of bacterial-style interconversion of form. The second tRNA is apparently cytoplasmic (nonchloroplastic) and is incapable of such an interconversion. It would be interesting to search for this property in other organisms.

4.3.5 RNA Molecular Weights

Sedimentation coefficients of RNA molecules in gradients of cesium chloride or sucrose have been used to indicate their molecular weights. In turn, these molecular weights have been examined for information concerning phylogeny. The basis of the phylogenetic significance of molecular weights is, however, unclear. All prokaryotes, chloroplasts, and possibly mitochondria, on one hand, possess RNA species sedimenting at 23 S and 16 S; the nuclei and extraorganellar cytoplasm, on the other hand, possess heavier species sedimenting at 25 S–28 S and 18 S (Table 5). The 16 S and 23 S units have molecular weights of 0.56 and 1.1 $\times$ 10^6 daltons, respectively, whereas the eukaryotic 18 S has a molecular weight of about 0.70 $\times$ 10^6 daltons. The molecular weight of the 28 S species varies from 1.30 to 1.75 $\times$ 10^6 daltons, depending on the organism.

More subtle differences among molecular weights of the 28 S rRNA subunit have been shown to support the thesis that the "fungi" (*sensu lato*) are at least diphyletic. The molecular weights of this subunit from members of the Chytridiomycetes, Zygomycetes, Ascomycetes, and Basidiomycetes fell within the range 1.30–1.36 $\times$ 10^6 daltons, but molecular weights were markedly larger among the Oomycetes (1.40–1.43 $\times$ 10^6 daltons) and the myxomycetes *Physarum polycephalum* and *Dictyostelium discoideum* (1.42–1.45 $\times$ 10^6 daltons). The phylogenetic position of the hyphochytridiomycete *Rhizidiomyces apophysatus* (1.36 $\times$ 10^6 daltons) was not clarified by this approach (Lovett and Haselby, 1971).

Eukaryotic algae and the higher plants resemble the fungi in possessing 28 S rRNA subunits of approximately 1.3 $\times$ 10^6 daltons, whereas among the higher animals the trend toward increasing morphological complexity is paralleled with an increase in the molecular weight of the 28 S subunit. *Euglena gracilis* is an exception among photosynthetic organisms by possessing a heavier (1.3–1.5 $\times$ 10^6 daltons) subunit. *Acanthamoeba castellani* (1.55 $\times$ 10^6 daltons) and *Entamoeba* sp. (1.6 $\times$ 10^6 daltons), but not *Naegleria* sp. (1.3 $\times$ 10^6 daltons), possess 28 S subunits uncharacteristic of other protozoa. Loening (1968, 1973) has explored the possible phylogenetic significance of variation in the molecular weights and stability of this subunit among the protozoa.

TABLE 5

Distribution of the Two Patterns of rRNA Sedimentation[a]

Organism	Reference
Prokaryotic pattern	
Bacteria (5 spp.)	Loening, 1968; Thompson *et al.*, 1971
Anabaena sp.	Loening, 1968
Nostoc sp.	Loening, 1968
Oscillatoria sp.	Loening and Ingle, 1967
Phormidium persicinium	Howland and Ramus, 1971
Euglena gracilis var. *bacillaris* chloroplasts	Van Pel and Cocito, 1973
Griffithsia pacifica chloroplasts	Howland and Ramus, 1971
Porphyridium aerugineum chloroplasts	Howland and Ramus, 1971
Volvox carteri chloroplasts	Kochert and Sansing, 1971
Higher plant chloroplasts	Loening, 1968; Whitton *et al.*, 1971
Crithidia oncopelti "bipolar body"	Spencer and Cross, 1975
Eukaryotic pattern	
Acanthamoeba castellani[b]	Loening, 1973
Chlamydomonas sp.	Loening, 1968
Chlorella sp.	Loening, 1968
Crithidia oncopelti	Spencer and Cross, 1975
Dictyostelium discoideum	Lovett and Haselby, 1971
Entamoeba sp.[b]	Loening, 1973
Euglena gracilis[b]	Loening, 1968
Griffithsia pacifica[c]	Howland and Ramus, 1971
Gyrodinium cohnii	Rae, 1970
Naegleria sp.	Loening, 1973
Paramecium sp.	Loening, 1968
Porphyridium aerugineum[c]	Howland and Ramus, 1971
Physarum polycephalum	Lovett and Haselby, 1971
Tetrahymena pyriformis W	Loening, 1968
Volvox carteri	Kochert and Sansing, 1971
Hyphochytridiomycetes (1 sp.)	Lovett and Haselby, 1971
Oomycetes (2 spp.)	Lovett and Haselby, 1971
Chytridiomycetes (4 spp.)	Lovett and Haselby, 1971
Zygomycetes (2 spp.)	Lovett and Haselby, 1971
Ascomycetes (2 spp.)	Lovett and Haselby, 1971
Imperfect fungi (2 spp.)	Lovett and Haselby, 1971
Basidiomycetes (3 spp.)	Lovett and Haselby, 1971
Higher plants (6 spp.)	Loening, 1968; Lovett and Haselby, 1971
Higher animals[b]	Loening, 1968

[a] All data are for cytosoluble rRNA unless otherwise specified.

[b] Heavy "28 S" component.

[c] Lower molecular weight (1.21 million daltons) than in most other eukaryotes but definitely not prokaryotic and not "truly transitional" (Howland and Ramus, 1971).

4.3.6 Maturation of rRNA

Even more subtle information concerning properties of these rRNA molecules may be found by examining the biosynthetic steps following their transcription from the DNA; these steps are collectively termed "maturation" or "processing." Basically, two types of maturation sequences have been observed in the protists. In prokaryotes, the observed products of transcription are molecules of RNA slightly larger than the mature 16 S and 23 S forms; the precursor to 23 S, or p23, is about 8% larger than the mature 23 S (m23) (Sogin *et al.*, 1973). In eukaryotes, however, a 45 S precursor is first transcribed, and later dissociates into the 18 S and 28 S species (Grierson *et al.*, 1970).

The prokaryote type of rRNA maturation has been demonstrated in bacteria (Pace, 1973) and in chloroplasts (Table 6), but the situation in mitochondria may (Sanders *et al.*, 1975) or may not (Kuriyama and Luck, 1973) be similar. The Cyanophyceae appear to possess at least a p16 form (Seitz and Seitz, 1973), and possibly, in addition, a p23 form (Doolittle,

TABLE 6

Distribution of the Two Patterns of rRNA Maturation

Organism	Reference
Prokaryotic pattern	
Enteric bacteria	Pace, 1973
Anacystis nidulans	Doolittle, 1972; Szalay *et al.*, 1972; Seitz and Seitz, 1973
Chlamydomonas reinhardi chloroplasts	Miller and McMahon, 1974
Chlorella fusca chloroplasts	Galling, 1974
Euglena gracilis Z chloroplasts	Heizmann, 1974; Scott, 1974
Higher plant chloroplasts	Detchon and Possingham, 1973; Munsche and Wollgiehn, 1973
Eukaryotic pattern	
Amoeba proteus	Craig and Goldstein, 1969
Blastocladiella emersonii	Murphy and Lovett, 1966
Chlamydomonas reinhardi	Miller and McMahon, 1974
Euglena gracilis Z	Scott, 1974
Gyrodinium cohnii	Rae, 1970
Schizosaccharomyces pombe	Taber and Vincent, 1969
Tetrahymena pyriformis	Leick, 1969
Higher animals	Grierson *et al.*, 1970; Perry *et al.*, 1970a,b
Higher plants	Loening, 1967

1972; Szalay *et al.*, 1972). The difference in maturation sequences observed between prokaryotes and eukaryotes probably reflects the presence or absence, respectively, of RNase III (Ginsberg and Steitz, 1975; Hayes *et al.*, 1975; Nikolaev *et al.*, 1975).

In plants and in animals up to and including reptiles, the "45 S" precursor rRNA has a molecular weight of 2.4 to 2.8×10^6 daltons; in higher animals a molecular weight of 4.0 to 4.2×10^6 daltons is transcribed (Perry *et al.*, 1970a,b). The former is sometimes called the "38 S" type, in contrast to the "true 45 S" molecule. The "38 S" molecule has been reported in *Tetrahymena pyriformis, Amoeba proteus, Schizosaccharomyces pombe, Gyrodinium cohnii, Chlamydomonas reinhardi,* and *Blastocladiella emersonii.*

Methylation of the four common ribonucleotide bases to form the so-called unusual bases occurs in the final stages of maturation, or after maturation, in prokaryotes (Seitz and Seitz, 1973) and in chloroplasts (Rijven and Zwar, 1973; Posner *et al.*, 1974). In eukaryotes, however, the precursors as well as the mature rRNA's are methylated (Grierson *et al.*, 1970; Detchon and Possingham, 1973). Methylation of chloroplast rRNA has been found to be less extensive (less than 25%) than in cytoplasmic rRNA (80%), and again the chloroplast rRNA resembles that of prokaryotes in this regard. *Escherichia coli,* for instance, has 15% to 20% of its rRNA bases methylated (Rijven and Zwar, 1973). Methylases responsible for some of these modifications have been isolated (Burkard *et al.*, 1973; Pace, 1973), and methylated ribose has also been reported (Smith and Dunn, 1959).

Initiation of translation by N-formylated methionyl-tRNA is well known in prokaryotes, including *Anacystis nidulans* and *Nostoc* sp. (Leach and Herdman, 1973), and is also found in chloroplasts of higher plants (Burkard *et al.*, 1973), *Euglena gracilis* Z (Schwartz *et al.*, 1967), *Acetabularia mediterranea* (Bachmayer, 1970), and in some mitochondria (references in Burkard *et al.*, 1973). A chloroplastic enzyme has been found which formylates chloroplast methionyl-tRNA but is inactive with extrachloroplastic methionyl-tRNA.

4.4 RIBOSOMES

RNA functions in the biosynthesis of proteins in the cell as a template (mRNA), as the amino acid carriers (tRNA), and in an as yet poorly understood but important (Woese, 1973) manner in the ribosomes (rRNA). Ribosomes appear to be involved in protein synthesis in all living systems, although a few highly specialized proteins, such as cyanophycin in blue-green algae, are not formed on ribosomes (Simon, 1973). Ribosome-like particles have been found in some mammalian viruses as well (Farber and Rawls, 1975).

4.4.1 Sedimentation Coefficients and Subunits

Two major patterns of ribosome sizes are seen in organisms, corresponding to the dichotomy between prokaryotes and eukaryotes. The former organisms possess ribosomes sedimenting in sucrose or cesium chloride gradients at approximately 70 S, varying slightly with the size, weight, conformation, composition, and degree of hydration of the ribosome. The range of reported sedimentation values for prokaryotic ribosomes extends from about 62 S to about 72 S, but this range does not overlap that of eukaroytic ribosomes (79 S to 87 S); the latter are termed "80 S" ribosomes (Taylor and Storck, 1964; Boulter *et al.*, 1972). The distribution of these two sedimentation types is listed in Table 7. It is interesting but probably

TABLE 7

Distribution of Sedimentation Patterns in Ribosomes

Organism	Sedimentation (S)	Reference
Prokaryotic pattern		
Bacillus licheniformis	72	Taylor and Storck, 1964
Escherichia coli BB	69	Taylor and Storck, 1964
Rhodopseudomonas spheroides	66	Reisner *et al.*, 1968
Rhodospirillum rubrum	65	Taylor and Storck, 1964
Staphylococcus epidermidis	67	Taylor and Storck, 1964
Streptomyces griseus	69	Taylor and Storck, 1964
Anabaena cylindrica	72	Taylor and Storck, 1964
A. variabilis	70	Craig and Carr, 1968
Anacystis montana	70	Rodríguez-López and Vasquez, 1968
A. nidulans	69	Taylor and Storck, 1964
Oscillatoria sp.	[a]	Loening and Ingle, 1967
Acetabularia mediterranea chloroplasts	70[b]	Apel and Schweiger, 1973
Chlamydomonas reinhardi chloroplasts	68	Hoober and Blobel, 1969
Chlorella pyrenoidosa chloroplasts	70	Green, 1974; Yurina and Odintsova, 1974
C. fusca chloroplasts	70	Galling and Ssymank, 1970

TABLE 7 (*Continued*)

Organism	Sedimentation (S)	Reference
Euglena gracilis var. *bacillaris* chloroplasts	68	Van Pel and Cocito, 1973; Hecker *et al.*, 1974
Griffithsia pacifica chloroplasts	70	Howland and Ramus, 1971
Porphyridium aerugineum chloroplasts	70	Howland and Ramus, 1971
Higher plant chloroplasts	62–70	Detchon and Possingham, 1973; Leach and Herdman, 1973; Yurina and Odintsova, 1974
Crithidia oncopelti "bipolar body"	67	Spencer and Cross, 1975
Eukaryotic pattern		
Acanthamoeba castellani	[c]	Loening, 1973
Acetabularia mediterranea	80	Apel and Schweiger, 1973
Chilomonas paramecium	80	Rodríguez-López and Muñoz Calvo, 1971
Chlamydomonas reinhardi	80	Hoober and Blobel, 1969
Chlorella fusca	80	Galling and Ssymank, 1970
C. protothecoides	80	Green, 1974
C. pyrenoidosa	80	Green, 1974
Crithidia oncopelti	87	Spencer and Cross, 1975
Cryptomonas ovata	80	Rodríguez-López and Muñoz Calvo, 1971
Euglena gracilis var. *bacillaris*	87	Van Pel and Cocito, 1973
Griffithsia pacifica	[c]	Howland and Ramus, 1971
Paramecium aurelia	80	Reisner *et al.*, 1968
Porphyridium aerugineum	[c]	Howland and Ramus, 1971
Chytridiomycetes (1 sp.)	83	Lovett, 1963
Zygomycetes (7 spp.)	80–84	Taylor and Storck, 1964
Ascomycetes (8 spp.)	80–83	Taylor and Storck, 1964
Imperfect fungi (4 spp.)	79–81	Taylor and Storck, 1964
Basidiomycetes (7 spp.)	79–82	Taylor and Storck, 1964
Mammals	80	Van Pel and Cocito, 1973
Higher plants	78–80	Ellis *et al.*, 1973; Van Pel and Cocito, 1973

[a] Sedimentation value not given, but prokaryotic pattern.

[b] An 80 S ribosome type was also found in chloroplast preparations (Apel and Schweiger, 1973).

[c] Sedimentation value not given, but molecular weight in eukaryotic range.

coincidental that the 80 S ribosomes of *Tetrahymena pyriformis* have been converted to a 70 S form by the heat shocks used to synchronize growth of the organism (Hill, 1972).

Chloroplasts possess ribosomes with the typical prokaryotic type of sedimentation (Table 7), although there are differences between the composition of chloroplastic and bacterial ribosomes (Yurina and Odintsova, 1974). Mitochrondria do not possess this pattern, instead having ribosomes ("mitoribosomes") which may be classified in two, or possibly three, categories according to sedimentation values. Fungal and *Euglena gracilis* mitoribosomes sediment at about 72–75 S, whereas mitoribosomes of *Tetrahymena pyriformis* sediment at 80 S, and those of higher animals at 55 S to 60 S. Further data concerning mitoribosomes are available in Borst and Grivell (1971), Raff and Mahler (1972), Cohen (1973), and Curgy *et al.* (1974).

Subunits from ribosomes can themselves be dissociated to yield the various component nucleic acids and proteins. 70 S ribosomes (from prokaryotes) are composed of two main subunits, of approximately 50 S and 30 S; the heavier of these contains the 23 S and 5 S rRNA, and the 30 S subunit contains the 16 S rRNA. In eukaryotes, the 80 S ribosomes are composed of 60 S and 40 S subunits, which in turn are again composed of analogous rRNA species (28 S, 5 S; 18 S).

Although the sedimentation values are of interest, they are limited to supporting the prokaryote–eukaryote divergence and to providing circumstantial evidence on the origin of chloroplasts. These data possess insufficient detail to provide further input into protistan phylogeny. Neither amino acid sequences of ribosomal proteins nor ribonucleotide base sequences of rRNA are known, so researchers have turned to other types of data for phylogenetically interesting detail.

The formation of functional hybrid ribosomes—composed of subunits from different sources—has been explored in this regard. It would be expected *a priori* that closely related organisms would possess sufficiently similar ribosomes to allow the formation of functional hybrids, whereas less proximate organisms should be less able to do so. Assuming this to be true, chloroplasts from *Euglena gracilis* and from spinach are related to *Escherichia coli.* Interestingly, mitoribosomes from yeast do not appear capable of forming functional hybrids with the appropriate subunits from *E. coli.* Functional ribosomes can be formed using pairs of ribosome subunits from rat liver, pea seedlings, and *Tetrahymena pyriformis*, although these organisms have significantly different rRNA GC ratios and electrophoretically dissimilar ribosomal proteins, and certainly are not closely allied phylogenetically (Lee and Evans, 1971; Grivell and Walg, 1972; Cammarano *et al.*, 1972).

4.4.2 Ribosomal Proteins

Ribosomes contain numerous proteins, mostly of unknown functions, many of which can be separated by two-dimensional electrophoresis on polyacrylamide gels. By this method some fifty-five proteins have been observed from prokaryotic ribosomes, and some seventy to eighty from eukaryotic ribosomes (Yurina and Odintsova, 1975). The two-dimensional patterns, or "fingerprints," observed after electrophoresis can be subjected to statistical analysis and thus can be used in a comparison of ribosomes from different species.

Among higher animals, the ribosomal protein fingerprints reinforce the morphologically based taxonomy; fingerprint patterns from mammals are statistically less similar to crustacean or molluscan fingerprints than to those from reptiles, birds, fishes, or amphibians. The similarity is still less between mammals and higher plants, and is extremely low between eukaryotes and bacteria. Within the bacteria there is great variability in ribosomal protein fingerprint patterns (Wittmann, 1970; Delaunay *et al.*, 1973; Delaunay and Schapira, 1974a,b). The extent of similarity between prokaryotic and eukaryotic organellar ribosomes has not been investigated by the electrophoretic fingerprint technique, although Wittmann (1970) was unable to demonstrate serological similarities between ribosomes of higher plant chloroplasts and prokaryotes.

4.4.3 Other Parameters

A test commonly used to characterize ribosomes involves their sensitivities *in vitro* to various antibiotics. Ebringer (1972) has listed 144 antibiotics that have been used in the study of transcription and translation; many of these compounds act to interfere with ribosome function. Although in general the sensitivity of chloroplast ribosomes parallels the sensitivity of bacterial ribosomes to the same antibiotics, exceptions are known, as with virginiamycin M (Van Pel and Cocito, 1973). Exceptions such as this either could arise from evolutionary divergence of the two ribosomes since their origin from a common ancestral set of genes, or could reflect fundamental dissimilarities. Although the sheer number of similarities in their reactions with antibiotics is suggestive of common origin of the ribosomes, there is little indication of the biophysical basis for most of the antibiotics' action.

Finally, Kurtz (1974) has reported that ribosomes from bacteria and from mitochondria are similarly prone to committing errors of translation, whereas the cytoplasmic eukaryotic ribosomes examined are not.

5

Proteins

PART I: GENERAL CONSIDERATIONS

5.1 POLYPEPTIDES, PROTEINS, AND ENZYMES

It has been noted earlier (Section 4.1) that the evolution of DNA or of RNA cannot be considered independently of the evolution of other biological macromolecules. The same applies to a consideration of proteins. On one level, a functional enzyme may be composed of several polypeptides, and examination of the genes coding for these polypeptides may provide an interesting insight into enzyme evolution. On another level, the appearance of coordinated groups of enzymes was a significant event in the evolution and adaptation of living systems.

It would be ideal to consider biosynthetic enzymes and biosynthetic pathways concurrently with the metabolites they produce; unfortunately the logistics of this approach would be cumbersome, and an overview of entire biosynthetic pathways might be obstructed. Consequently, the discussion of proteins has been segregated from the discussion of their products in the following pages. It is hoped that the insights into the special problems associated with phylogenies using biochemical characters from each of these levels will compensate for the inconveniences of this approach.

Not all proteins are enzymes, and although much of the subsequent discussion will deal with enzymes, it is important to lcavc thc impression that enzymes are not the only proteins of any importance either to the cell or to phylogenetic reconstruction. Relatively small oligopeptides serve as hormones and as toxic factors in several organisms, including some protists. Cyanophyceae produce a storage material and possible nitrogen reserve, "cyanophycin," by nonribosomal condensation of aspartic acid and arginine in an equimolar ratio (Simon, 1971, 1973). "Structural proteins" such as collagen and collagen-like polymers have been reported from a wide range of protists (Aaronson, 1970), and the "cuticle" of algae is apparently proteinaceous as well (Hanic and Craigie, 1969; McGregor-Shaw, 1971). Microtubules, histones, and lectins are also proteinaceous but are apparently not enzymatic. Protistan flagella are composed of proteins (Bouck, 1972) as are the chloroplast pyrenoids (Holdsworth, 1971). Glycoproteins appear to serve important nonenzymatic functions in biomembranes. Polypeptide units are common as constituents of glycoproteins and mucopolysaccharides that are found in all algae as structural or mucilage components. In all of these aspects, ignorance, often arising from experimental difficulties, precludes any phylogenetic considerations. In the case of the proteins that are functional at the cellular level we are more fortunate. Information, admittedly sparse but nevertheless phylogenetically significant, is available from such functional proteins as ferredoxins, plastocyanins, chlorophyll proteins, biliproteins, and cytochromes. These are discussed in Part II of this chapter.

5.2 PHYLOGENY AND THE STRUCTURE OF ENZYMES: GENERAL AND THEORETICAL CONSIDERATIONS

Evolutionary modification tends to obscure many features by which homologies among biomolecules can be recognized. Other features, however, may be strongly conserved among molecules in many groups of organisms, and are consequently more valuable in phylogenetic reconstruction. In the case of proteins, it can be asked whether common ancestry can be recognized only by similarities in amino acid sequences, or whether other parameters have been at the focus of selective conservation. These other features could include tertiary and subunit conformations, solubility and transport properties, geometry of the catalytic site, and characteristics of allosteric binding sites.

It seems almost self-evident that the geometry of the enzyme's catalytic site, at the very least, will have been conserved to a significant degree. Properties such as size and solubility may also be important for some enzymes, but might be less easily modified by individual mutational events. However, as non-Darwinians argue, it seems likely that some amino acid substitutions would have virtually no effect upon all of the above properties of enzymes. It has been argued (e.g., Fitch and Margoliash, 1967, 1971) that for a given set of proteins a minimal number of residues must be invariant in order for the proteins to retain the same general biological function. In the case of cytochrome *c* this number appears to be between 27 and 30 (out of a total of 104 residues). For those classes of enzymes or proteins (orthologous proteins: Fitch and Margoliash, 1971) for which sufficient sequence data are available, it does appear that nearly identical sequences around the active site are quite common. If this indicates, as it seems to do (see Swain, 1974, for references), that an orthologous set of proteins has evolved from a single ancestral type, the phylogeneticist has at his disposal a powerful tool for determining phylogenies and the time course of evolution (Section 13.2). This idea of a minimal number of invariant residues can also aid in the vexing problem of distinguishing between ancestral homology and convergent evolution (Nolan and Margoliash, 1968). These concepts have, in turn, led to arguments on the subject of whether or not molecular evolution is neutral. Neutral evolution (Kimura and Ohta, 1971; Ohta and Kimura, 1971) is interpreted as a "random fixation of neutral mutants rather than by natural selection." Swain (1974) suggests that this may be true for cytochrome *c* (see Section 2.4.3), but that it is unwise to extrapolate that the "whole genome or even the structural DNA undergoes only neutral changes during the course of evolution." If a change in enzyme activity requires additional DNA in the genome, or gene duplication (Watts and Watts, 1968a,b) or gene fusion (Yourno *et al.*, 1970), molecular evolution is probably not neutral. Amino acid sequence data have of course played an important role in phylogenetic studies on higher organisms. These investigations are currently being expanded to include several protists. It is important to realize that most of the data and ideas in the development of the concepts, and the appropriate arguments pro or con, originated with proteins from higher organisms, especially animals, and to a lesser extent higher plants. Methodological and interpretational problems inherent in studies of this nature have been treated by Boulter (1972), Dayhoff (1972a,b), Holmquist and Jukes (1972), and Peacock and Boulter (1975).

In all of these studies, however, the overriding concern is the detection of homology. Needless to say, the amino acid sequence is important for determining protein homologies.

Perhaps the best available approach to recognizing homology is the determination of protein tertiary structures by X-ray crystallography. Homologous proteins, shown by amino acid sequence data to differ by more than 50%, can still retain remarkably similar tertiary conformations (Blake,

1974; Hartley, 1974). By X-ray crystallography, "families" of homologous enzymes can be recognized in spite of the differing catalytic specificities of members of each "family," and strongly conserved or highly variable amino acid sequences can usually be correlated with tertiary features (Wootton, 1974). There is of course little assurance that enzyme crystal structures bear significant resemblance to their *in vivo* active conformations, but the method is nonetheless relatively direct and seems promising. Difficulties lie not only in performing the necessary experiments and in extrapolation to the living cell, but in quantification and treatment of the data.

Although constraints upon various parameters of enzyme structure doubtless exist, the fact remains that the primary structure is apparently the only parameter encoded directly in the DNA. This has given rise to the hope that techniques can be developed for the prediction of tertiary structure (and thus of catalytic properties?) directly from protein primary structure. There has in fact been some progress in this endeavor, although the day seems far off when the entire calculation can actually be made for a newly isolated protein. Whether or not scientists will ever be able to perform this calculation to any degree of certainty, the fact that it is done routinely in the living cell points out the subtle relationships that are the basis not only of day-to-day life but also of evolution (Conrad, 1974).

In the meantime, however, one must rely upon sequence data (amino acid changes) or minimal mutation distances (Nolan and Margoliash, 1968). These minimal mutation distances are the minimum number of single nucleotide changes needed to alter "the gene or gene segment coding for one protein segment to that coding for the other sequence" (Nolan and Margoliash, 1968). Swain (1974) has summarized the cytochrome *c* data for higher plants, and the different phylogenetic trees that have been constructed. Although these data have been used in different ways, the phylogenetic trees of angiosperm families are quite similar. There are, however, significant differences from the phylogenetic schemes constructed from the classic or morphological approach. This would reinforce the comments of Nolan and Margoliash (1968), echoed by Sussman (1974), that amino acid sequences "could be an independent test of the concepts of evolution because this process at the molecular level need not proceed by the same rules as gross morphology or even micro-molecules (micro-metabolites)," since they are direct translations of the nucleic acid code. Whether or not sequence data are more suited than simple chemical compounds (Nolan and Margoliash, 1968; Turner, 1969) is a matter for conjecture. In our current state of knowledge, however, we still believe that metabolites (and "nonsequence" data) must play an important role in phylogeny construction.

5.3 ENZYME AGGREGATES

It has been found that certain groups of enzymes occur together in the cell as enzyme aggregates. In many cases, it is suspected that the genes for these aggregated enzymes occur adjacently on the DNA, but this is not an absolute requirement for aggregation. The evolutionary significance of aggregation may lie in the increased efficiency of such aggregates, as the entropic factor ("searching out" the proper substrate) is greatly ameliorated.

An explanation for some types of aggregation could be offered by the theory of Horowitz (1945, 1965; see following Section). Gene duplication could have given rise to similar genes adjacent on the chromosome, one of which could eventually be modified by mutation to acquire a related synthetic activity. Cotranscription of these two genes could then presumably result in a functional aggregate. There are, however, some difficulties with this hypothesis: First, not all aggregates are formed from adjacent gene products. Second, this hypothesis does not explain why enzyme aggregates, although known in prokaryotes, are considerably more frequent in eukaryotes. It is not unreasonable to assume that the situation in prokaryotes could be a derived condition, since unique circumstances arising from the nonorganelle unicellular arrangement could have provided less of an evolutionary advantage for the maintenance of such aggregates.

Because the molecular basis for enzyme aggregation is not well understood, it is difficult to assess just how significant a biochemical character it may be. It is likely that it would have been the result of numerous modifications in the DNA, and that there could have been a large number of paths to similar aggregates. For these reasons it would be unwise to assume that apparently similar aggregates in different organisms are necessarily homologous.

5.4 EVOLUTION OF BIOSYNTHETIC PATHWAYS

The organization of individual enzymes into biosynthetic pathways capable of synthesizing complex metabolites is an important feature of modern living systems. Obviously, earlier organisms were on the whole less complex than most modern organisms, and several hypotheses have been offered to explain the evolution of complex biosynthetic pathways.

Perhaps the first such hypothesis was that of Horowitz (1945), who postulated that enzymatic steps in individual pathways have evolved in reverse order of their temporal biosynthetic sequence. The final enzyme in the pathway would then have been the first to have appeared; as its substrate was exhausted from the surrounding "primordial soup," another enzyme would appear that would increase the intracellular levels of that substrate, either by actively transporting it into the cell or by catalyzing its formation from another available substance. Those organisms that were not so able to increase the levels of this important compound or to utilize another would have been severely selected against. This hypothesis appears to be very reasonable for relatively simple molecules in a primordial soup.

If enzymes evolving in this fashion were to utilize related compounds available from the environment, it might further be postulated that the structures of the active sites of these new enzymes would resemble those of the original enzymes in the pathway. Hence the process of gene duplication might well be involved in Horowitzian evolution of biosynthetic pathways. Amino acid sequence homologies have been reported among histones, ferredoxins, cytochromes *c*, haptoglobin β-chains, chymotrypsinogen A, and immunoglobin H-chains (Bauer, 1971), and internal sequence homologies have been noted for numerous proteins (Dayhoff, 1972a; Hall *et al.*, 1972). Unfortunately, insufficient numbers of amino acid sequences from enzymes of a single biosynthetic pathway are available for a better test of the hypothesis. Gene duplication would also be consistent with topological relationships seen in some genes of enzymes from single biosynthetic pathways (Horowitz, 1965).

A somewhat extended version of the above hypothesis allows for the participation of modified genes coding in other pathways to participate in the evolution of a new biosynthetic pathway. In the well-known research of Lerner *et al.* (1964) and Wu *et al.* (1968), a modified D-arabitol transport system was produced that was able to actively transport xylitol into the cells of *Aerobacter aerogenes*. Xylitol metabolism was brought about by modifications in the control and probably the structure of a ribitol dehydrogenase. Modification of the control of histidinol utilization has been reported in *Pseudomonas aeruginosa*, and apparent gene duplication and modification have been found to operate in histidinol utilization by *Arthrobacter histidinolovorans* (Dhawale *et al.*, 1972).

In a recent discussion, Hartman (1975) has commented that the Horowitzian viewpoint, the origin of autotrophy from heterotrophy (or that the first organism was simple in a complex environment), is the dominant view held today. Hartman (1975) has argued for an autotrophic origin that requires clays, transition metals, disulfide, dithiols, cyanide, and ultraviolet light. One of the objections to the Horowitzian viewpoint is the possibility that a complex primordial soup did not exist. Building on this idea, and the suggestion that life might have evolved in a secondary atmosphere of CO_2,

N_2, and H_2O which replaced the primary atmosphere of CH_4, NH_3, and H_2, Hartman has proposed an origin of metabolism built around the citric acid cycle.

The data so far collected suggest that Horowitzian evolution of biosynthetic pathways may explain some observations, but that apparently unrelated enzymes may also participate. Hegeman and Rosenberg (1970), Clarke (1974), and Hartley (1974) have reviewed many of these investigations. It is possible that Horowitzian evolution was more prevalent during the very early evolutionary stages, when the environment was relatively rich in organic molecules and when there were relatively few other enzymes to call upon. Later evolutionary events would then be less likely to have involved this mechanism, although there are perhaps certain resemblances between modern parasitic and saprobic ways of life and the situation in the primitive soup.

The role of gene duplication in this evolution can be accommodated by either the Horowitzian or the cooperative type of evolution. In the former, gene duplication would have been a more direct influence, whereas in the latter, gene duplication might have given rise to enzymes whose functions could have diverged considerably before eventually being recruited at a later stage.

5.5 EVOLUTION OF METABOLIC ENERGY PRODUCTION

As described above, evolution of biosynthetic pathways after the model of Horowitz (1945, 1965) seems to provide a rational method by which primitive forms of life could have extracted energy from a carbon-rich environment. Events following the depletion of the primordial soup are more difficult to envision, and phylogeneticists usually attempt to reconstruct likely series of events by arranging hypothetical ancestors to various modern groups of protists in a scheme of increasing metabolic complexity (Hall, 1971). Thus anaerobic organisms capable of deriving energy only from substrate-level phosphorylation (as is true of some modern clostridia and lactobacilli) are considered to have been ancestral to anaerobes capable of electron transport as well (as are other clostridia and modern streptococci), which in turn may have been ancestral to cytochrome-containing anaerobes utilizing simple electron acceptors such as sulfate, carbonate, or nitrate. Broda (1970, 1971a,b, 1975a), however, has added a few cautions. Although a clostridium-type fermentation may be the oldest, these organisms are occasionally capable of glycolysis, whereas lactic acid bacteria can be microaerophilic. The earliest cells probably obtained their energy through redox fermentations of materials in the primordial soup. Depletion of these substrates brought about the development of photosynthesis. This would have coincided with the elaboration of the anaerobes' cytochromes into bacterial chlorophylls. However, in connection with these cytochrome-containing anaerobes Broda (1970, 1971a,b, 1975a) has pointed out two weaknesses. He notes that sulfate (oxidized sulfur) probably existed only in minimal quantities prior to the appearance of the photosynthetic sulfur bacteria, and that it is almost inconceivable that nitrate (oxidized ammonia-nitrogen) antedated oxygen (Broda, 1975a–c). Hall (1971, 1973b) and Egami (1973, 1976) have argued, however, that nitrate preceded oxygen. Carbonate reducers have not been shown to possess cytochromes. By these arguments, therefore, anaerobic respirers could not occupy such an intermediate evolutionary position. As a result Broda argues for an evolutionary sequence of

Fermentation → photoorganotrophy → photolithotrophy → "plant" photosynthesis → aerobic respirers (or oxidative phosphorylators)

The first three transitions would have been brought about by a shortage of fermentable substrates, of organic carbon, and of inorganic electron donors (other than water). This scheme would surprisingly have the nonsulfur photosynthetic bacteria preceding the sulfur photosynthetic bacteria. Interestingly enough, the former can live aerobically in the dark, and *Chloroflexus*, with chemical and morphological features of green sulfur bacteria and blue-green algae (Pierson and Castenholz, 1974a,b), can grow aerobically in both light and dark (chemoorganotrophically) or anaerobically in the light (photoorganotrophically). Even more recently (Cohen *et al.*, 1975), certain blue-green algae have been shown capable of adapting to an anaerobic photosynthesis using sulfide as the electron donor, thus resembling the photolithotrophic photosynthetic sulfur bacteria. One might expect the photosynthetic sulfur bacteria, as obligate anaerobes, to have preceded the photoorganotrophs. *Chlorobium* is nominally an obligate photolithotroph, although Kelly (1974) has suggested it may be photomixotrophic in nature. Data from metabolite considerations (Sections 9.4 and 10.2) would suggest that the green sulfur bacteria at least are a dead end offshoot from the mainstream. Hartman (1975), however, suggested that the "early metabolism" deriving from an autotrophic origin would have involved CO_2 and nitrogen fixation and ultraviolet light. A possible source of free energy for metabolism would have been a ferrous/ferric and sulfhydryl/disulfide oxidation–reduction. Sulfide, ferredoxins, and nitrogen fixation figure prominently in prokaryote modes of life.

Organisms capable of aerobic respiration may have evolved directly from nitrate reducers,* or from photosynthetic bacteria, long before the atmosphere became oxygen-rich from biological photolysis of water. Many modern aerobes, including organisms classified in the genera *Bacillus, Corynebacterium, Escherichia, Micrococcus, Proteus, Pseudomonas, Staphylococcus,* and *Thiobacillus,* are facultative nitrate respirers. This may be an adaptive feature because of the question of the presence of nitrate prior to oxygen production. As the oxygen content of the atmosphere increased further, major modern groups of aerobes and autotrophs could have evolved in relation to a variety of local environmental exigencies.

Finally, one should note Hartman's comments about coenzymes. These are an essential consideration for the evolution of metabolism, and Hartman (1975) has speculated that coenzymes came before enzymes and, in fact, must have had their own evolutionary history.

5.6 BIOCHEMICAL METHODS IN THE PHYLOGENETIC STUDY OF PROTEINS

Different parameters have been used in the study and comparison of proteins; some of these parameters may be more phylogenetically useful than others. Biochemical data concerning proteins are subject to certain limitations imposed by the experimental techniques used in protein biochemistry. Limitations other than those listed here will become evident upon examination of the primary literature.

1. Demonstration of an enzymatic *activity* in an extract is not as dependable a biochemical datum as is the purification and characterization of the actual enzyme (or enzymes) responsible for this catalytic activity.
2. Many enzymes are present in amounts insufficient for amino acid sequencing; therefore, it may be necessary to characterize the enzymes in less elegant manners. Amino acid sequences published on the basis of microgram quantities of enzymes are more susceptible to experimental errors than are sequences derived from milligram or centigram quantities of purified enzyme (this is particularly a problem in certain higher plant cytochromes).
3. The characteristics of a given protein may depend upon the age, condition, or developmental stage of the cells from which it was extracted. Different isoenzymes of a given enzyme may be present at different develop-

* Oxygen is a competitive inhibitor of nitrate reductase, and the respiration of oxygen and of nitrogen involves many of the same electron carriers (Hall, 1971).

mental stages, or in different regions of the organism, or in certain subcellular organelles, or throughout the entire cytoplasm.

4. The experimental determination of the positions of glutamine and asparagine residues in proteins is particularly difficult, as these residues are subject to change into glutamate and aspartate, respectively, during certain extraction and handling procedures. The statistical treatment of amino acid sequences necessitates the proper assignment of these residues before firm phylogenetic facts can be derived.

5. Because examined enzymes are often derived from cultured organisms or from mutant strains of bacteria, there is the possibility that slight differences exist between the determined protein characteristics and those of the wild-type organisms.

In all, protein biochemistry is in a relatively advanced state compared to the situation for other biological molecules, especially for other biopolymers. But doubtless some of the reports found in the literature are erroneous, clouding the conclusions reached here and elsewhere.

PART II: HEME PROTEINS, METALLOPROTEINS, AND HISTONES

5.7 CYTOCHROMES

Cytochromes comprise a class of moderate-sized heme proteins (molecular weights about 10,000 daltons) that function in electron transport and in oxidation-reduction reactions both in prokaryotes* and in eukaryotes. In view of their molecular variability, moderate size, and widespread distribution, they may prove to be of considerable interest in protistan phylogeny, as they have in the phylogeny of higher organisms.

Because they are involved in a variety of reactions in organisms, it is not surprising that cytochromes are structurally diverse. This diversity was first noticed in their spectral properties, molecular weights, and redox potentials. More recently, amino acid sequence studies have borne out these early conclusions. Nonetheless, all investigated cytochromes possess strong tertiary structural similarities, and about 30 invariant amino acids (Section 5.2).

In order to organize the data, cytochromes are usually classified into four major categories on the basis of their spectral characteristics: cytochromes

* Cytochromes appear to be absent from a few gram-positive bacteria, e.g., *Lactobacillus* spp., *Clostridium* spp. (Bishop *et al.*, 1962), although some *Lactobacillus* species can utilize exogenously supplied heme (Bryan-Jones and Whittenbury, 1969).

a, cytochromes *b*, cytochromes *c*, and cytochromes *d*. The cytochromes *c* have been most intensely studied, and will be discussed here. Cytochromes *b*, at first thought to be confined to oxygen-evolving organisms and Athiorhodaceae (Olson, 1970), have recently been detected in green and purple sulfur bacteria as well (Fowler, 1974; Knaff and Buchanan, 1975). Further information on other cytochromes is available in recent books and reviews by Okunuki *et al.* (1968), Keilin (1970), Kamen (1973), and Lemberg and Barrett (1972).

The cytochromes *c* have themselves been subdivided into many categories on the basis of physical and chemical properties. Although some of these subdivisions represent groups of homologous proteins, other subdivisions doubtless contain proteins of no detectable primary sequential homologies. The classification of Lemberg and Barrett (1972), similar to one proposed by Kamen (1973), will be followed here.

1. Cytochromes *c*: the classic, soluble mitochondrial cytochromes of fungi and animals.
2. Cytochromes c_1: mitochondrial membrane-bound cytochromes of animals, possibly present in higher plants and in some bacteria. Closely homologous with cytochromes *c*.
3. Cytochromes c_2: soluble enzymes of nonsulfur purple bacteria, highly purified from *Rhodospirillum rubrum* (cytochrome c_{550}). Homologous with cytochrome *c* (see Ambler *et al.*, 1976).
4. Cytochrome c_3: soluble cytochrome from *Desulfovibrio desulfuricans*.
5. Cytochrome c_4: soluble enzyme from *Azotobacter vinelandii* (cytochrome c_{551}) and *Pseudomonas* spp.
6. Cytochrome c_5: soluble enzyme from *A. vinelandii* (cytochrome c_{555}).
7. Cytochromes c_6: "cytochromes *f*" in Cyanophyceae, eukaryotic algae, and higher plants. Chloroplastic in higher plants and probably in eukaryotic algae; only weakly homologous with cytochrome *c* by amino acid sequences, and with somewhat similar tertiary structures (Timkovich and Dickerson, 1973).
8. Cytochromes c' and c'': soluble cytochromes from some bacteria, including purple photosynthetic bacteria (Meyer *et al.*, 1975).
9. Flavocytochromes *c*: in *Chromatium* spp. and in *Chlorobium thiosulfatophilum*.
10. Other cytochromes, including cytochromes from *Mycobacterium* spp., cytochromes c_{551} from *Pseudomonas* spp., and other poorly characterized bacterial cytochromes.

Comparative studies on these cytochromes have centered on the available amino acid sequence data, but other characteristics have also been utilized for phylogenetic studies. These include redox potentials, methylation of lysine residues (Ramshaw *et al.*, 1974), molecular weights, and electronic properties of the chelated iron atoms (see A-2 in Kamen, 1973). The three-dimensional geometry of the active site could also be of considerable interest in phylogenetics. Yet the parameter most easily treated in a quantitative sense, and the one that has actually been examined most closely for phylogenetic information, is the primary structure. Amino acid sequences for the cytochromes *c* have been compiled by Dayhoff (1972a), and new ones are regularly listed in the *Journal of Molecular Evolution*.

The moderate rate of amino acid substitutions in the cytochromes *c* allows phylogenetic relationships among representatives of the biological kingdoms (*sensu* Whittaker, 1969) to be reconstructed from amino acid sequence data (Boulter, 1972; McLaughlin and Dayhoff, 1973; Dayhoff *et al.*, 1975; Fitch, 1976). Of the 105 possible phylogenetic permutations of the prokaryotes, fungi, *Crithidia* spp., *Euglena gracilis* Z, higher plants, and higher animals, the permutation requiring the fewest amino acid substitutions is shown in Figure 1. One difficulty with this approach to phylogeny is that the evolution of the cytochromes may not have been parsimonious with respect to amino acid (or codon) replacements. On a more practical level, the data base for "moneran" and "protistan" sequences is very limited.

The proposed phylogenetic relationship of *E. gracilis* Z, *C. fasciculata*, and *C. oncopelti* cytochromes *c* (Figure 1) deserves special comment. Amino acid sequence data suggest that these flagellates have diverged from a common ancestral protist more recently than from lines leading to other kingdoms. Assuming an extrachloroplastic location for the cytochrome *c* gene, these data require either that there were at least two independent origins of the chloroplast (see Section 3.3) or that *Crithidia* spp. arose from a photosynthetic ancestor. Interestingly, *E. gracilis* Z and *C. oncopelti* possess further, unusual *c*-type cytochromes with only a single cysteine residue each (Pettigrew *et al.*, 1975). Additional cytochrome *c* amino acid sequences from other euglenoids and from protozoa would be most desirable.

Among the fungi, the cytochrome *c* sequence from the basidiomycete *Ustilago sphaerogena* appears equally dissimilar to sequences from five ascomycetes and imperfect fungi. The latter five sequences can be further divided into two groups. Ascomycetes are represented in each group, indicating that there probably was no phylogenetic divergence *en masse* between ascomycetes and imperfect fungi. Instead, it is more likely that individual fungi independently lost sexuality, becoming "imperfect" (as we chauvinistically term the condition). It must be remembered, however, that

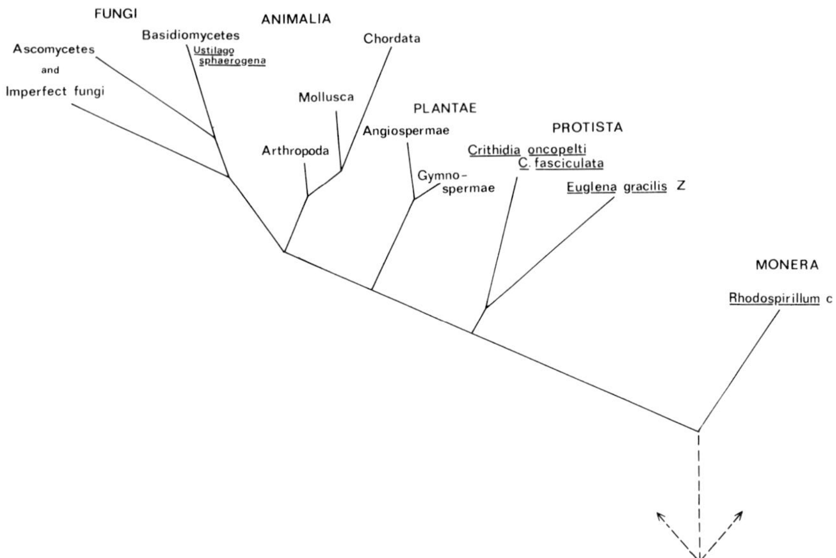

Figure 1. Phylogenetic relationships among protists, based on amino acid sequence data from cytochromes *c*. G. E. Tarr (personal communication) has recently determined the amino acid sequences in cytochrome *c* from *Tetrahymena pyriformis* and *Physarum polycephalum*. The latter cytochrome is more closely related to the *Crithidia–Euglena* cytochromes than to other known cytochromes. The *T. pyriformis* cytochrome appears to have diverged somewhat earlier than the *Crithidia–Euglena–Physarum* cytochromes. After Figure 4 from Dayhoff *et al.* (1975). Reproduced by permission of Springer-Verlag, Inc., and the authors.

the imperfect fungi are grouped in a form taxon, the Deuteromycetes, because sexual reproduction is unknown. This could be due to a "lost sexuality," in which case the above cytochrome data would suggest an ascomycete origin. In contrast to this, there is always a possibility that sexual reproduction simply has not been observed. Although there may be a morphological resemblance to the Ascomycetes in asexual reproduction, this is no guarantee that the individuals concerned are in fact Ascomycetes. Amino acid sequence data of cytochromes, and other proteins, may well be the solution to the systematic problem posed by the imperfect fungi. Errors in fungal sequence data collected by Dayhoff (1972a) have been corrected by Lederer and co-workers (Lederer, 1972; Lederer *et al.*, 1972; Lederer and Simon, 1974).

Cytochromes *f* (cytochromes c_6) have been sequenced in a few protists, although not all these sequences have been finalized. There appears to be homology among amino acid sequences from *Anabaena variabilis*, *Plec-*

tonema boryanum, Synechococcus (Aitken, 1976), *Spirulina maxima* (Ambler and Bartsch, 1975), *Porphyra tenera* (R. P. Ambler and R. G. Bartsch, unpublished), *Monochrysis lutheri* (Laycock, 1972), *Alaria esculenta* (Laycock, 1975), and *Euglena gracilis* Z (Pettigrew, 1974), but a smaller cytochrome *f* has been reported from *Bumilleriopsis filiformis* (Lach *et al.*, 1973). The sequence divergence between cytochromes *f* from *A. esculenta* and *M. lutheri* is greater than might be expected; this may represent a very ancient phylogenetic divergence, or may reflect differences in selective pressures encountered by a unicellular organism (*M. lutheri*) and a macrophyte (*A. esculenta*), or both. Although cytochrome *f* functions in photosynthetic electron transport in many protists, the location of its gene is not known.

5.8 HEME PROTEINS OTHER THAN CYTOCHROMES

Catalase (EC 1.11.1.6) and peroxidase (EC 1.11.1.7) are widely distributed enzymes containing cyclic tetrapyrroles as prosthetic groups. Neither of these enzymes has been reported from lactobacilli or clostridia, organisms that also appear to lack ubiquinones, vitamin K_2, and cytochromes (Margulis, 1969; De Ley and Kersters, 1975). If these absences are not due to secondary losses in biosynthetic capacity nor to insensitivity in the methods of assay, this distribution could indicate the divergence of all other organisms from a common ancestor existing after the subsequent appearance of later stages in the δ-aminolevulinic acid pathway. Secondary losses of catalase activity have presumably occurred among some Charophyceae (Silverberg and Sawa, 1973).

Interestingly, clostridia do possess a sulfite reductase (EC 1.8.99.2), hence presumably a siroheme prosthetic group synthetic capability. Siroheme, a derivative of uroporphyrinogen-III, is presumably present in all sulfite reductases (Murphy and Siegel, 1973; Peck *et al.*, 1974). Thus, at least the early steps in porphyrin biosynthesis occur in clostridia.

Hemoglobin, with an iron-containing porphyrin group, exhibits a limited distribution among the protists. It has been reported in some Ascomycetes, *Paramecium caudatum, Tetrahymena pyriformis*, and in root nodules containing *Rhizobium* spp. (Keilin and Ryley, 1953; Willmer, 1974).

5.9 PHYCOBILIPROTEINS

These proteins are photoreactive accessory pigments found in the Cyanophyceae, Rhodophyceae, and Cryptophyceae, where they comprise a major part of the "antenna" pigment system of the photosynthetic apparatus. The term phycobiliprotein is perhaps an arbitrary designation. Phytochrome, also a biliprotein like the above, is also found in (at least) the Rhodophyceae and Chlorophyceae (Bennett and Siegelman, 1977). Some workers exclude it from the phycobiliproteins because it is not an energy trapping pigment, but rather a "trigger" or photomorphogenic pigment. Very little is known about the chemistry of this protein. In the absence of data indicating a homology with the phycobiliproteins, only limited consideration will be given to this type of biliprotein.

The distinguishing feature of the biliproteins is the presence of a linear, or open chain tetrapyrrole (Siegelman *et al.*, 1968; Rüdiger, 1971, 1975; Bennett and Siegelman, 1977), covalently bound (cf. Chapman, 1973; Köst *et al.*, 1975) to the protein (Table 8). A second distinguishing feature of these proteins is their location on the chloroplast thylakoid in distinct macromolecular aggregates, or phycobilisomes (Gantt and Conti, 1966). The Cryptophyceae are an exception in that the biliproteins appear to be located in intrathylakoidal spaces (Gantt *et al.*, 1971).

Phycobiliproteins are usually compared on the basis of their absorption spectra, which are related to the number, nature, and environments of the bilin prosthetic groups in the molecule. On this basis at least six types of phycobiliproteins have been recognized in the Cyanophyceae and the Rhodophyceae. The Cryptophyceae contain as many as six further phycobiliproteins, all of which differ from those of other algae in spectral, aggregational, and immunological properties (Table 9). Despite this multiplicity of phycobiliproteins, only two distinct bilin chromophores, phycocyanobilin and phycoerythrobilin, have been identified. The problem of "phycourobilin" as a third discrete chromophore or a bilin–protein interac-

TABLE 8

Bile Pigment Nomenclature

Common groupings	Groupings by numbers of methine bridges	Examples
Verdins	Bilatrienes	Phytochrome (phytobilin); phycocyanobilin; pigments in invertebrates, shells, and eel blood
Violins	Biladienes	Bilirubin; phycoerythrobilin; some mollusk shell pigments; *Aplysia* ink, *Haliotis* shell pigments
Urobilins	Bilenes (monoenes)	Jaundice pigment
Inogen	No methine bridges	Human bile pigment excretion

TABLE 9

Distribution of Phycobiliproteins in the Algae[a]

Organism	Phycobiliproteins[b]								
	1	2	3	4	5	6	7	8	9
Cyanophyceae	+	[c]	+			+	+	[d]	
Cyanellae[e]	+		+						
Rhodophyceae									
Bangiophycidae	+	+	+	+	+			+	
Florideophycidae	+	+	+	+				+	
Cryptophyceae									+

[a] Data for cyanellae: Chapman (1966, 1973); Cryptophyceae: O'hEocha (1965); Cyanophyceae and Rhodophyceae: Chapman (1973), Chapman *et al.* (1968), O'hEocha (1965), Glazer (1976), Bogorad (1975), Bennett and Siegelman (1977). Older references are not treated (e.g., Allen, 1959).

[b] Phycobiliprotein types: 1, C-phycocyanins; 2, R-phycocyanins; 3, *allo*-phycocyanins; 4, B-phycoerythrins; 5, b-phycoerythrins; 6, C-phycoerythrins; 7, *allo*-phycocyanins B; 8, R-phycoerythrins; 9, phycocyanins and phycoerythrins different from those in Cyanophyceae and Rhodophyceae (see text).

[c] A phycoerythrobilin-phycocyanobilin mixed biliprotein has recently been isolated from *Anabaena* (Bryant *et al.*, 1976).

[d] Recently, Fujita and Shimura (1974) have claimed the presence of R-phycoerythrin in *Trichodesmium thiebautii*, based on spectral assignment. Phycoerythrins (especially R-phycoerythrin) are very labile, and the fact that these cells were stored at −20°C prior to extraction raises questions about the degree of denaturation. In the absence of conclusive proof to the contrary, we will regard this as a C-phycoerythrin type molecule.

[e] Cyanellae in *Cyanophora paradoxa* and *Glaucocystis nostochinearum*.

tion has not been resolved (Chapman, 1973; O'Carra and O'hEocha, 1976). The chromophores do not reveal any phylogenetic implications, since both types are found in all three classes. The appearance of these tetrapyrroles in aerobic photosynthesizers undoubtedly relates to the need for oxygen in the cleavage of a porphyrin ring by heme oxygenase to produce the "linear" bilin. Amino acid sequences of the apoprotein moieties are beginning to provide the molecular detail necessary for phylogenetic comparison of phycobiliproteins. Sequence homology has been observed among partial amino-terminal sequences of both α and β subunits of phycobiliproteins from a number of strains of Cyanophyceae, two species of Rhodophyceae, and one species of Cryptophyceae (Williams *et al.*, 1974; Brown *et al.*, 1975; Frank *et al.*, 1975; Harris and Berns, 1975; Troxler *et al.*, 1975; Glazer *et al.*, 1976). Unfortunately, primary sequence data are not sufficiently complete to provide details of the phylogeny of these organisms, but nevertheless some general observations can be made. Within the Cyanophyceae and

Rhodophyceae, the α-chain of C-phycocyanin shows very little variation and is very conservative with regard to substitutions, suggesting a common evolutionary origin of cyanophycean and rhodophycean C-phycocyanin (Williams *et al.*, 1974; Harris and Berns, 1975; Troxler *et al.*, 1975). These workers (see Glazer, 1976) have also commented on the similarity between the α- and β-chains, suggesting that one may have arisen from the other by gene duplication from a common ancestral gene. Comparisons of homologies have also been extended to C-phycoerythrin and B-phycoerythrin, and the similarities (admittedly based on limited data) would suggest that both phycoerythrins and phycocyanins evolved from a common gene source (Harris and Berns, 1975; Troxler *et al.*, 1975). Similarities also occur between the α- and β-chains of *allo*-phycocyanin and in turn to C-phycocyanin (Brown *et al.*, 1975). These observations, in concert with the known facts that all Cyanophyceae and Rhodophyceae appear to possess *allo*-phycocyanin (Chapman, 1973) which mediates the transfer of energy from the other biliproteins to chlorophyll *a* (Gantt and Lipschultz, 1973; Lemasson *et al.*, 1973), would suggest that *allo*-phycocyanin may be the antecedent to phycocyanin and the phycoerythrins. In the sequence

allo-phycocyanin → C-phycocyanin → phycoerythrins

there is an increase in the number of chromophores per subunit. This sequence parallels the energy transfer sequence

Phycoerythrin → phycocyanin → *allo*-phycocyanin → chlorophyll *a*

and the increasing complexity and arrangement of the phycobilisome (Glazer and Hixson, 1975). This would appear to parallel the evolutionary sequence of the phycobiliproteins. Chapman (1973) has commented that the simple coccoid Cyanophyceae are generally characterized by C-phycocyanin and *allo*-phycocyanin and the lack of C-phycoerythrin, whereas the advanced filamentous Oscillatoriaceae are frequently characterized by a preponderance of C-phycoerythrin, and the Rhodophyceae in turn by a preponderance of phycoerythrins and R-phycocyanin (with the exception of the non-phycoerythrin-containing *Cyanidium caldarium, Porphyridium aerugineum*, and *Asterocytis ornata*). It is very tempting to propose the phylogenetic sequence given in Figure 2 and to suggest a direct relationship between the Cyanophyceae and the red algal chloroplast.

The Cryptophyceae are a phylogenetic enigma. The presence of the bilin chromophores themselves probably has little significance beyond indicating perhaps a distant connection to the Cyanophyceae. Immunological comparisons of cryptophycean biliproteins reveal that these proteins are quite different from those of the Cyanophyceae and Rhodophyceae (Berns, 1967; Bennett and Bogorad, 1973; Glazer *et al.*, 1971), while these same studies

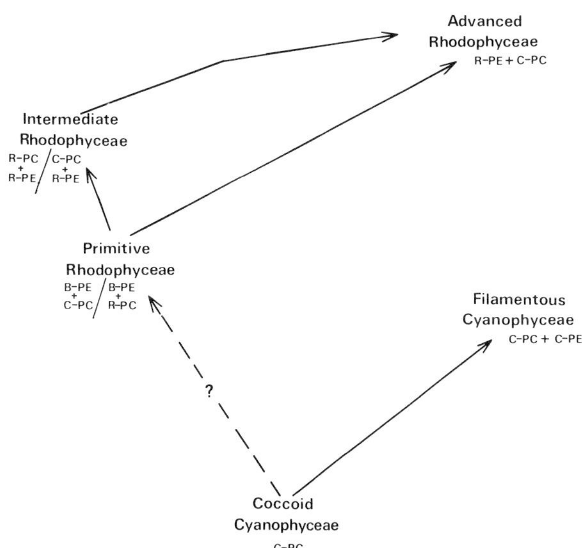

Figure 2. Phylogenetic relations of Cyanophyceae and Rhodophyceae, based on phycobiliprotein biochemistry. C-PC, C-phycocyanins; C-PE, C-phycoerythrins; B-PE, B-phycoerythrins; R-PC, R-phycocyanins; R-PE, R-phycoerythrins. *allo*-Phycocyanin is considered to be present in all biliprotein-containing Cyanophyceae and Rhodophyceae.

reveal extensive homologies between biliproteins of the same class from both eukaryotes and prokaryotes. The minimal sequence data (Harris and Berns, 1975) of a cryptomonad phycocyanin would support this evolutionary isolation and suggests that the cryptophycean biliproteins may have had a different evolutionary origin or that they developed independently of any constraints imposed by the structure and function of biliproteins in phycobilisomes. In this regard, one should remember that cryptophycean biliproteins appear to be located within the thylakoid (Gantt *et al.*, 1971). In the absence of comparable data on biochemistry and taxonomic distributions, we will avoid any phylogenetic speculation based on phytochrome.

5.10 FERREDOXINS

Ferredoxin (Fd) is an iron–sulfur oxidoreductase that functions in bacterial hydrogenase systems, nitrogen reduction, hydroxylations, photosynthetic electron transport, and respiratory electron transport. Although Fd activity is often associated with the chloroplast in photosynthetic eukaryotes, the gene coding for Fd is located in the nuclear genome, at least in tobacco (Kwanyuen and Wildman, 1975). This observation is consistent with either an endosymbiotic or a gradual origin of tobacco chloroplasts, but is more easily rationalized by (and is indeed predicted by) a gradualistic origin of eukaryotes from a cyanophycean-like ancestor.

Amino acid sequences are available for soluble ferredoxins from the protists *Clostridium acidi-urici, C. butyricum, C. pasteurianum, C. tartarivorum, C. thermosaccharolyticum, Clostridium* strain M-E, *Desulfovibrio gigas, Peptostreptococcus elsdenii, P. aerogenes* (= *Micrococcus aerogenes*), *Chromatium* D, *Chlorobium limicola, Spirulina platensis, S. maxima, Scenedesmus quadricauda,** and several higher plants (Hall *et al.*, 1972, 1973a,b, 1975a,b; Tanaka *et al.*, 1974, 1975a,b,c, 1976; Yasunobu and Tanaka, 1974; Wada *et al.*, 1975a,b), and partial sequences are available for Fd from *Aphanothece sacrum* (Hase *et al.*, 1976; Wada *et al.*, 1974) and *Porphyra umbilicalis* (Andrews *et al.*, 1976). Membrane-bound Fd's are present in many protists, but have received relatively little attention.

Ferredoxins are largely composed of only nine amino acids, all of which have been synthesized in Miller-type "primitive earth" experiments (Section 2.3), and most of which have been found in meteorites and in lunar soil (Hall *et al.*, 1973b). These facts, coupled with the low redox potentials of Fd's (near that of molecular hydrogen), the widespread distribution of Fd's, and the involvement of Fd in basic cellular energetics, have been taken to suggest that Fd's played an important part in the origin of living systems.

Homology among the ferredoxins has been shown by comparison of their primary structures. Moreover, amino acid sequences of clostridial Fd's give strong evidence for two internal homologous regions of twenty-five or twenty-six residues, suggesting that these proteins were formed by gene duplication. Lipmann (1971) has suggested that each of these twenty-five-residue segments may itself have arisen from smaller units, a conclusion that may be supported by the observed molecular weight distribution of Fd's (Table 10). The *Chromatium* D Fd could have arisen by a triplication of the gene coding for the twenty-five residue polypeptide, and algal and higher-plant Fd's could have evolved by gene duplication from a fifty-residue molecule.

The blue-green algal Fd's differ from Fd's in other prokaryotes in their amino acid compositions, ultraviolet absorption spectra, circular dichroism spectra (dependent upon the tertiary structure of the protein in the region of the chromophore), electron paramagnetic resonance spectra (also determined by fine structural characteristics), and Mössbauer spectra

* Species identified by one of us (M.A.R.) from description and photograph kindly supplied by Dr. Hiroshi Matsubara.

TABLE 10

Properties of Protistan Ferredoxins[a]

Bound Fe + S per molecule	Apoprotein molecular weight (approximate) in daltons							
	6,000	7,000	8,000	9,000	10,500	12,500	14,500	24,000
8 Fe + 8 S	1[b]	2		3			4	
4 Fe + 4 S	5		6	7			8	
2 Fe + 2 S				9	10	11		12

[a] We are grateful to Drs. D. O. Hall and K. K. Rao for advice concerning preparation of this table.

[b] Numbers 1 through 12 refer to the following groups of enzymes: (1) *Clostridium* spp., *Peptococcus aerogenes, P. elsdenii, Veilonella alcalescens,* etc., ferredoxins (Fd's); (2) *Chlorobium limicola* Fd; (3) *Chromatium* D (light-grown) Fd, *Rhodospirillum rubrum* Type I Fd (membrane-bound); (4) *Azotobacter vinelandii* Type I Fd; (5) *Desulfovibrio gigas, Spirochaeta auriantia* Fd's; (6) *Bacillus stearothermophilus* Fd, higher plant chloroplastic membrane-bound Fd's; (7) *Bacillus polymyxa* Types I and II Fd's; "high potential iron protein" (HiPIP) from *Chromatium* D and *Thiocapsa pfennigii*; (8) *Rhodospirillum rubrum* Type II Fd; (9) *Rhizobium japonicum* Fd; (10) soluble Fd's from Cyanophyceae (*Aphanothece sacrum, Microcystis flos-aquae, Nostoc* strain MAC (Types I and II Fd's), *Spirulina maxima, S. platensis,* and species of *Anabaena, Anacystis, Phormidium,* and *Tolypothrix*), eukaryotic algae (species of *Botrydiopsis, Bumilleriopsis, Chlamydomonas, Chlorella, Cladophora, Cyanidium, Euglena* (Johnson *et al.*, 1968), *Navicula, Porphyridium. Porphyra,* and *Scenedesmus*). and all investigated higher plants; a similar Fd has been reported in *Agrobacterium* sp. (Hall *et al.*, 1975c); immunologically similar where investigated (Tel-Or *et al.*, 1975); (11) *Escherichia coli* and *Pseudomonas putida* Fd's; adrenodoxin; (12) *Clostridium pasteurianum* "EPR Protein."

(examining the electronic state of the bound iron-57 atoms). In all these characteristics the cyanophycean protein resembles those of higher plants. However, the primary structues of blue-green algal Fd's show considerable divergence not only with respect to Fd's from higher plants but also among themselves (Wada *et al.*, 1975a). Using the accumulated data, Hall *et al.* (1975b) proposed an evolutionary development of the ferredoxins from

Anaerobic heterotrophs → green sulfur bacteria → red sulfur bacteria → sulfate-reducing bacteria → "plant" photosynthesizers and aerobic respirers

This scheme is somewhat different from that proposed by Broda (1971a,b, 1975a) (Section 5.5) and that indicated by micrometabolites. However, the ferredoxin data for the photosynthetic bacteria are very meager, and more information may well bring about revisions, especially with regard to the positioning of one or more groups of photosynthetic bacteria as evolutionary dead ends.

5.11 METALLOPROTEINS OTHER THAN FERREDOXINS

Although a number of metalloproteins are known, only two have been examined to an extent that allows any phylogenetic discussion. These are plastocyanin, a key Cu-protein constituent of the photosynthetic electron transport chain; and superoxide dismutase, a Mn- or Cu-Zn enzyme responsible for the dismutation of the hydroxyl radical. Although the study of these two proteins is in its infancy, it is to be expected that the future will see the evolutionary "use" of these proteins to the same extent that cytochromes, biliproteins, and ferredoxins are being used.

5.11.1 Plastocyanin

Although most of the work has involved higher plants, plastocyanin from *Anabaena variabilis* (Aitken, 1975), *Chlorella fusca* (Kelly and Ambler, 1974), and *Plectonema boryanum* (Aitken, 1976) has been isolated and sequenced. When compared with plastocyanin from such higher plants as spinach (Scawen *et al.,* 1975) or French bean (Milne *et al.,* 1974), considerable homologies are found. Admittedly the sequence similarity decreases as one advances from prokaryote through *Chlorella* to higher plant (variation dependent upon starting alignment of sequence):

Anabaena: *Chlorella*	52 or 55	Amino acids similar
Chlorella: Higher plant	52–58	Amino acids similar
Anabaena: Higher plant	41–45	Amino acids similar

However, Aitken (1975) has pointed out that the sequence is nonetheless highly conserved and that the "similarity in amino acid sequence between the prokaryote plastocyanin and eukaryote plastocyanin is too great to reasonably suggest convergent evolution."

5.11.2 Superoxide Dismutase

This enzyme (Fridovich, 1974b,c), which appears to be quite universal, has been examined in a number of protists (Asada *et al.,* 1975; Lumsden and Hall, 1974, 1975a; Misra and Keele, 1975) and has been found even in obligate anaerobes (Hewitt and Morris, 1975) and in facultative anaerobes (Lindmark and Müller, 1974) and yeast (Ravindrath and Fridovich, 1974). This enzyme is either an Fe-Mn type (prokaryote) or Cu-Zn type (eukaryote). Lumsden and Hall (1975b) have suggested a dual role as water splitter in aerobic photosynthesis and hydroxyl radical detoxifier for a prokaryote Mn-enzyme. In an interesting proposal, they suggest that first a symbiosis between a primitive aerobic bacterium and a blue-green alga occurred to give a cyanella-like organism, which was later followed by a second symbiosis between the cyanome and an aerobic prokaryote to give a rhodophycean-like alga. Following from Hall (1971) they argue that herein lies an explanation for the nuclear coding for mitochondrial proteins.

5.12 HISTONES

Histones comprise a class of incompletely characterized proteins present in the nuclei of many, but not all, eukaryotes (Table 11). Along with other nuclear proteins, they are thought to be involved in the control of gene transcription, a field under active investigation at the present time.

Because histones contain numerous basic amino acid residues, they can be extracted at low pH's (usually under pH = 2.1). Although this method provides relative ease of extraction, it is also likely to compromise some of the tryptophan, glutamine/glutamate, and asparagine/aspartate assignments. Methionine, cysteine, and tryptophan residues are scarce in these extracted histone fractions. Five major classes of histones have been recognized by Dayhoff (1972a); these differ in amino acid composition, electrophoretic mobilities, evolutionary amino acid substitution rates, and taxonomic distributions. Amino acid sequences are not known for any protistan histone fractions, but they are likely to be of interest in protistan phylogeny. More information on histones is available in the recently published monograph by Hnilica (1973).

It has been suggested that the transition from prokaryotes to eukaryotes coincided with the evolution of histones or related basic nuclear proteins. Histones are indeed absent from investigated bacteria, Cyanophyceae, and chloroplasts, although basic proteins not homologous with known histones have been found in *Thermoplasma acidophilum* (Searcy, 1975). Moreover, neutral or weakly acidic proteins are bound to the DNA of *Spirulina platensis* (Levitina and Pinevich, 1974) and *Anabaena cylindrica* (Makino and Tsuzuki, 1971).

No histones have been found in certain groups of eukaryotic protists. Among the protozoa, the absence of histones is often correlated with the parasitic condition. Most Eumycota have nuclear proteins apparently unrelated to histones, whereas histones comprise only a very small portion of the nucleoproteins of *Gyrodinium cohnii*. The potential contribution of histone research to phylogenetics is likely to arise from comparison of primary structures and in elucidation of different functional roles for histones in various protists, rather than in matching the presence or absence of histones with groups of prokaryotes, "mesoprotists," or eukaryotes. Rizzo (1976) has pointed out that histones are a highly conserved group of proteins (0.06 mutations per 100 residues per 100 million years, Dayhoff, 1972b; see Section 13.2) and that a more rigorous definition of histones should be used.

TABLE 11

Distribution of Histones

Organism	Present (+) or absent (−)	Identification[a]	Reference
Bacteria			
Escherichia coli	−	E	Wilkins and Zubay, 1959; Zubay and Watson, 1959
Cyanophyceae			
Anabaena cylindrica	−	E	Makino and Tsuzuki, 1971
Anabaena sp.	−	S	De and Ghosh, 1965
Aphanocapsa sp.	−	S	De and Ghosh, 1965
Oscillatoria sp.	−	S	De and Ghosh, 1965
Polycystis sp.	−	S	De and Ghosh, 1965
Rhodophyceae			
Rhodymenia palmata	+	E	Duffus *et al.*, 1973
Dinophyceae			
Gyrodinium cohnii	[b]	E	Rizzo and Noodén, 1972, 1974a,b
Peridineum trachoideum	[b]	E	Rizzo and Noodén, 1972, 1974a,b
Oxyrrhis marina	+	A	Stewart and Beck, 1967
Peridinium spp. (2 spp.)	+	A	Stewart and Beck, 1967
Prorocentrum spp. (2 spp.)	+	A	Stewart and Beck, 1967
Cryptophyceae			
Chilomonas spp. (2 spp.)	+	A	Stewart and Beck, 1967
Cyathomonas truncata	+	A	Stewart and Beck, 1967
Chlorophyceae			
Chlamydomonas angulosa	+	A	Stewart and Beck, 1967
Chlorella ellipsoidea	+	E	Iwai, 1964
C. vulgaris	+	E	Levitina and Pinevich, 1974
Polytoma uvella	+	A	Stewart and Beck, 1967
Polytomella agilis	+	A	Stewart and Beck, 1967
Volvox carteri	+	E	Bradley *et al.*, 1974
Euglenophyceae			
Astasia longa	+	E	Levitina and Pinevich, 1974
Euglena deses var. *major*	+	A	Stewart and Beck, 1967
E. gracilis var. *bacillaris*	+	E	Netrawali, 1970
E. gracilis Z	+	E	Leedale, 1970; Levitina and Pinevich, 1974
Peranema trichophorum	+	A	Stewart and Beck, 1967
Higher plants	+	E	Bonner and Ts'o, 1964
Sarcodina			
Actinophrys sp.	+	A	Stewart and Beck, 1967
Amoeba spp. (4 spp.)	+	A	Stewart and Beck, 1967
Endolimax spp. (9 str.)	−	A	Stewart and Beck, 1967

TABLE 11 *(Continued)*

Organism	Present (+) or absent (−)	Identification[a]	Reference
Hartmannella spp. (2 spp.)	+	A	Stewart and Beck, 1967
Mayorella palestinensis	+	A	Stewart and Beck, 1967
Pelomyxa carolinensis	+	A	Stewart and Beck, 1967
Mastigophora			
(Kinetoplastida; Bodonina)			
Bodo saltans	+	A	Stewart and Beck, 1967
Oikomonas sp.[c]	+	A	Stewart and Beck, 1967
Kinetoplastida; Trypanosomatina			
Crithidia fasciculata	−	A	Stewart and Beck, 1967
C. oncopelti	+	E	Leaver and Ramponi, 1971
Endotrypanum schaudinni	−	A	Stewart and Beck, 1967
Herpetomonas muscarum	−	A	Stewart and Beck, 1967
Leishmania spp. (10 str.)	−	A	Stewart and Beck, 1967
Leptomonas collosoma	−	A	Stewart and Beck, 1967
Strigomonas oncopelti	−	A	Stewart and Beck, 1967
Trypanosoma spp. (17 str.)	−	A	Stewart and Beck, 1967
T. lewisi	+	E	Levitina and Pinevich, 1974
Rhizomastigida			
Heteramoeba clara	+	A	Stewart and Beck, 1967
Histomonas meleagridis (2 str.)	−	A	Stewart and Beck, 1967
Naegleria gruberi	+	A	Stewart and Beck, 1967
Tetramitus rostratus	+	A	Stewart and Beck, 1967
Trichomonadida			
Trichomonas spp. (2 spp.)[c]	−	A	Stewart and Beck, 1967
Tritrichomonas spp. (2 spp.)	−	A	Stewart and Beck, 1967
Ciliatea			
Oxytrichia sp.	+	*E*	Rizzo, 1976
Paramecium aurelia	+	*E*	Rizzo, 1976
Stylenychia mytilus	+	*E*	Rizzo, 1976
Tetrahymena pyriformis	+	*E*	Iwai *et al.*, 1965, 1970 Johmann and Gorovsky, 1976
Myxomycetes			
Physarum polycephalum	+	E	Mohberg and Rusch, 1969; Bradbury *et al.*, 1973
Plasmodiophoromycetes			
Dictyostelium discoideum	+	E	Charlesworth and Parish, 1975
Polysphondylium pallidum	+	E	Horgen and O'Day, 1973

TABLE 11 (*Continued*)

Organism	Present (+) or absent (−)	Identification[a]	Reference
Oomycetes			
Achlya bisexualis	+	E	Horgen *et al.*, 1973
Chytridiomycetes			
Allomyces arbuscula	−	E	Stumm and van Went, 1968
Blastocladiella emersonii	−	E	Horgen *et al.*, 1973
Zygomycetes			
Phycomyces blakesleeanus	−	E	Leighton *et al.*, 1971
	−	E	Cohen and Stein, 1975
Ascomycetes			
Cordyceps militaris	−[d]	E	Indik *et al.*, 1975
Baker's yeast	−[d]	E	Tonino and Rozijn, 1966; Franco *et al.*, 1974
Neurospora crassa	+	E	Hsiang and Cole, 1973
Neurospora spp. (2 spp.)	−	E	Leighton *et al.*, 1971
Saccharomyces cerevisiae	+[d]	E	Wintersberger *et al.*, 1973
Schizosaccharomyces pombe	−[d]	E	Duffus, 1971
Imperfect fungi			
Aspergillus nidulans	+	E	Felden *et al.*, 1976
Microsporum gypseum	−	E	Leighton *et al.*, 1971
Higher animals	+	E	Bonner and Ts'o, 1964

[a] A, antigenic or immunofluorescence test; E, acid extraction and chemical characterization; S, cytochemical staining method. A critical discussion of these methods has been presented by Rizzo (1976).

[b] Dinophyceae have very low nucleoprotein contents. Histones are apparently present, but may be functionally different from histones in other eukaryotes (Rizzo and Noodén, 1974a,b). Rizzo (1976) discusses histones, *sensu stricto*, in Dinophyceae.

[c] Sometimes considered to be a colorless chrysophyte.

[d] There is disagreement whether histone-like proteins isolated from Eumycota are homologous with histones from higher plants and higher animals.

This is particularly important in considering the presence or absence of these basic proteins in protists and the phylogenetic implications thereof.

In a very recent publication, Rizzo (1976), using acid extraction techniques and chemical characterization has established the presence of histones in *Paramecium aurelia, Oxytricha* sp., and *Stylenychia mytilus* (all Ciliatca), and histone-like compounds in the dinophycean *Peridinium trochoideum*. Horgen *et al.* (1973) also recorded histones in *Blastocladiella emersonii* (Chytridiomycetes).

6

Proteins: Enzymes

6.1 EMBDEN-MEYERHOF-PARNAS GLYCOLYSIS

The Embden–Meyerhof–Parnas (EMP) glycolytic pathway, responsible for the conversion of glucose derivatives into pyruvic acid, is one of the most central of the cellular metabolic pathways. Many of the enzymes responsible for individual glycolytic steps* have been isolated and characterized, although only rarely from a single protist. All EMP enzyme activities have been found in *Euglena gracilis* (although not all activities have been reported in any one strain of *E. gracilis*) (Smillie, 1968) and in *Tetrahymena pyriformis* (Hill, 1972), and most EMP enzyme activities have been reported in *Trichomonas vaginalis* (Danforth, 1967), Cyanophyceae (Wolk, 1973), streptomycetes, and fungi (Blumenthal, 1965). Even so, it must be borne in mind that the EMP, hexose monophosphate, and Entner–Doudoroff† pathways have many enzymes in common; both enzyme purification and labeling experiments are required before truly unambiguous identification of any metabolic pathway is possible.

More commonly, evidence for the operation of the EMP glycolytic pathway consists of (a) the ability of cell extracts to metabolize EMP pathway intermediates; (b) the distribution of carbon-14 from exogenously supplied precursors in accordance with the operation of the EMP pathway;

* Hexokinase, glucosephosphate isomerase, phosphofructokinase, fructose-1,6-diphosphate aldolase, triosephosphate isomerase, triosephosphate dehydrogenase, phosphoglycerate kinase, phosphoglycerate mutase, enolase, and pyruvate kinase. Lactic dehydrogenase, glucose-6-phosphatase, and diphosphofructose phosphatase are sometimes included as well.

† The latter pathway has been found almost exclusively among flagellated gram-negative bacteria (De Ley and Kersters, 1975).

or (c) the presence of "key" enzyme activities in cell extracts. Evidence of these three types has been used to suggest the operation of the EMP pathway in one or more organisms from the bacteria (including photosynthetic bacteria) (Klein and Cronquist, 1967; Doelle, 1969); actinomycetes (Blumenthal, 1965; Bradley and Bond, 1974); Ascomycetes and Basidiomycetes (Blumenthal, 1965); photoautotrophic Cyanophyceae (Gapochka and Danilova, 1973; Wolk, 1973); Phaeophyceae (Johnston and Davies, 1969); Chrysophyceae (Reazin, 1956); Chlorophyceae (Lloyd, 1974); Euglenophyceae (Hurlbert and Rittenberg, 1962; Danforth, 1967); higher plants; cellular slime molds (Wright, 1964); Sarcomastigophora (Shorb, 1964; Danforth, 1967); Sporozoa (Danforth, 1967); Ciliatea (Hill, 1972); and higher animals. Fructose-1,6-diphosphate aldolase, an EMP pathway enzyme, has been reported in all classes of eukaryotic algae examined.

Diversity in EMP pathway reactions in certain protists (Horecker, 1963) is largely limited to the terminal steps, as in the production of ethanol, lactic acid, acetylmethyl carbinol, acetone, or butyric acid. The Rhodophyceae are somewhat anomalous in their metabolic utilization of galactose derivatives, similar to the many ways that glucose is utilized in other organisms. Nevertheless, these algae probably possess a perfectly normal EMP pathway (Jacobi, 1962; Klein and Cronquist, 1967). The distribution of the central EMP pathway enzymes is very suggestive of a derivation of all these forms of life from a common ancestor, but homology among individual EMP pathway enzymes has yet to be established for the majority of protists.

6.1.1 Fructose-1,6-diphosphate Aldolases

The EMP pathway enzyme fructose-1,6-diphosphate aldolase (FDP aldolase: EC 4.1.2.7) catalyzes the conversion of FDP to glyceraldehyde 3-phosphate and dihydroxyacetone phosphate. Owing to a degree of nonspecificity it will catalyze the cleavage of 5-deoxyketopentose 1-phosphates to dihydroxyacetone phosphate and acetaldehyde. It also functions in the reversal of glycolysis and in the photosynthetic reductive pentose phosphate reactions of some protists. Aldolase activity is apparently absent from most lactobacilli and bifidobacteria (De Ley and Kersters, 1975).

The research of Rutter (1964, 1965) has established the existence of two general classes of FDP aldolases. Type I aldolases are usually constructed of four subunits totaling 150,000 daltons, and their catalytic action involves a lysine residue (Schiff base intermediate), the carboxy-terminal tyrosine residue, and probably a histidine residue (Rutter, 1965; Kaklij and

Nadkarni, 1974a,b; London, 1974). Type II aldolases usually possess only two subunits totaling about 70,000 daltons, have a catalytically active thiol group, and require divalent cations. As a result, the two enzyme types can be readily differentiated by their sedimentation coefficients and by their activity or inactivity in the presence of the chelating agent ethylenediaminetetraacetate (EDTA) (Table 12). These two enzymes are dissimilar in amino acid composition, but no sequences have been published. The characterization of aldolases is usually done on the basis of the above parameters, on their electrophoretic mobilities, and on immunological cross-reactivities (London and Kline, 1973). At least within the lactic acid bacteria, these different types of biochemical characters are not always in good agreement with each other.

Within each of the two major aldolase classes, multiple forms differing in minor respects have been reported. As many as four variants of the Type I aldolase have been found in vertebrates, and are usually localized in different tissues. Three forms were reported in *Lactobacillus casei* (Kaklij and Nadkarni, 1974a,b). The distribution of these multiple forms may be of phylogenetic significance, particularly if there is a correlation with intracellular localization. No multiple forms of a Type II aldolase have been reported from within an individual organism.

Perhaps the only conclusion immediately clear from the distributional data (Table 12) is that no clear phylogenetic dichotomy is obvious. A more puzzling fact is that the predominant aldolase in the Cyanophyceae is Type II, whereas chloroplasts in green algae, three strains of *Euglena gracilis*, and higher plants possess a Type I aldolase. In pea, the chloroplastic Type I enzyme differs in physical properties but not in chemical (catalytic) properties from the cytoplasmic aldolase; this has been interpreted to indicate that these two enzymes possess the same function *in vivo* (Anderson, 1972). Although there is no indication that the chloroplastic aldolase is encoded in chloroplast DNA, it does however appear to be biosynthesized on chloroplast ribosomes (Mo *et al.*, 1973).

It is probably possible to construct a phylogenetic scheme consistent with all these data, showing the evolutionary similarity of a few gram-positive bacteria, protozoa, higher animals, chloroplasts, red algae, brown algae, chrysophycean algae, pennate diatoms, prasinophycean algae, and higher plants. The usefulness of such an exercise seems dubious at best. A more promising approach, given this distribution, would be to make the speculative but reasonable assumption that the transition from Type II to Type I enzyme has often been evolutionarily advantageous, or at least evolutionarily likely. Such a transition, involving a further dimerization, would provide a degree of independence from divalent metal requirements and would allow enzyme activity throughout a wider pH range (Rutter, 1965).

TABLE 12

Distribution of FDP Aldolase Activities

Organism	Aldolase	References
Bacteria and Actinomycetes		
Brucella suis	II	Doelle, 1969
Clostridium perfringens	II	Doelle, 1969
Escherichia coli	I + II	Stribling and Perham, 1973
Lactobacillus acidophilus	II	London, 1974
L. bifidus	II	Doelle, 1969
L. casei (3 strains)	I + II[a]	Kaklij and Nadkarni, 1970, 1974a,b; London, 1974
L. delbrueckii	II	London, 1974
L. jensenii	II	London, 1974
L. salivarius	II	London, 1974
L. xylosus	II	London, 1974
Leptospira spp. (2 spp.)	nd[b]	Baseman and Cox, 1969
Micrococcus aerogenes	I	Lebherz and Rutter, 1969
Mycobacterium smegmatus	I	Jayanthi Bai *et al.*, 1975
M. tuberculosis (2 strains)	II	Jayanthi Bai *et al.*, 1975
Pediococcus cerevisiae	II	London, 1974
Peptococcus aerogenes	I[a]	Lebherz and Rutter, 1973
Rhodopseudomonas spheroides	II	Willard and Gibbs, 1968a
Saprospira thermalis	II	Willard and Gibbs, 1968a
Streptococcus faecalis	II	London, 1974
Streptomyces spp. (2 spp.)	nd	Blumenthal, 1965
Cyanophyceae (10 spp.)[c]	II	Antia, 1967; Willard and Gibbs, 1968a; Doelle, 1969; Ikawa *et al.*, 1972; Smith, 1973
Eumycota		
Rhizopus MX	nd	Blumenthal, 1965
Saccharomyces cerevisiae	II	Rutter, 1965
Aspergillus niger	II	Rutter, 1965
Epidermophyton sp.	nd	Kozarenko and Sukhenko, 1971
Fusarium lini	nd	Blumenthal, 1965
Microsporum spp. (3 spp.)	nd	Blumenthal, 1965; Kozarenko and Sukhenko, 1971
Penicillium chrysogenum	nd	Blumenthal, 1965
Tilletia caries	nd	Blumenthal, 1965
Ustilago maydis	nd	Blumenthal, 1965
Protozoa	I	Rutter, 1965
Euglenophyceae		
Euglena gracilis[d]		
Cytoplasm (str. Z)	II	Mo *et al.*, 1973
Chloroplast (str. Z)	I	Mo *et al.*, 1973

TABLE 12 (*Continued*)

Organism	Aldolase	References
Photoautotrophically grown (str. B, Z, *bacillaris*)	I + II	Willard and Gibbs, 1968b; Mo *et al.*, 1973; Bukowiecki and Anderson, 1974
Dark-grown (str. B, Z, *bacillaris*)	II	Willard and Gibbs, 1968b; Mo *et al.*, 1973
Rhodophyceae (14 spp.)	I + II	Willard and Gibbs, 1968b; Ikawa *et al.*, 1972
Phaeophyceae (10 spp.)	I + II	Ikawa *et al.*, 1972
Chrysophyceae		
Monochrysis lutheri	I	Antia, 1967
Ochromonas danica	I + II	Willard and Gibbs, 1968b
Haptophyceae (3 spp.)	II	Antia, 1967
Bacillariophyceae		
Centrales (2 spp.)	II	Antia, 1967
Pennales (1 spp.)	I	Antia, 1967
Cryptophyceae (3 spp.)	II	Antia, 1967
Dinophyceae (1 sp.)	II?	Antia, 1967
Prasinophyceae (1 sp.)	I	Antia, 1967
Chlorophyceae		
Chlamydomonas mundana		
Photoautotrophically grown	I	Russell and Gibbs, 1967; Willard and Gibbs, 1968b
Dark grown	II	Russell and Gibbs, 1967; Willard and Gibbs, 1968b
Chlorella pyrenoidosa		
Photoautotrophically grown	I	Russell and Gibbs, 1967; Willard and Gibbs, 1968b
Dark grown	II	Russell and Gibbs, 1967; Willard and Gibbs, 1968b
Cladophora japonica	I	Ikawa *et al.*, 1972
Codium latum	I	Ikawa *et al.*, 1972
Dunaliella tertiolecta	I	Antia, 1967
Ulva pertusa	I	Ikawa *et al.*, 1972
Higher plants		
Leaves, seeds, chloroplasts, and cytoplasm	I	Stumpf, 1948; Willard and Gibbs, 1968b; Lebherz and Rutter, 1969; Rapoport *et al.*, 1969

[a] Variants of Type I present.

[b] nd, Present but class not determined (*i.e.*, I vs. II).

[c] Possible traces of Type I in *Anacystis marina* (Antia, 1967).

[d] Frequently quoted but incorrect data have been reported by Rutter (1964, 1965). Consult references listed for discussion.

The above assumption, although unproved, is not without some degree of support. If the Type I aldolase were to have been independently derived from Type II enzymes on several occasions, this polyphyletic origin might be reflected in differing properties of enzymes from various organisms. Mo *et al.* (1973) have reported that the isoelectric point of Type I aldolases from pea chloroplasts (pI = 4.5–4.75) and from *Euglena gracilis* strain Z chloroplasts (pI = 4.60) differs markedly from that of animal Type I aldolases (pI = 9.1–9.7). Bukowiecki and Anderson (1974) have extended this pattern to other chloroplast-cytoplasm systems.

In the final analysis, the data on distribution of types of FDP aldolase, based on molecular weights, cofactor requirements, and pI ranges, are too weak to be used as an unimpeachable argument for or against any given phylogeny. There has been much opportunity for convergence of properties and for multiple ancestry: In how many ways can an enzyme acquire an increased pI range? Moreover, environmental conditions may be responsible for the accumulation of a particular type of aldolase, as seen in *Chlamydomonas mundana* and *Chlorella pyrenoidosa* (Table 12). Amino acid sequences, tertiary structures, and gene localization will be necessary to explain the distributional patterns and establish any phylogenetic significance.

The distinction between covalent cation-requiring and Schiff base-type aldolases seems to be of general occurrence. Rutter (1965) has listed seventeen analogous aldolase-like enzymes falling into one of the above two categories. The subunit structures of most of these enzymes are unknown.

6.1.2 Triosephosphate Isomerases

Triosephosphate isomerase (D-glyceraldehyde-3-phosphate ketol-isomerase, EC 5.3.1.1) is responsible for the conversion of dihydroxyacetone phosphate to D-glyceraldehyde 3-phosphate during EMP glycolysis. Anderson (1971) has suggested that there is a specific interaction between this enzyme and FDP aldolase that influences the direction of the aldolase reaction. As with FDP aldolase, there appear to be two different forms of triosephosphate isomerase in *Euglena gracilis* Z and in the higher plants. Both of these forms appear to function in glycolysis or, alternatively, in gluconeogenesis; the early suggestion that one of these forms might be specialized or compartmentalized exclusively for pentose phosphate metabolism has received no experimental support (Mo *et al.*, 1973).

The "Type A" isomerase is found in chloroplasts of *E. gracilis* Z and higher plants; a second type, "Type B," is cytoplasmic. The chloroplastic isomerase is translated on chloramphenicol-sensitive (presumably chloroplastic) ribosomes, but the location of its gene(s) is not known. The two forms can be distinguished by their differing isoelectric points, molecular weights, Michaelis constants, reactivity to inhibitors, and pH activity curves. No differences in catalytic activities have been observed (Anderson, 1971, 1972). Some hybridization (AA,AB,BB) has been reported for isomerases from animals, but no such hybridization has been found in *E. gracilis* Z. Both forms of isomerase from higher plants differ from both animal forms by possessing much lower pI values; the pattern in *E. gracilis* Z resembles that in higher plants.

6.1.3 Glyceraldehyde-3-phosphate Dehydrogenases

D-Glyceraldehyde-3-phosphate (G3P) dehydrogenase [triosephosphate dehydrogenase; D-glyceraldehyde-3-phosphate: NAD^+ ($NADP^+$) oxido-reductase; NAD-linked, EC 1.2.1.12; NADP-linked, 1.2.1.13] is the enzyme responsible for the conversion of D-glyceraldehyde 3-phosphate to D-1,3-diphosphoglyceric acid. It has been studied both at the primary structural level (e.g., Jones and Harris, 1972) and at the secondary and tertiary levels. Not only have NAD-requiring G3P dehydrogenases in diverse taxa retained remarkably similar tertiary structures, but clear resemblances have also been observed among nucleotide-binding regions in these enzymes and in other nucleotide-binding proteins such as mammalian liver alcohol dehydrogenase (Brändén *et al.*, 1973), soluble malate dehydrogenase, lactate dehydrogenase, glutamate dehydrogenase (Wootton, 1974), and possibly even bacterial flavodoxin (Buehner *et al.*, 1973). In these cases, similar folding patterns have been retained despite large-scale divergence at the primary structural level (Rossman, 1974).

Although few protistan G3P dehydrogenases have been studied at either the primary or tertiary structural levels, it appears from the distributional data (Table 13) that many protists possess an NAD-linked G3P dehydrogenase activity encoded in the nuclear DNA. The appearance of an NADP-linked activity may have coincided with the appearance of photosynthetic capacity in the bacteria. Thus it is interesting that chloroplasts may possess a single enzyme with both NAD- and NADP-linked activities, which is immunologically dissimilar to extrachloroplastic G3P dehydrogenases from the same species (McGowan and Gibbs, 1974). Nonetheless, part or all of the chloroplastic activity may be coded in nuclear DNA in *Euglena gracilis* var. *bacillaris* (Bovarnick *et al.*, 1974).

6.2 PENTOSE PHOSPHATE PATHWAY

The pentose phosphate pathway (hexose monophosphate shunt, Warburg–Lipmann–Dickens pathway, phosphogluconate oxidative pathway)

TABLE 13

Distribution of G3P Dehydrogenase Activities

Organism	Cofactor(s)	Reference
Bacteria		
Thiobacillus denitrificans	NAD	Trudinger, 1956
Rhodospirillum rubrum	NAD,NADP	Fuller and Gibbs, 1959
Cyanophyceae		
Anabaena variabilis	NAD,NADP	Pearce and Carr, 1969
Anacystis nidulans	NAD,NADP	Fuller and Gibbs, 1959
Nostoc muscorum	NAD,NADP	Fewson *et al.*, 1962
Tolypothrix tenuis	NAD,NADP	Latzko and Gibbs, 1969
Rhodophyceae (13 spp.)	NAD,NADP	Ziegler and Ziegler, 1967
Phaeophyceae (8 spp.)	NAD,NADP	Ziegler and Ziegler, 1967
Chlorophyceae		
Bryopsis plumosa	NAD,NADP	Jacobi, 1962
Chaetomorpha linum	NAD,NADP	Jacobi, 1962
Chlorella pyrenoidosa		
Chloroplast	NAD,NADP	Willard and Gibbs, 1968a,b
Cytoplasm	NAD	Willard and Gibbs, 1968a,b
C. variegata		
Autotrophically grown	NAD,NADP	Fuller and Gibbs, 1959
Heterotrophically grown	NAD,NADP[a]	Fuller and Gibbs, 1959
Ulva lactuca	NAD,NADP	Jacobi, 1962
Euglenophyceae		
Astasia sp.	NAD	Fuller and Gibbs, 1959
Euglena gracilis (str. B, Z, *bacillaris*)		
Green	NAD,NADP	Fuller and Gibbs, 1959
Streptomycin bleached	NAD	Fuller and Gibbs, 1959
Dark grown	NAD,NADP[a]	Fuller and Gibbs, 1959
Chloroplast	NAD,NADP	Willard and Gibbs, 1968a,b
Cytoplasm	NAD	Willard and Gibbs, 1968a,b
Higher plants		
Spinach, pea		
Chloroplast	NAD,NADP	Willard and Gibbs, 1968a,b
Cytoplasm	NAD	Willard and Gibbs, 1968a,b
Barley seedlings		
Green	NAD,NADP	Fuller and Gibbs, 1959
X-ray induced albino	NAD,NADP	Fuller and Gibbs, 1959
Protozoa		
Tetrahymena gelii	NAD	Fuller and Gibbs, 1959
Eumycota	NAD	Blumenthal, 1965
Animals	NAD	Buehner *et al.*, 1973

[a] A minor amount of NADP-linked activity was also observed.

was discovered in bacterial cells, but evidence is accumulating for its operation in many other organisms. Reports have suggested its operation in most bacteria (Klein and Cronquist, 1967) and actinomycetes (Bradley and Bond, 1974); photosynthetic bacteria (Kondrat'eva, 1965); Cyanophyceae (Fewson *et al.*, 1962; Wolk, 1973); chloroplasts (Schnarrenberger and Oeser, 1974); Rhodophyceae (Cohen, 1950; Bean and Hassid, 1956); Chrysophyceae (Reazin, 1956); Phaeophyceae (Johnston and Davies, 1969); Chlorophyceae (Cohen, 1950; Lloyd, 1974; Neilson and Lewin, 1974); Euglenophyceae (Hurlbert and Rittenberg, 1962; Smillie, 1968); Ascomycetes and Basidiomycetes (Blumenthal, 1965); *Dictyostelium discoideum* (Wright, 1964); protozoa (van Wagtendonk, 1963; Shorb, 1964; Wright, 1964; Honigberg, 1967; Hill, 1972); higher animals; and higher plants. This pathway may be entirely replaced by the Entner–Doudoroff pathway in *Entamoeba histolytica* (Danforth, 1967), and modified forms may exist in *Leuconostoc mesenteroides* (Horecker, 1963), *Pseudomonas methanica* (Doelle, 1969), methane-utilizing bacteria and yeasts (Strøm *et al.*, 1974), and *Caldariomyces fumago* (Ramachandran and Gottlieb, 1963).

Some pentose phosphate pathway enzymes* are shared with the EMP pathway. In *C. fumago*, enzymes from both of these pathways and from the Entner–Doudoroff shunt are utilized in glucose oxidation, but no single pathway has been shown to be totally operative. The red algal hexose oxidase (EC 1.1.3.5) appears to differ from the glucose oxidase of other organisms by utilizing a wide range of substrates (Bean and Hassid, 1956; Sullivan and Ikawa, 1973).

In certain groups of organisms (e.g., some Athiorhodaceae, the Cyanophyceae, and the Rhodophyceae) the pentose phosphate reactions may be of greater physiological importance than are those of the EMP pathway. Nonetheless, the presence of genes for these enzymes is the basic factor in biochemical phylogenies. It appears that the distribution of the pentose phosphate pathway as a whole is of little phylogenetic significance, except to indicate a likely common ancestor to all these protists.

6.3 TRICARBOXYLIC ACID CYCLE

The tricarboxylic acid cycle (TCA cycle, Krebs cycle, citric acid cycle) is important in the oxidative metabolism of glucose derivatives. As such it could be expected to be functional in all aerobic organisms, and as a rule

* Glucose-6-phosphate dehydrogenase, phosphogluconate dehydrogenase, pentose epimerase, pentose isomerase, transketolase, transaldolase, phosphatase, and hexose isomerase are pentose phosphate pathway enzymes. Triosephosphate isomerase and FDP aldolase are used in both EMP glycolysis and pentose phosphate metabolism.

this is indeed what has been observed. The cycle might be expected to be inoperative in anaerobic organisms, but the question of whether individual TCA cycle enzymes* are present in anaerobes has no *a priori* answer. α-Ketoglutarate dehydrogenase often cannot be detected in anaerobes (nor can it be found in Cyanophyceae: Smith, 1973). Its absence would restrict the function of the rest of the TCA enzymes to providing carbon skeletons for amino acid biosynthesis (Stanier *et al.*, 1970).

Only in some enteric bacteria and in mitochondria of higher animals have individual TCA cycle enzyme activities been thoroughly studied. Various individual enzyme activities have been examined among protozoa (Danforth, 1967), Euglenophyceae (Smillie, 1968), and eukaryotic algae (Mehler, 1950; Jacobi, 1962; Johnston and Davies, 1969), but the presence of all TCA cycle enzymes has been demonstrated in only a few eukaryotic protists (Danforth, 1967; Lloyd, 1974). More commonly, patterns of carbon-14 incorporation into TCA cycle intermediates, inhibition studies, and growth stimulation by TCA cycle intermediates are used as evidence for the operation of the TCA cycle in protists. Microorganisms grown heterotrophically in darkness usually metabolize acetate by the TCA cycle; such growth has been recorded in members of the Cryptophyceae, Dinophyceae, Bacillariophyceae, Chrysophyceae, Xanthophyceae, Euglenophyceae, and Chlorophyceae (Neilson and Lewin, 1974). Nonetheless, due to the potential multiplicity of interlocking biosynthetic pathways and the limited specificity of many inhibitors, it is not possible to establish the presence of the TCA cycle from such studies.

The distribution of the TCA cycle (Table 14) is suggestive of a common ancestry of all these organisms, but requires confirmation by demonstration of homology among the enzymes involved. The absence of a functional TCA cycle could obviously be due to mutation or to repression of any of a large number of genes, and could have evolved several times by dissimilar events in different protists.

6.3.1 Isocitrate Dehydrogenases

Isocitrate dehydrogenase is one of the few TCA cycle enzymes that has been further investigated for phylogenetic information. It occurs in two different forms in the protists, one form linked *in vivo* to NAD (EC 1.1.1.41), the other linked to NADP (EC 1.1.1.42). Physical properties of the enzyme

* Citrate synthetase, aconitase, isocitrate dehydrogenase, α-ketoglutarate oxidase, succinyl thiokinase, succinate dehydrogenase, fumarase, and malic dehydrogenase. Some of these same enzymes are involved in the "reductive carboxylic acid cycle" accompanying some bacterial photosynthesis (Buchanan, 1972).

TABLE 14

Distribution of the TCA Cycle

Organism	Cycle complete?[a]	Reference
Bacteria and Actinomycetes		
Actinomycetes	Complete	Kornberg, 1959
Athiorhodaceae (aerobic)	Complete	Krebs and Lowenstein, 1960
Chlorobium thiosulfatophilum	Incomplete	Smillie and Evans, 1963
Clostridium spp.	Incomplete	Klein and Cronquist, 1967
Desulfovibrio sp.	Incomplete	Klein and Cronquist, 1967
Escherichia coli		
Aerobically grown	Complete	Amarasingham and Davis, 1965
Anaerobically grown	Incomplete	Amarasingham and Davis, 1965
Hydrogenomonas H16G	Complete	Doelle, 1969
H. facilis	Incomplete	Doelle, 1969
Leptospira spp. (2 spp.)	Complete	Baseman and Cox, 1969
Methylococcus capsulatus	Incomplete	Smith, 1973; Patel *et al.*, 1975
Methylotrophic bacteria		
With "Type I" membranes	Incomplete	Anthony, 1975
With "Type II" membranes	Complete	Anthony, 1975
Micrococcus spp. (2 spp.)	Complete	Klein and Cronquist, 1967; Doelle, 1969
Mycobacterium spp.	Complete	Kornberg, 1959; Bradley and Bond, 1974
Mycoplasma hominis	Complete	Doelle, 1969
Nitrifying bacteria	Incomplete	Klein and Cronquist, 1967
Nitrosocystis oceanus	Complete	Smith, 1973
Nitrosomonas europaea	Complete	Smith, 1973
Propionibacterium pentosaceum	Complete	Doelle, 1969
Pseudomonads	Complete	Doelle, 1969
Thiorhodaceae	Incomplete	Klein and Cronquist, 1967; Doelle, 1969
Cyanophyceae		
Anabaena variabilis	Incomplete	Pearce and Carr, 1969
Anacystis nidulans	Incomplete	Smith *et al.*, 1967; Pearce and Carr, 1969
Chlorogloea fritschii	Complete	Miller and Allen, 1972
Coccochloris peniocystis	Incomplete	Smith *et al.*, 1967
Gloeocapsa alpicola	Incomplete	Smith *et al.*, 1967

TABLE 14 (*Continued*)

Organism	Cycle complete?[a]	Reference
Protozoa		
Crithidia spp. (2 spp.)	Complete	Honigberg, 1967
Entamoeba histolytica	Incomplete	Danforth, 1967
Leishmania donovani	Complete	Honigberg, 1967
Paramecium spp. (2 spp.)	Complete	van Wagtendonk, 1963; Danforth, 1967
Plasmodium berghei	Incomplete	Danforth, 1967
P. gallinaceum	Complete	van Wagtendonk, 1963; Danforth, 1967
Tetrahymena pyriformis	Complete	van Wagtendonk, 1963; Hill, 1972
Trichomonads	Incomplete	Shorb, 1964
Trypanosoma spp.	[b]	Danforth, 1967; Honigberg, 1967
Plasmodiophoromycetes		
Dictyostelium discoideum	Complete	Wright, 1964
Chytridiomycetes (4 spp.)	Complete	Niederpruem, 1965
Zygomycetes (4 spp.)	Complete	Niederpruem, 1965
Ascomycetes		
Hemiascomycetidae (5 spp.)	Complete	Niederpruem, 1965
Euascomycetidae (5 spp.)	Complete	Niederpruem, 1965
Imperfect fungi (22 spp.)	Complete	Niederpruem, 1965
Basidiomycetes (9 spp.)	Complete	Niederpruem, 1965
Oomycetes (2 spp.)	Complete	Niederpruem, 1965
Bacillariophyceae (1 sp.)	Complete?	Cooksey, 1974
Cryptophyceae (1 sp.)	Complete	Danforth, 1967
Chlorophyceae (5 spp.)	Complete	Jacobi, 1962; Marsh *et al.*, 1965; Lloyd, 1974
Euglenophyceae		
Astasia longa	Complete	Danforth, 1967
Euglena gracilis var. *bacillaris*	Complete	Hurlbert and Rittenberg, 1962
Higher plants	Complete	Beevers, 1961

[a] Complete cycle: all enzyme activities found, where investigated, in members of the taxon; or labeling pattern consistent with operation of the cycle. Incomplete cycle: at least one enzyme activity totally absent from all members of the taxon; or labeling pattern inconsistent with operation of the cycle.

[b] Depends on species and life form.

have for the most part remained uninvestigated. The NAD-linked activity has been found in Chytridiomycetes and in Zygomycetes, but has not been found in Oomycetes and Hyphochytridiomycetes examined. NADP-linked activity has been reported in members of both the Eumycota and the Oomycota; in the Chytridiomycetes it catalyzes an easily reversible reaction, whereas in the Oomycetes the reaction is not easily reversible (LéJohn, 1971a).

6.3.2 Citrate Synthase

Citrate synthase (EC 4.1.3.7), another TCA cycle enzyme, has been found to occur either as a tetrameric (molecular weight 250,000 daltons) or a dimeric (molecular weight 100,000 daltons) enzyme. Gram-positive bacteria possess the dimeric form, whereas gram-negative bacteria contain the tetramer.* In all bacteria so far investigated, the former is insensitive to NADH, but the tetrameric form is inhibited by NADH (De Ley and Kersters, 1975). Investigated eukaryotes contain the NADH-insensitive dimer.

It has been demonstrated that five species of Cyanophyceae contain a citrate synthase of molecular weight 250,000, thus sharing this character with other gram-negative prokaryotes (Greenblatt and Sarkissian, 1973; Lucas and Weitzman, 1975). However, the blue-green algal synthase is not affected by NADH. This difference presumably reflects both the differing physiological environment of the blue-green algal enzyme and the phylogenetic distinctiveness of the Cyanophyceae as a group.

6.4 HATCH–SLACK PATHWAY

The Hatch–Slack pathway (C_4 dicarboxylic acid pathway) is found in at least six orders of higher plants, both monocotyledonous and dicotyledonous, where it is to some degree an auxiliary pathway for carbon dioxide fixation and reduction. In these reactions, oxaloacetate and/or malate are the major products of photosynthetic carbon fixation instead of 3-phosphoglycerate. Certain cells in the plant leaf are specialized to carry out these reactions, and both chloroplasts and mitochondria are involved (Hatch and Slack, 1966; Moore, 1974; Kagawa and Hatch, 1975).

There are strong indications, both biochemical and morphological, that higher plants and green algae have evolved from a common ancestor. It

* Massarini and Cazzullo (1975) have found both dimeric and tetrameric citrate synthases in a marine *Pseudomonas* sp.

would therefore be of interest to determine whether the green algae possess the Hatch–Slack pathway. Gibbs *et al.* (1970) and Codd and Merrett (1971) were unable to find significant levels of incorporation of labeled carbon dioxide into dicarboxylic acids by green algae.* The simplest explanation is that the higher plants evolved these enzymes at a somewhat later stage than their divergence from a common ancestor with the green algae; indeed "intermediate" higher plants have been described (Kennedy and Laetsch, 1974). This "late evolution" of the Hatch–Slack pathway is in keeping with its occurrence principally in tropical plants, growing in high light intensities, and which, because of the pathway, show a more efficient CO_2 uptake and incorporation. In addition to this low CO_2 compensation point there appears to be an absence of photorespiratory capacity. Thus pathways dealing with fundamental metabolic processes can be evolved by morphologically advanced groups of organisms.

6.5 GLYCOLATE OXIDATION

Plants utilizing the Calvin cycle (Section 6.6) possess a light-dependent metabolism of glycolic acid into glyoxylate, then glycine, serine, hydroxypyruvate, glycerate, and finally a hexose; this process is termed photorespiration, and results in a physiologically (and economically) significant reduction in the net efficiency of carbon fixation (Merrett and Lord, 1973). Seven enzymes† are involved in the photorespiratory reactions. Their subcellular localization is not known for many protists, although glycolate-formation often appears to be associated with the chloroplast. The glycolate oxidizing enzyme is mitochondrial in diatoms (Paul *et al.*, 1975), and catalase is characteristically found in the peroxisomes of *Euglena gracilis* Z and of higher plants. The status of peroxisomes in most algae is uncertain (Frederick *et al.*, 1973).

Glycolate is involved in other reactions in the cell, including a modification of the TCA cycle (Müller *et al.*, 1968; Stanier *et al.*, 1970), and is excreted by certain marine algae of the classes Chlorophyceae, Dinophyceae, Chrysophyceae, and Bacillariophyceae (Hellebust, 1965) and by *Rhodomicrobium vannielii, Rhodospirillum rubrum,* and *Chromatium* D (Codd and Turnbull, 1975).

* Döhler (1974a,b) has recently presented evidence that may be interpreted as showing C_4-carboxylation in *Chlorella vulgaris* and in *Anacystis nidulans* during the induction of photosynthesis.

† The glycolate-oxidizing enzyme, glycolate-glutamate aminotransferase, serine hydroxymethyltransferase, serine-α-oxoglutarate aminotransferase, hydroxypyruvate reductase, and glycerate kinase in the pathway proper, and catalase in the destruction of the hydrogen peroxide formed during glycolate oxidation.

A major variation in the pathway, which appears to be of phylogenetic interest, concerns the nature of the enzyme responsible for glycolate oxidation. In most higher plants so far investigated, glycolate oxidase (glycolate: oxygen oxidoreductase, EC 1.1.3.1) is the major oxidizing enzyme. This enzyme can oxidize L-lactate (but not D-lactate) and is insensitive to inhibition by cyanide ion. A different enzyme activity, "glycolate dehydrogenase," is found in many green algae. The dehydrogenase oxidizes only the D-isomer of lactate, is usually inhibited by cyanide (but see Paul *et al.*, 1975), and does not transfer electrons directly to molecular oxygen as does the oxidase (Frederick *et al.*, 1973; Gruber *et al.*, 1974). Some higher plants grown under specialized conditions may also possess some glycolate dehydrogenase activity (Roth-Bejerano and Lips, 1973). Serological tests have provided evidence for the similarity of the oxidase and the dehydrogenase (Codd and Schmid, 1972).

The distribution of these two enzyme activities (Table 15) is consistent with Pickett-Heaps' reassessment of green algal phylogeny (Pickett-Heaps

TABLE 15

Distribution of Glycolate Oxidase and Glycolate Dehydrogenase Activities

Organism	Activity	Reference
Cyanophyceae		
Anabaena cylindrica	Dehydrogenase	Codd and Steward, 1974
A. flos-aquae	Oxidase?	Grodzinski and Colman, 1970
Lyngbya sp.	Dehydrogenase	Tolbert, 1976
Oscillatoria sp.	Oxidase?	Grodzinski and Colman, 1970
Bacillariophyceae		
Cylindrotheca fusiformis	Dehydrogenase	Paul *et al.*, 1975
Nitzschia alba	Dehydrogenase	Paul *et al.*, 1975
Thalassiosira pseudonana	Dehydrogenase	Paul and Volcani, 1974
Chlorophyceae		
Ankistrodesmus braunii	Dehydrogenase	Hess and Tolbert, 1967
Boergesenia forbesii	Dehydrogenase	Tolbert, 1976
Caulerpa spp. (3 spp.)	Dehydrogenase	Tolbert, 1976
Chaetomorpha crassa	Dehydrogenase	Tolbert, 1976
Chlamydomonas reinhardi	Dehydrogenase	Codd *et al.*, 1969
Chlorella spp. (2 spp.)	Dehydrogenase	Frederick *et al.*, 1973
Cladophora fascicularis	Dehydrogenase	Tolbert, 1976
Codium sp.	Dehydrogenase	Frederick *et al.*, 1973
Coleochaete scutata	Oxidase	Frederick *et al.*, 1973
Dictyosphaeria versluysii	Dehydrogenase	Tolbert, 1976
Dunaliella tertiolecta	Dehydrogenase	Frederick *et al.*, 1973

TABLE 15 *(Continued)*

Organism	Activity	Reference
Enteromorpha flexuosa	Dehydrogenase	Tolbert, 1976
Eremosphaera viridis	Dehydrogenase	Frederick *et al.*, 1973
Halimeda spp. (2 spp.)	Dehydrogenase	Tolbert, 1976
Klebsormidium flaccidum	Oxidase	Frederick *et al.*, 1973
Microspora sp.	Dehydrogenase	Frederick *et al.*, 1973
Neomeris vanbossae	Dehydrogenase	Tolbert, 1976
Netrium digitus	Oxidase	Frederick *et al.*, 1973
Oocystis polymorpha	Dehydrogenase	Frederick *et al.*, 1973
Protosiphon botryoides	Dehydrogenase	Frederick *et al.*, 1973
Scenedesmus obliquus	Dehydrogenase	Hess and Tolbert, 1967
Spirogyra varians	Oxidase	Frederick *et al.*, 1973
Stigeoclonium helveticum	Dehydrogenase	Frederick *et al.*, 1973
Udotea argentea	Dehydrogenase	Tolbert, 1976
Valonia sp.	Dehydrogenase	Tolbert, 1976
Charophyceae		
Nitella sp.	Oxidase	Frederick *et al.*, 1973
Higher plants		
Most investigated species	Oxidase	Frederick *et al.*, 1973; Tolbert, 1976
Cymodocea rotundata	Dehydrogenase	Tolbert, 1976
Thalassia hemprichii	Dehydrogenase	Tolbert, 1976
Euglenophyceae		
Euglena gracilis str. Z	Dehydrogenase	Codd *et al.*, 1969
Protozoa		
Acanthamoeba sp.	Not detected	Müller *et al.*, 1968
Tetrahymena pyriformis	"Oxidase"	Müller *et al.*, 1968
Eumycota		
Investigated Eumycota	"Oxidase"	Gibbs *et al.*, 1970
Higher animals		
Investigated animals	"Oxidase"	Müller *et al.*, 1968

and Marchant, 1972; Pickett-Heaps, 1975), and supports the evolution of higher plants from an ancestral line common to certain green algae (*Klebsormidium*, the Zygnematales) and the Charophyceae. This evolutionary line is characterized by a "phragmoplast" system in which microtubules are oriented perpendicular to the plane of cytokinesis. At the biochemical level this would be the glycolate oxidase line. The other principal phylogenetic line in the evolutionary dichotomy is characterized by a system ("phycoplast") in which the microtubules are oriented in the plane of cytokinesis. This evolutionary line appears to be characterized by the glycolate dehydrogenase system. However, the appearance of the dehydrogenase system in the higher plants *Cymodocea* and *Thalassia* (Tolbert, 1976) indicates further work is needed.

Extracts of *Rhodomicrobium vannielii* and *Rhodospirillum rubrum* contain a glycolate-oxidizing enzyme that is not stimulated by FMN, as are the enzymes of some Cyanophyceae, Chlorophyceae, and all investigated higher plants (Codd and Turnbull, 1975). The glycolate-oxidizing enzyme of *Anabaena cylindrica* appears to be of the dehydrogenase type, but may be a facultative oxidase as well (Codd and Stewart, 1974); such an explanation would also be consistent with the data of Grodzinski and Colman (1970) from two other Cyanophyceae. In *A. cylindrica,* glycolate metabolism involves tartronic semialdehyde as at least a partial alternative to glycine and serine (Codd and Stewart, 1973). Gruber *et al.* (1974) have suggested that D-lactate dehydrogenase (EC 1.1.1.28) activity is found in organisms possessing glycolate dehydrogenase activity (which itself can oxidize D-lactate), whereas L-lactate dehydrogenase activity occurs in organisms possessing glycolate oxidase (which oxidizes L-lactate). Both lactate dehydrogenase activities have been reported in prokaryotes (Sanchez *et al.*, 1975).

Two interesting facts derive from Pickett-Heaps' (1975) proposals:

1. The *Chlamydomonas*-type does *not* represent the primitive ancestral stock from which all other Chlorophyceae derived.
2. The Prasinophyceae now become a key group in the search for the ancestral stock of the green algae.

It will be very interesting to see if these phylogenetic proposals hold up and if they are substantiated by independent biochemical data. Whatever the outcome, this situation emphasizes the problem (Section 1.4) of what is a representative alga. Furthermore, there is an important caveat for the comparative biochemist. *Chlamydomonas* and/or *Chlorella,* two favorites of the biochemist, may not be the best organisms to use in phylogenetically based experiments involving the full evolutionary line from prokaryote (or primitive eukaryote) through to higher plant. This is particularly true when a green alga is used as a hypothetical or presumptive ancestor of higher plants, be they bryophytes, gymnosperms, or angiosperms.

6.6 CALVIN CYCLE

The Calvin (Calvin–Benson, reductive pentose phosphate) cycle is the sequence of reactions by which photosynthetic organisms reduce carbon dioxide, with the resultant incorporation of the carbon atom in their photosynthate (usually hexoses). These reactions have been summarized numerous times (e.g., Walker, 1974) and will not be repeated here. As a rule, the Calvin cycle enzymes* have not been thoroughly studied individually, although FDP aldolase and ribulose-1,5-diphosphate carboxylase are exceptions. In all, some twelve enzymes are more or less directly involved in the cycle, and it is unlikely that this could be the result of convergent evolution in many different protists.

Most enzyme activities of the Calvin cycle have been established in *Tolypothrix tenuis* (Latzko and Gibbs, 1969), *Chlorella pyrenoidosa* (Peterkofsky and Racker, 1961; Latzko and Gibbs, 1969), *Euglena gracilis* Z (Peterkofsky and Racker, 1961; Smillie, 1968), and *E. gracilis* var. *bacillaris* (Latzko and Gibbs, 1969). Labeling experiments have given an indication of the presence of the cycle in many other protists, including *Anacystis nidulans, Nostoc* spp., *Phormidium* sp., another *Synechococcus* sp., *Chlorococcum* sp., *Haematococcus* sp., *Scenedesmus obliquus, Spirogyra* sp., and *Porphyridium* sp. (Norris *et al.*, 1955; Kindel and Gibbs, 1963; Russell and Gibbs, 1968). Similar labeling patterns have also been observed in the cyanellae of *Cyanophora paradoxa* and *Glaucocystis nostochinearum* (Schenk and Hofer, 1972). It is probable that the distribution of the Calvin cycle enzymes will be found to correspond very closely with the distribution of photosynthetic capability. Hurlbert and Lascelles (1963), Kondrat'eva (1965), Latzko and Gibbs (1969), McFadden (1973), and Tabita *et al.* (1974a) have presented evidence consistent with its operation in various photosynthetic bacteria, although recent research has cast doubt upon its presence in *Chlorobium thiosulfatophilum* (Sirevåg, 1974). In contrast, Smillie *et al.* (1962) claimed that *Chlorobium* showed all the activities of the Calvin cycle enzymes. Ribulose-1,5-diphosphate carboxylase activity has not been demonstrated in *Chloropseudomonas* (Callely *et al.*, 1968; Fuller, 1971). This organism may have a special CO_2 fixation mechanism, since there is an obligate requirement for a 2-carbon source for growth in addition to CO_2. The Chlorobacteriaceae may well utilize the reductive carboxylic acid cycle (Sirevåg and Ormerod, 1970a,b; Sirevåg, 1974) for CO_2 fixation. De Ley and Kersters (1975) have commented that the Chlorobacteriaceae appear special and may well be a separate phylogenetic line. The Calvin cycle may also operate in nonphotosynthetic organisms of the genera *Hydrogenomonas, Micrococcus, Nitrobacter,*

* Including ribulose-1,5-diphosphate carboxylase, phosphoglycerate kinase, phosphoglyceraldehyde dehydrogenase, triosephosphate isomerase, FDP aldolase, phosphatase, transketolase, transaldolase, epimerase, and phosphoribulokinase. Some of these enzymes are shared with EMP glycolysis or with pentose phosphate metabolism.

Nitrosocystis, Nitrosomonas, Pseudomonas, and *Thiobacillus* (McFadden, 1973).

Photoreduction (anaerobic hydrogen assimilation under the influence of light) and the "oxyhydrogen reaction" (carbon dioxide and hydrogen assimilation in the dark at low partial pressures of oxygen) may be considered as auxiliary processes to the Calvin cycle. Hydrogenase activity, catalyzing the first step of the photoreductive reactions, has been found in some species of Cyanophyceae, Rhodophyceae, Phaeophyceae, Chlorophyceae, Euglenophyceae (Healey, 1970a,b; Kessler, 1974) and photosynthetic bacteria (Kondrat'eva, 1965), although it appears to be coupled to carbon dioxide reduction only in the Chlorophyceae. These auxiliary processes have not been reported among the higher plants.

6.6.1 Ribulose-1,5-diphosphate Carboxylase

D-Ribulose-1,5-diphosphate carboxylase (carboxydismutase; Fraction I Protein; EC 4.1.1.39) is the enzyme directly involved in the assimilation of molecular carbon dioxide by autotrophic organisms.* In eukaryotes and in some prokaryotes, the enzyme is constructed of eight heavy (molecular weight 51,000–58,000 daltons) and eight lighter (molecular weight 12,000–18,000 daltons) subunits. In tobacco, the codons for the heavy subunits are located in the chloroplast genome (Singh and Wildman, 1973), and these polypeptides are synthesized on chloroplastic ribosomes (Blair and Ellis, 1973). The smaller subunits are encoded in nuclear DNA (Kawashima and Wildman, 1972) and are synthesized on extrachloroplastic ribosomes. The spinach leaf carboxydismutase also catalyzes the oxygenation of ribulose 1,5-diphosphate to 3-phosphoglycerate and phosphoglycolate (Ryan and Tolbert, 1975).

It is likely that carboxydismutase did not originate with the early photosynthetic organisms, because it is present in chemolithotrophic bacteria and is inducible in *Escherichia coli* grown on carbon dioxide (Fuller and Gibbs, 1959; McFadden, 1973). Addition of ribulose 1,5-diphosphate to cultures of *Astasia* sp. also induces carboxydismutase activity.

* Carboxydismutase is not the only enzyme by which carbon dioxide can be assimilated. The Hatch–Slack (Section 6.4) and crassulacean acid pathways are alternatives in some organisms (Walker, 1974). The photosynthetic bacteria incorporate carbon dioxide partly by a reductive TCA cycle, of which there are several modifications. Sirevåg (1974) has reviewed reports of other mechanisms in photosynthetic bacteria. Lynch and Calvin (1952) have studied carbon dioxide fixation in bacteria, *Tetrahymena geleii, Physarum polycephalum, Allomyces arbuscula, Blastocladia pringsheimii,* and *Hansenula anomala.* Higher animals also fix carbon dioxide into the carboxyl position of oxaloacetate, but cannot utilize the product in hexose formation.

Due to the large size of the enzyme, comparative studies have involved only relatively indirect biochemical methods. Immunological tests have demonstrated the similarity of bacterial carboxydismutases, on one hand, and green algal and higher plant enzymes on the other. The amino acid composition of the large subunit from a wide range of protists and higher plants is relatively similar, although there is less similarity among the small subunits where present (McFadden and Tabita, 1974). Furthermore, there appear to be at least four broad groups of carboxydismutases as judged by molecular weight: (a) molecular weight approximately 120,000 daltons (dimeric); (b) molecular weight approximately 360,000 daltons (hexameric); (c) molecular weight approximately 450,000 daltons (octameric); (d) molecular weight approximately 550,000 daltons (eight heavy and eight lighter subunits). The distribution of these forms is shown in Table 16.

6.7 POLYSACCHARIDE BIOSYNTHETIC ENZYMES

The usual biosynthetic route to polysaccharides requires at least three types of enzymes: phosphorylases, synthases, and branching enzymes.* Phosphorylases (EC 2.4.1.1) add glycosyl phosphates onto two- or three-unit primer molecules in a linear fashion. Synthases (EC 2.4.1.11) require long-chain primers, and add to these primers glycosyl units derived from adenosine diphosphoglycosyl or uridine diphosphoglycosyl donors. Both phosphorylases and synthases form α-1,4 linkages. Branching enzymes (EC 2.4.1.18) may be classified into two categories, those with true branching activity (branching enzymes, or BE's) and those with "Q" activity. The former are able to form branches on amylopectin (branched) polymers, whereas Q enzymes can branch only amylose (straight chain) molecules (Fredrick, 1970). Both BE and Q activities result in the formation of α-1,6 bonds.

Because other chemical linkages are found in protistan polysaccharides, it is clear that enzymes other than these three types must be active also. Moreover, although these enzymes are usually thought of as acting upon glucose monomers or polymers, polymers containing other sugars and sugar derivatives are common among the protists. Therefore the complete picture is not to be found in the examination of only these three enzyme types. Unfortunately, knowledge of other polysaccharide biosynthetic enzymes is minimal.

* For alternative routes see Lwoff (1951).

TABLE 16

Distribution of D-Ribulose-1,5-Diphosphate Carboxylases

Organism	MW type[a]	Purified enzyme?	Reference
Bacteria			
Chlorobium limicola	Absent		Chernyad'ev *et al.*, 1975
C. thiosulfatophilum	b	Yes	Tabita *et al.*, 1974a
Chromatium D	d	Yes?	McFadden, 1973
Ectothiorhodospira spp. (2 spp.)	+	No	McFadden and Tabita, 1974; Chernyad'ev *et al.*, 1975
Escherichia coli			
Glucose grown	Absent		Fuller and Gibbs, 1959
Xylose + CO_2 grown	+	No	Fuller and Gibbs, 1959
Hydrogenomonas spp. (2 spp.)	d	Yes	McFadden, 1973
Rhodopseudomonas spp. (2 spp.)	b	No	McFadden, 1973
Rhodospirillum rubrum	a	No	McFadden, 1973; Sirevåg, 1974
Thiobacillus denitrificans	b	No	McFadden, 1973
T. neopolitanus	+	No	MacElroy *et al.*, 1968
T. thioparus	+	No	MacElroy *et al.*, 1968
Thiopedia sp.	+	No	Hurlbert and Lascelles, 1963
Cyanophyceae			
Agmenellum quadruplicatum	c	Yes	Tabita *et al.*, 1974b
Anabaena cylindrica	+	No	Lord *et al.*, 1975
A. spiroides	+	No	Lord *et al.*, 1975
Anabaena sp.	+	No	Lord *et al.*, 1975
Anabaenopsis circularis	+	No	Lord *et al.*, 1975
Anacystis nidulans	d	No	McFadden, 1973
Plectonema boryanum	d	No	McFadden, 1973
Tolypothrix tenuis	+	No	Latzko and Gibbs, 1969
Dinophyceae			
Gonyaulax polyedra	+	No	Bush and Sweeney, 1972
Phaeophyceae			
Laminaria hyperborea	+	No	Weidner and Küppers, 1973
Chlorophyceae			
Chlamydomonas reinhardi	d	Yes	Givan and Criddle, 1972; Iwanij *et al.*, 1974
Chlorella ellipsoidea	d	Yes	McFadden, 1973
C. fusca	d	Yes	Lord and Brown, 1975

TABLE 16 (*Continued*)

Organism	MW type[a]	Purified enzyme?	Reference
C. pyrenoidosa			
Light grown	+	No	Gibbs *et al.*, 1970
Light + glucose grown	+	No	Gibbs *et al.*, 1970
Dark + glucose grown	+	No	Gibbs *et al.*, 1970
C. variegata			
Light grown		Absent	Fuller and Gibbs, 1959
Light + ribulose $diPO_4$	+	No	Fuller and Gibbs, 1959
Dark grown		Absent	Fuller and Gibbs, 1959
Dark + glucose grown		Absent	Fuller and Gibbs, 1959
Euglenophyceae			
Astasia sp.			
Normal medium		Absent	Fuller and Gibbs, 1959
Medium + ribulose $diPO_4$		Trace	Fuller and Gibbs, 1959
Euglena gracilis str. Z	d	Yes	McFadden, 1973; Lord and Brown, 1975
Photoautotrophic		Trace	Fuller and Gibbs, 1959
Green + ribulose $diPO_4$	+	No	Fuller and Gibbs, 1959
Streptomycin bleached		Absent	Fuller and Gibbs, 1959
Dark grown		Trace	Gibbs *et al.*, 1970
Dark + ribulose $diPO_4$		Trace	Fuller and Gibbs, 1959
Higher plants (8 spp.)	d	Yes	Fuller and Gibbs, 1959; McFadden, 1973
Ascomycetes (2 spp.)		Absent	Fuller and Gibbs, 1959
Protozoa (1 sp.)		Absent	Fuller and Gibbs, 1959

[a] a,b,c,d refer to molecular weight classes described in text. + indicates presence of carboxydismutase of uninvestigated molecular weight. Trace refers to very low activities of carboxydismutation. Absent denotes inability to find detectable levels of activity.

6.7.1 Phosphorylases

Phosphorylases have been studied in bacteria, several algae, and the higher plants. Although it now appears that their primary function is in carbohydrate degradation (Fredrick, 1970), they are probably also concerned with the early stages of starch formation. *Cyanidium caldarium* possesses only one phosphorylase isoenzyme (Fredrick, 1972), but two isoenzymes differing between themselves in substrate and primer specificities have been found in Cyanophyceae and Chlorophyceae. Other Rhodophyceae appear to possess three or more phosphorylase isoenzymes, none of which requires a primer molecule (Craigie, 1974). Phosphorylases have been less thoroughly examined in members of the Euglenophyceae, Chrysophyceae (Craigie, 1974), Phaeophyceae (Percival and McDowell, 1967), and Zygomycetes (Yin, 1948).

6.7.2 Synthases

Polysaccharide synthases (nucleotide glucosyltransferases) have been most extensively studied in the higher plants, although a few protists have also been examined. It is interesting that substrate specificities of these enzymes have been demonstrated to be very dependent upon how strongly the enzyme is adsorbed to the polymer it is synthesizing. A starch synthetase, for example, was mechanically released from the starch granules it was adsorbed to; when released, its affinity for adenosine diphosphoglucose increased immediately, and its affinity for uridine diphosphoglucose decreased simultaneously. Other such phenomena have been discussed by Fredrick (1970). Although most closely studied in *Dictyostelium discoideum,* they may be characteristic of other protists (and other enzyme systems) as well. Such findings point out that biochemical criteria should be thoroughly understood before phylogenies are constructed on them.

No significant differences have been found among polyglucoside synthases of Cyanophyceae and Rhodophyceae (including *Cyanidium caldarium*). All these organisms possess two isoenzymes acting on uridine diphosphoglucose and on adenosine diphosphoglucose (Fredrick, 1970, 1972). Although a chloroplast-associated synthase has been reported in red algae, there is no indication that it is coded in the chloroplast genome. Although there appear to be interesting modifications in the control of polyglucoside biosynthesis in Chlorophyceae, two synthase isoenzymes electrophoretically similar to those in blue-green and red algae are present in these organisms as well.

6.7.3 Branching Enzymes

It is the branching enzymes that show the most promise as a phylogenetically interesting group of polysaccharide biosynthetic enzymes. In bacteria, only a single branching enzyme is present, possessing both BE and Q activities (Fredrick, 1970). It is perhaps significant that *Cyanidium caldarium* produces branching enzymes closely resembling those of the Cyanophyceae (Fredrick, 1968b, 1972), since this organism has been regarded (Chapman, 1975) as a rhodophycean alga that is either primitive or reduced. In the red algae, three electrophoretically distinct isoenzymes have been found; only one of them is able to branch amylopectins, whereas the other two possess Q activity. Of the three isoenzymes discovered in green algae, none possesses BE activity (Fredrick, 1970). Fredrick (1970) has also pointed out that the relationship among the Cyanophyceae, Rhodophyceae, and Chlorophyceae (on the basis of the Q/BE ratio) closely parallels the phylogeny drawn by Klein and Cronquist (1967) on the basis of the distribution of carbohydrates in these taxa.

A distinction must be drawn at this point, however, between a trend evident in a group of biochemical (or other) data and a phylogeny. It is amply clear that the more morphologically primitive groups (prokaryotes) possess low Q/BE ratios whereas the more morphologically advanced algae have higher ratios. The fact that more isoenzymes of the branching enzymes are found in eukaryotes may constitute a fair case for evolution of isoenzymes by gene duplication. But it does not necessarily follow that the green algae have evolved from the red algae, nor that *Cyanidium caldarium* is phylogenetically intermediate between blue-green algae and eukaryotes, nor that the blue-green algae have evolved from the bacteria. Amino acid sequences and tertiary structures of the various isoenzymes might be better suited to phylogenetic reconstruction than isoenzyme ratios.

Branching enzymes have also been investigated in the higher plants (Fredrick, 1968a), Hemiascomycetidae (Kjölberg and Manners, 1963), and higher animals.

6.8 FATTY ACID SYNTHASES

Fatty acid synthases catalyze the formation of long-chain saturated fatty acids from malonyl coenzyme A and acetyl coenzyme A, and are probably found in all organisms except the viruses. Because they have been examined in a number of protists and higher organisms, they have the potential to provide an interesting biochemical character for phylogenetics.

These enzymes occur in nonmembranous aggregates in yeasts, although in bacteria and higher plants they are nonaggregated and soluble (Erwin, 1973). Because the multienzyme complexes of yeast fatty acid synthases have molecular weights of greater than one million daltons, it is unlikely that primary and tertiary structures will be available in the near future (Lynen, 1967). Fatty acid synthases have been classified into three groups on the basis of their molecular weights, requirements for substrates and primers, and enzymatic products.

Type I synthases occur as high molecular weight enzyme complexes that do not require acyl carrier protein (Volpe and Vagelos, 1973), do require NADH and NADPH, and produce a product (largely palmitate) bound to coenzyme A. Type II enzymes (which are ACP-dependent) are not aggregated into a high molecular weight complex and produce more stearate than palmitate. A third type of synthase has recently been described which is also ACP-dependent, but appears to require longer fatty acids as chain initiators (Goldberg and Bloch, 1972). It is not known whether all fatty acid synthases of a given type are homologous among themselves, although weak immunological cross-reactivity has been observed between Types I and II synthases from *Euglena gracilis* var. *bacillaris* (Ernst-Fonberg *et al.*, 1974).

There is circumstantial evidence that some Type II fatty acid synthases are homologous among themselves. Goldberg and Bloch (1972) noted that ACP from *Escherichia coli* was able to satisfy the ACP requirement of the *Euglena gracilis* Z Type II synthase; a similar phenomenon has been seen in the spinach chloroplast enzyme, which has many properties of Type II synthases (Kannangara *et al.*, 1973). Alternatively, this could reflect evolutionary conservation of features of ACP.

The distribution of these three classes of fatty acid synthases (Table 17) reveals similarity among enzymes from *Mycobacterium phlei,* yeast, *Euglena gracilis* strains Z and *bacillaris,* and higher animals. The other prokaryotes examined differ from *M. phlei* in possessing the nonaggregated Type II fatty acid synthases, and *Chlamydomonas reinhardi* and higher plants differ from *E. gracilis* similarly. The Type II synthase of *E. gracilis* is localized in the chloroplast and is translated on chloroplastic ribosomes (Goldberg and Bloch, 1972; Ernst-Fonberg *et al.*, 1974), but is probably encoded in the nuclear DNA, as in *Chlamydomonas reinhardi* (Sirevåg and Levine, 1972). This result does not support the endosymbiotic theory of chloroplast evolution. Fatty acid synthases of Cyanophyceae have not been closely studied.

Data on the distribution of the Type III synthase, only recently discovered, are too incomplete to warrant speculation on its significance.

6.9 LYSINE BIOSYNTHETIC PATHWAYS

One of the more interesting examples of biosynthetic variation among the protists is provided by the pathways of lysine biosynthesis. All organisms capable of lysine biosynthesis utilize either the *meso*-α,ϵ-diaminopimelic acid pathway (DAP pathway) or the L-α-adipic acid pathway (AAA pathway); no organism is known to utilize both sets of biosynthetic enzymes. Chloroplasts also biosynthesize lysine, but the pathway used is not known (Kirk and Leech, 1972), although a recent report (Mazelis *et ai.*, 1976) indicates the presence of DAP carboxylase in higher-plant chloroplasts.

TABLE 17

Distribution of Fatty Acid Synthases

Organism	Type	Reference
Bacteria and Actinomycetes		
Bacillus subtilis	II	Butterworth and Bloch, 1970
Clostridium kluyveri	II	Goldman *et al.*, 1963
Escherichia coli	II	Goldberg and Bloch, 1972
Mycobacterium phlei	I + III	Goldberg and Bloch, 1972
Ascomycetes		
Yeast	I	Lynen 1961, 1967
Higher animals		
Chicken, pigeon, rat, and rabbit tissues	I	Volpe and Vagelos, 1973
Euglenophyceae		
Euglena gracilis var. *bacillaris*		
Dark grown	I	Ernst-Fonberg *et al.*, 1974
Light grown	I + II	Ernst-Fonberg *et al.*, 1974
E. gracilis str. Z		
Dark grown	I + III	Goldberg and Bloch, 1972
Light grown	I + II + III	Goldberg and Bloch, 1972
Chloroplastic	II + III	Goldberg and Bloch, 1972
Cytoplasmic	I + III	Goldberg and Bloch, 1972
Chlorophyceae		
Chlamydomonas reinhardi		
Dark grown	II + III	Goldberg and Bloch, 1972; Sirevåg and Levine, 1972
Light grown	II + III	Goldberg and Bloch, 1972; Sirevåg and Levine, 1972
Chloroplastic	II	Goldberg and Bloch, 1972; Sirevåg and Levine, 1972
Higher plants		
Entire plant	II	Goldberg and Bloch, 1972
Chloroplastic	II	Kannangara *et al.*, 1973

The demonstration of these pathways in protists has involved the use of radioactive tracer labeling experiments (Vogel, 1965). Isolation of the intermediate compounds is difficult because they occur in tiny quantities. Although AAA itself has been found in higher plants (Fowden, 1965; Møller, 1974) and in the culture medium of a *Corynebacterium hydrocarboclastus* mutant (Coudert and Vandecasteele, 1975), it apparently does not participate in lysine biosynthesis in these organisms. Unfortunately, most lysine biosynthetic enzymes have not been isolated. The DAP pathway involves seven biosynthetic steps, and the AAA pathway eight (none identical with reactions of the other pathway). Thus it is likely that the dichotomy in distribution of lysine biosynthetic pathways represents a significant biochemical difference among the protists.

There is considerable disagreement in the literature whether protozoa (especially rumen ciliates) are capable of lysine biosynthesis. Most protozoa that have been successfully cultured in defined media require exogenously applied amino acids, including lysine, for growth (Kidder, 1967). The absence of a lysine requirement by *Crithidia oncopelti* was shown to be due to an endosymbiotic gram-positive bacterium that biosynthesized lysine via the DAP pathway (Gill and Vogel, 1963; Mandel, 1967). Nine other trypanosomes were found to grow in a lysine-free medium containing *meso*-DAP, presumably synthesizing their own lysine from the DAP (AAA was ineffective) (Guttman, 1967), although bacterial contamination was not ruled out.

In a recent series of papers, Onodera and Kandatsu (1973, 1974; Onodera *et al.*, 1974) have reported that rumen ciliates (largely *Entodinium* spp.) isolated from the digestive tract of a goat are capable of converting DAP into lysine, but cannot biosynthesize DAP itself. Instead they may obtain it from their diet of rumen bacterial cell walls. These results strongly suggest that an important biological relationship exists between rumen protozoa and rumen bacteria. Unfortunately these data shed little light upon the phylogeny of rumen protozoa.

Investigation of lysine biosynthesis by rumen protozoa is further complicated by the controversy concerning whether these protozoa usually contain bacterial endosymbionts. Hungate (1966) has raised this possibility for the biosynthesis of starch and cellulose by rumen protozoa. Many protozoa are known to harbor bacteria, both as endo- (Margulis, 1970; Bonhomme-Florentin, 1971; Stevenson, 1972) and as ectosymbionts (Cleveland and Grimstone, 1963). Guttman and Eisenman (1965) have reported that *Crithidia oncopelti*, when cured of its bacterial endosymbiont, was no longer able to biosynthesize lysine and did not contain DAP carboxylase. It is difficult to comment further on the phylogenetic repercussions of this research until the specialists involved reach some measure of consensus.

The presence of the AAA pathway in certain fungi (Eumycota) and in *Euglena gracilis* strain T argues strongly for the common origin of these two groups. Given the presence of the DAP pathway in the Oomycota, prokaryotes, algae, and possibly in some protozoa (Table 18), it may be necessary to derive the AAA pathway organisms from DAP pathway ones. This is in many respects somewhat unsavory to phylogeneticists: How could the loss of an entire (DAP) pathway have been of selective advantage, and

TABLE 18

Distribution of Lysine Biosynthetic Pathways

Organism	Reference
DAP PATHWAY	
Bacteria and Actinomycetes	
Aerobacter aerogenes	Vogel, 1965
Agrobacterium radiobacter	Vogel, 1965
Alcaligenes faecalis	Vogel, 1965
Arthrobacter globiformis	Vogel, 1965
Azotobacter agilis	Vogel, 1965
Bacillus subtilis	Vogel, 1965
Beggiatoa alba[a]	Holm-Hansen *et al.*, 1965
Erwinia carotovora	Vogel, 1965
Escherichia freundii	Vogel, 1965
Flexibacter litorale[a]	Holm-Hansen *et al.*, 1965
F. rubrum[a]	Holm-Hansen *et al.*, 1965
Hydrogenomonas facilis	Vogel, 1965
Micrococcus denitrificans	Vogel, 1965
M. lysodeikticus	Vogel, 1965
Nocardia spp.[a]	Aaronson and Hutner, 1966; Lechevalier *et al.*, 1971
Proteus rettgeri	Vogel, 1965
Pseudomonas fluorescens	Vogel, 1965
Rhodopseudomonas spheroides	Vogel, 1965
Salmonella typhimurium	Vogel, 1965
Saprospira grandis[a]	Holm-Hansen *et al.*, 1965
S. thermalis[a]	Holm-Hansen *et al.*, 1965
Serratia marcescens	Vogel, 1965
Streptococcus bovis	Vogel, 1965
Cyanophyceae	
Anabaena variabilis[a]	Holm-Hansen *et al.*, 1965
?*Anacystis nidulans*[a]	Holm-Hansen *et al.*, 1965
Nostoc muscorum	Vogel, 1965
Nostoc sp.[a]	Holm-Hansen *et al.*, 1965
Plectonema boreyanum	Vogel, 1965
Tolypothrix tenuis[a]	Holm-Hansen *et al.*, 1965
Rhodophyceae	
Cyanidium caldarium[a]	Hirose, 1958
Chlorophyceae	
Chlorella ellipsoidea[a]	Fujiwara and Akabori, 1954
C. pyrenoidosa	Vogel, 1965
C. vulgaris	Vogel, 1965

TABLE 18 (*Continued*)

Organism	Reference
Oomycetes	
Achlya bisexualis	Vogel, 1965
Pythium ultimum	Vogel, 1965
Sapromyces elongatus	Vogel, 1965
Sirolpidium zoophthorum	Vogel, 1965
Thraustotheca clavata	Vogel, 1965
Hyphochytridiomycetes	
Hyphochytrium catenoides	Vogel, 1965
Rhizidiomyces sp.	Vogel, 1965
Protozoa	
Blastocrithidia leptocoridis	Guttman, 1967
Crithidia spp. (6 spp.)	Guttman, 1967
Entodinium and others	Onodera and Kandatsu, 1973
Leptomonas spp. (2 spp.)	Guttman, 1967
Trypanosoma mega	Guttman, 1967
Higher plants (10 spp.)	Vogel, 1965; Vogel *et al.*, 1970; Møller, 1974
AAA PATHWAY	
Euglenophyceae	
Euglena gracilis str. T	Vogel *et al.*, 1970
Chytridiomycetes	
Allomyces macrogynus	Vogel, 1965
Monoblepharella laurei	Vogel, 1965
Phlyctochytrium punctatum	Vogel, 1965
Rhizophlyctis rosea	Vogel, 1965
Zygomycetes	
Cunninghamella blakesleeana	Vogel, 1965
Rhizopus stolonifer	Vogel, 1965
Syncephalastrum racemosum	Vogel, 1965
Ascomycetes	
Dipodascus uninucleatus	Vogel *et al.*, 1970
Gibberella fujikuori	Vogel *et al.*, 1970
Morchella crassipes	Vogel *et al.*, 1970
Neurospora crassa	Vogel *et al.*, 1970
Saccharomyces cerevisiae	Kurtz and Bhattacharjee, 1975
Sclerotinia fructicola	Vogel, 1965
Taphrina deformans	Vogel *et al.*, 1970
Venturia inaequalis	Vogel *et al.*, 1970
Imperfect fungi	
Penicillium chrysogenum	Vogel *et al.*, 1970

TABLE 18 (*Continued*)

Organism	Reference
Basidiomycetes	
Calvatia gigantea	Vogel *et al.*, 1970
Coprinus radians	Vogel *et al.*, 1970
Polyporus tulipiferus	Vogel *et al.*, 1970
Rhodotorula glutinis	Kurtz and Bhattacharjee, 1975
Ustilago maydis	Vogel *et al.*, 1970

[a] DAP has been reported but the DAP–lysine pathway has not properly been established.

how were all the new AAA pathway enzymes recruited? LéJohn (1974) has considered this problem and has pointed out that, surprisingly, lysine degradation appears to proceed by a reversed AAA pathway, with no evidence for a reversed DAP pathway. If one accepts the idea that catabolism preceded biosynthesis in evolution, one is then left with an ancestral group that possessed an AAA degradative pathway. LéJohn (1974) then proposes that three phylogenetic lines emerged.

1. Development of AAA degradation and then DAP biosynthesis replacing it (prokaryotes, algae, Oomycota, and higher plants).
2. Retention of only the AAA degradation pathway (higher animals and possibly protozoa).
3. A third line, which evolved from the second, in which an AAA biosynthetic pathway emerged, replacing the degradative one.

This is an intriguing proposal since it places the DAP as the later development. It does provide a solution for some problems, such as the failure of higher animals to biosynthesize lysine. Higher animals do, however, possess catalytic activity corresponding to the final AAA pathway enzyme, saccharopine dehydrogenase (Fellows and Lewis, 1973). It still leaves unanswered the problem of the sudden appearance of the DAP biosynthetic pathway. However one regards this problem, there is still the significant dichotomy that separates the Euglenophyceae and the Eumycota. Lysine biochemistry is obviously a prime candidate for further work. Investigation of the primary and tertiary structures of the enzymes may provide some answers. The range of organisms (and classes) examined should obviously be extended and in view of LéJohn's (1974) suggestion it would be profitable to examine the lysine biochemistry of some primitive anaerobes (e.g., clostridia).

A further use of lysine biosynthesis in phylogenetics is seen among the actinomycetes; although all utilize the same (DAP) pathway, some actinomycetes accumulate *meso*-2,6-diaminopimelate in their cell walls. Others preferentially accumulate its direct precursor, LL-diaminopimelate. The mycobacteria, which accumulate the latter compound, are morphologically distinguishable from some members of the genus *Nocardia*, which accumulate *meso*-DAP (Thiemann *et al.*, 1969; Lechevalier *et al.*, 1971). Such differences could result from small differences in the terminal lysine biosynthetic enzymes of these organisms, or could be caused by enzymes of cell wall biosynthesis.

6.10 ORNITHINE BIOSYNTHESIS AND THE ORNITHINE–CITRULLINE CYCLE

Ornithine is a widely distributed amino acid that is not normally incorporated into proteins. Most investigated protists biosynthesize ornithine from glutamic γ-semialdehyde in a reaction catalyzed by ornithine δ-transaminase. However, a few bacteria (and possibly a yeast) have been shown to possess a pathway leading through N-acetylated intermediates (Table 19). Scher and Vogel (1957) were unable to detect ornithine δ-transaminase activity in any of twelve gram-negative bacteria or in *Anacystis nidulans*. Apparently there have been no reports of ornithine biosynthesis by isolated chloroplasts.

Most protists are able to metabolize ornithine to citrulline, then to arginosuccinate, and finally to arginine. Some or all of the three enzyme activities required for this conversion (ornithine transcarbamylase, arginosuccinate synthetase, arginosuccinase) have been demonstrated in *Streptococcus faecalis* (Knivett, 1954), pleuropneumonia-like organisms (Smith, 1957), *Nostoc muscorum* (Holm-Hansen and Brown, 1963), *Gracilaria confervoides*, *Neomonospora furcellata* (Reuter, 1961), *Polysiphonia lanosa* (Citharel, 1971), *Nitzschia closterium* (Rho, 1957), *Chlorella* spp. (Walker and Myers, 1953; Naylor, 1959), *Cladophora prolifera*, *Ulva lactuca* (Reuter, 1961), *Micrasterias crux-melitensis* (Citharel and Gueune, 1974), yeast (Ratner and Petrack, 1953), *Neurospora crassa* (Fincham and Boylen, 1955), *Torulopsis utilis* (Vogel, 1955), *Agaricus bisporus* (Levenberg, 1962), *Ustilago cynodontis* (Prieur and Tavlitzki, 1974), higher plants (Naylor, 1959; Reuter, 1961), and higher animals (Cohen and Brown, 1960). They have not been detected in *Taonia* sp. (Reuter, 1961), *Crithidia fasciculata* (Kidder, 1967), or *Tetrahymena pyriformis* (Hill, 1972).

Some organisms are able to convert arginine to urea, regenerating ornithine (the ornithine–citrulline cycle or Krebs–Henseleit urea cycle). Arginase, the enzyme responsible for this conversion, has been detected in relatively few algae (Citharel and Gueune, 1974; Morris, 1974), although it

TABLE 19

Pathways of Ornithine Biosynthesis in Protists

Organism	Reference
Biosynthesis from N-acetylated intermediates	
Bacillus subtilis	Vogel and Vogel, 1963
Escherichia coli	Vogel, 1953
Micrococcus glutamicus	Udaka and Kinoshita, 1958
Saccharomyces cerevisiae?	DeDeken, 1963
Biosynthesis involving ornithine δ-transaminase activity	
Arthrobacter histidinolovorans	Scher and Vogel, 1957
Bacillus spp. (4 spp.)	Scher and Vogel, 1957
Micrococcus lysodeikticus	Scher and Vogel, 1957
Mycobacterium ranae	Scher and Vogel, 1957
Coprinus lagopus	Scher and Vogel, 1957
Neurospora crassa	Scher and Vogel, 1957
N. sitophila	Scher and Vogel, 1957
Saccharomyces cerevisiae	Scher and Vogel, 1957
Torulopsis utilis	Scher and Vogel, 1957
Chlamydomonas moewusii	Scher and Vogel, 1957
C. reinhardi	Scher and Vogel, 1957
Tetrahymena pyriformis	Scher and Vogel, 1957
Spinach leaf	Scher and Vogel, 1957
Calf liver	Scher and Vogel, 1957

is common in fungi and higher animals (Cohen and Brown, 1960). Urea, whether arising from arginase activity or from the environment, can be assimilated by many protists. Two separate enzymatic mechanisms for urea metabolism have been identified: (a) urease (EC 3.5.1.5), and (b) "ATP: urea amidolyase," composed of urea: carbon dioxide ligase (ADP-linked) (EC 6.3.4.6) and aryl-acylamide amidohydrolase (EC 3.5.1.13). The taxonomic distribution of these two metabolic routes is shown in Table 20. Neither pathway could be found in *Cyanidium caldarium, Porphyridium aerugineum* (Berns *et al.*, 1966), *Trypanosoma cruzi, Trichomonas* (two species), and *Leishmania* (three species) (Seneca *et al.*, 1962).

6.11 ISOLEUCINE BIOSYNTHESIS PATHWAYS

Isoleucine biosynthesis in bacteria and blue-green algae (Hood and Carr, 1972), and presumably in many other protists, involves the enzymes threonine deaminase, α-acetohydroxyacid synthase, isopropylmalate synthase, and transaminase B. Remarkably, many of these enzymes are also

TABLE 20

Distribution of Pathways of Urea Assimilation

Organism	Urease activity	"ATP:urea amidolyase" activity	Reference
Bacteria (34 spp.)	+		Seneca *et al.*, 1962; Reithel, 1971; but see Brierley and Brierley, 1968
Cyanophyceae			
Phormidium luridum	+		Berns *et al.*, 1966
Plectonema calothricoides	+		Berns *et al.*, 1966
Synechococcus lividus	+		Berns *et al.*, 1966
Rhodophyceae			
Porphyridium cruentum	±		Berns *et al.*, 1966
Bacillariophyceae			
Phaeodactylum tricornutum	+		Leftley and Syrett, 1973
Chrysophyceae			
Monochrysis lutheri	+		Leftley and Syrett, 1973
Ochromonas malhamensis	+		Lui and Roels, 1970
Xanthophyceae			
Monodus subterraneus	+		Leftley and Syrett, 1973
Chlorophyceae			
Ankistrodesmus braunii		+	Leftley and Syrett, 1973
Asterococcus superbus		+	Leftley and Syrett, 1973
Chlamydomonas reinhardi		+	Leftley and Syrett, 1973
Chlorella ellipsoidea		+	Roon and Levenberg, 1968
C. fusca var. *vacuolata*		+	Morris, 1974
C. pyrenoidosa		+	Roon and Levenberg, 1968
C. vulgaris var. *viridis*		+	Thompson and Muenster, 1971
Dunaliella primolecta		+	Leftley and Syrett, 1973
Micrasterias crux-melitensis	+		Citharel and Gueune, 1974

TABLE 20 *(Continued)*

Organism	Urease activity	"ATP:urea amidolyase" activity	Reference
Nannochloris coccoides		+	Leftley and Syrett, 1973
N. oculata[a]	+		Leftley and Syrett, 1973
Scenedesmus basilensis		+	Leftley and Syrett, 1973
S. obliquus		+	Leftley and Syrett, 1973
Stichococcus bacillaris		+	Leftley and Syrett, 1973
Prasinophyceae			
Tetraselmis subcordiformis	+		Leftley and Syrett, 1973
Tetraselmis sp.	+		Leftley and Syrett, 1973
Higher plants	+		Reithel, 1971
Ascomycetes			
Aspergillus tamarii	+		Zawada and Sutcliffe, 1974
Saccharomyces cerevisiae		+	Roon and Levenberg, 1968
Imperfect fungi			
Candida utilis		+	Roon and Levenberg, 1970

[a] This organism does not contain chlorophyll *b* and is probably not a member of the Chlorophyceae.

used in the biosynthesis of leucine and valine. Antia and Kripps (1973) have compiled properties of threonine deaminase in eighteen taxonomically diverse species of protists. Alternative pathways of isoleucine biosynthesis have been found in mutants of yeast (Vollbrecht, 1974).

In the gram-positive bacterium *Leptospira interrogans*, a pathway bypassing threonine appears to be operative; after acetyl coenzyme A and pyruvate are converted to β-methyloxaloacetate, α-ketobutyrate is formed with the loss of carbon dioxide. α-Ketobutyrate is then converted to L-isoleucine (Charon *et al.*, 1974). This biosynthetic pathway has been found in a very limited number of other bacteria of divergent species. At this early stage, it can hardly be predicted whether this character will be of interest in biochemical phylogenetics.

6.12 TRYPTOPHAN BIOSYNTHESIS AND THE POLYAROMATIC BIOSYNTHETIC PATHWAY

Although tryptophan biosynthesis appears to involve the same complement of enzymes in all protists, specific properties of the enzymes themselves are sufficiently variable to be of use in phylogenetic studies. Amino acid sequence studies have been carried out on only one of these enzymes (the α-chain of tryptophan synthase), demonstrating considerable sequence homology among seven bacteria (Crawford, 1975).

The pathway from erythrose 4-phosphate and phosphoenolpyruvate to tryptophan is detailed in Figure 3. Enzymes leading from erythrose 4-phosphate and phosphoenolpyruvate to chorismate are known as the "polyaromatic biosynthetic enzymes," or "shikimic acid pathway enzymes." The steps from chorismate to tryptophan are called the "tryptophan biosynthetic pathway."

6.12.1 The Polyaromatic Biosynthetic (Shikimate) Pathway

In enteric bacteria, the enzyme dehydroquinate synthase, dehydroquinase, dehydroshikimate reductase, shikimate kinase, and 3-enolpyruvylshikimate-5-phosphate synthase (see Figure 3) appear to be coded for by genes not organized into a single operon. In some fungi, however, these five enzymes occur together in a cluster having "operon-like characteristics" (Berlyn and Giles, 1969). Although the organization of the genes for these enzymes is not known for a wide range of protists—mutants for shikimate kinase have only rarely been reported, even in *Escherichia coli* and *Salmonella typhimurium*—physical properties of the enzymes and of their aggregates have been extensively studied. Again, these data possess certain weaknesses that amino acid sequence data might not have, but nonetheless phylogenetically interesting trends have emerged.

In bacterial extracts, all five enzymes are separable on sucrose gradients after mild extraction, strongly suggesting that they are not aggregated *in vivo* (Berlyn and Giles, 1969, 1973). In three species of Zygomycetes, one of Ascomycetes, and two of Basidiomycetes, all five enzymes consistently appeared in a single aggregate (Ahmed and Giles, 1969). Much the same situation was found in extracts of three Oomycetes, except that partial aggregates were also noticed. In Cyanophyceae, all five enzymes were separable, as in bacteria (Berlyn *et al.*, 1970), but in *Chlamydomonas reinhardi,* in a moss, and in tobacco plants, dehydroshikimate reductase and dehydroquinase activities could not be separated. The situation in *Euglena gracilis* differed considerably from that in the green algae and higher plants,

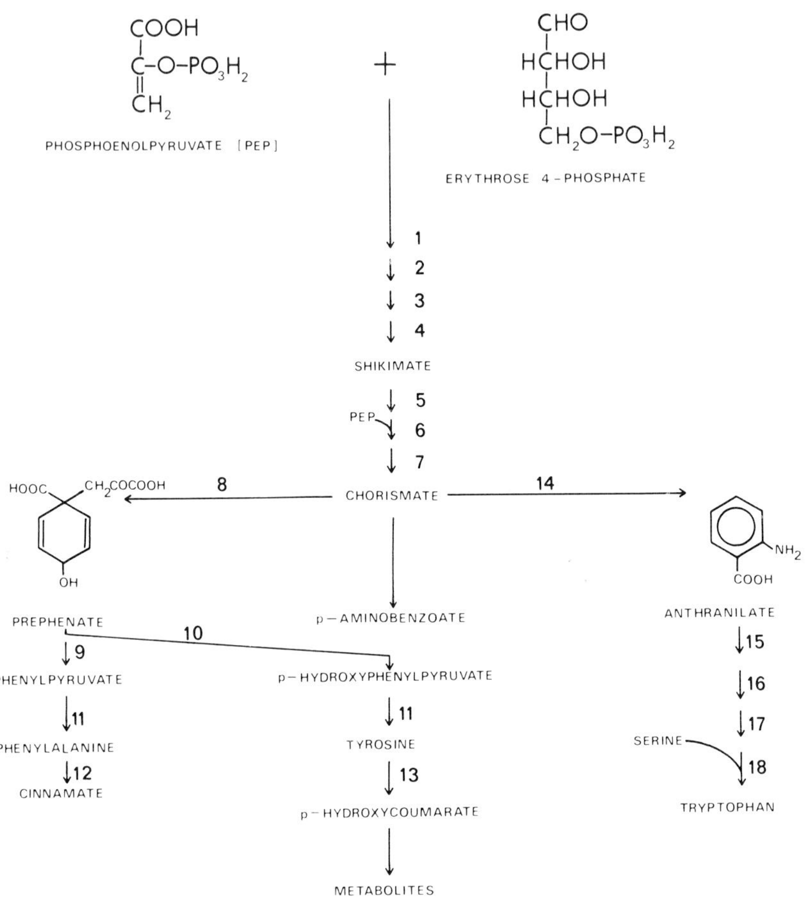

Figure 3. Aromatic shikimate and tryptophan biosynthetic pathways. Enzymes: (1) 3-deoxy-D-arabinoheptulosonate-7-phosphate (DAHP) synthase; (2) dehydroquinate synthase; (3) dehydroquinase; (4) dehydroshikimate reductase; (5) shikimate kinase; (6) 3-enolpyruvylshikimate-5-phosphate synthase; (7) chorismate synthase; (8) chorismate mutase; (9) prephenate dehydratase; (10) prephenate dehydrogenase; (11) transaminases; (12) phenylalanine ammonia-lyase; (13) tyrosine ammonia-lyase; (14) anthranilate synthase; (15) phosphoribosyltransferase; (16) *N*-(5′-phosphoribosyl)anthranilate isomerase; (17) indole-3-glycerophosphate synthase; (18) tryptophan synthase.

as a high molecular weight aggregate was found (as were smaller aggregates reminiscent of those in the Oomycetes) (Berlyn *et al.*, 1970).

Conclusions that can be drawn from these and other data presented by Giles's group are as follows:

1. The organization of these five enzymes in an aggregate appears to be similar in bacteria and Cyanophyceae, further reinforcing the view that these two groups are phylogenetically related.
2. The true fungi examined appear to be closely related to each other (on the basis of these data), but are less similar to the Oomycetes studied (see Ahmed, 1973).
3. The higher plants and a green alga possess similar aggregation patterns of these enzymes, reinforcing the view that they possess a common ancestor.
4. *Euglena gracilis* is in these regards more similar to the fungi than to green algae. It is of interest that the molecular weight of the five-enzyme aggregate in *E. gracilis* (115,000 daltons) is very similar to each of the two identical subunits of the aggregate in *Neurospora crassa* (total molecular weight 230,000 daltons). Moreover, quinic acid induced dehydroquinase, which functions in the anabolism of dehydroquinate to 5-dehydroshikimate (and on to protocatechuic acid), was found in *E. gracilis* and in several Ascomycetes; it has not been found in bacteria (Ahmed and Giles, 1969; Berlyn *et al.*, 1970).

6.12.2 The Tryptophan Biosynthetic Pathway

The L-tryptophan biosynthetic pathway consists of the enzymes anthranilate synthase, phosphoribosyltransferase, *N*-(5′-phosphoribosyl)anthranilate isomerase, indole-3-glycerophosphate synthase, and tryptophan synthase (Figure 3). In many protists, some or all of these enzyme activities occur together in one or more discrete aggregates. It is thought that organization of these enzymes into aggregates reflects clustering and cotranscription of the genes for these enzymes; however, the genetics of tryptophan biosynthesis has been well studied only in some enteric bacteria, where gene transduction can be carried out routinely (Hütter and DeMoss, 1967; Crawford, 1975). In all, eight sedimentation patterns can be discerned among the protists so far investigated (Table 21).

Several observations can be made from an examination of the distribution of these sedimentation patterns (Table 22 and Figure 4).

1. The two Oomycetes examined possess a different organization of tryptophan biosynthetic enzymes than does any of the true fungi (Eumycota) examined.

TABLE 21

Patterns of Tryptophan Pathway Enzyme Sedimentation[a]

Type 0
Separate: 14, 15, 16, 17, 18
Type I
Coprecipitated: 14 + 16 + 17
Separate: 15, 18
Type II
Coprecipitated: 14 + 17
Separate: 15, 16, 18
Type III
In the presence of L-glutamine and EDTA
Coprecipitated: 14 + 16 + 17
Separate: 15, 18
In the absence of L-glutamine and EDTA
Unaffected: 14
Coprecipitated but at a lower ammonium sulfate concentration: 16 + 17
Separate: 15, 18
Type IV
Coprecipitated: 16 + 17
Separate: 14, 15, 18
Type V
Coprecipitated: 14 + 15
Coprecipitated: 16 + 17
Separate: 18
Type VI
Coprecipitated: 14 + 15 + 16 + 17
Separate: 18
Type VII
Coprecipitated: 15 + 16 + 17 + 18
Separate: 14

[a] Enzymes: (14) anthranilate synthase; (15) phosphoribosyltransferase; (16) *N*-(5′-phosphoribosyl)anthranilate isomerase; (17) indole-3-glycerophosphate synthase; (18) tryptophan synthase.

2. The vast majority of the Eumycota examined possess either a Type I or a Type III sedimentation pattern. Since the difference between these types of sedimentation involves only the stability of anthranilate synthase in the absence of L-glutamine and EDTA, the organization of these aggregates could be very similar. The instability of anthranilate synthase could result from a number of factors that may or may not have derived from a common mutational event.

TABLE 22

Distribution of Tryptophan Pathway Sedimentation Patterns

Organism	Pattern type[a]	Reference
Bacteria		
Aerobacter aerogenes	V	Egan and Gibson, 1967
Acinetobacter calco-aceticus (= *Bacterium anitratum*)	0	Twarog and Liggins, 1970
Aeromonas formicans	IV	Largen and Belser, 1975
A. formicans	V	Crawford *et al.*, 1967
Citrobacter ballerupensis	V	Largen and Belser, 1975
C. freundii	V	Largen and Belser, 1975
Enterobacter aerogenes	V	Crawford, 1975
E. cloacae	V	Largen and Belser, 1975
E. hafniae (= *Hafnia alvei*)	IV	Largen and Belser, 1975
E. liquifaciens	IV	Largen and Belser, 1975
Erwinia carotovora	IV	Largen and Belser, 1975
E. dissolvens	V	Largen and Belser, 1975
Escherichia coli	V	Hütter and DeMoss, 1967
Proteus morganii	IV	Largen and Belser, 1975
P. vulgaris	IV	Largen and Belser, 1975
Pseudomonas putida	0	Enatsu and Crawford, 1968
Salmonella typhimurium	V	Crawford, 1975
Serratia marcescens	IV	Largen and Belser, 1975
S. marinorubra	IV	Largen and Belser, 1975
Cyanophyceae		
Anabaena variabilis	IV	Hütter and DeMoss, 1967
Euglenophyceae		
Euglena gracilis str. G	VII	Lara and Mills, 1972
Higher plants	0	Crawford, 1975
Plasmodiophoromycetes		
Physarum polycephalum	I	Hütter and DeMoss, 1967
Oomycetes		
Pythium sp.	IV	Hütter and DeMoss, 1967
Saprolegnia sp.	IV	Hütter and DeMoss, 1967
Chytridiomycetes		
Allomyces macrogynus	I	Hütter and DeMoss, 1967
Rhizophlyctis rosea	I	Hütter and DeMoss, 1967
Zygomycetes		
Mucor hiemalis	III	Hütter and DeMoss, 1967
Phycomyces blakesleeanus	III	Hütter and DeMoss, 1967
Rhizopus arrhizinus	III	Hütter and DeMoss, 1967

TABLE 22 *(Continued)*

Organism	Pattern type[a]	Reference
Ascomycetes		
Aspergillus nidulans	I	Hütter and DeMoss, 1967
Byssochlamys nivea	I	Hütter and DeMoss, 1967
Gibberella fujikuori	I	Hütter and DeMoss, 1967
Morchella esculenta	I	Hütter and DeMoss, 1967
Neurospora crassa	I	Hütter and DeMoss, 1967
Dipodascus uninucleatus	II	Hütter and DeMoss, 1967
Endomyces bisporus	II	Hütter and DeMoss, 1967
Saccharomyces cerevisiae	II	Hütter and DeMoss, 1967
Schizosaccharomyces pombe	II	Crawford, 1975
Basidiomycetes		
Coprinus lagopus	VI	Crawford, 1975
Polyporus circinatus	I	Hütter and DeMoss, 1967
Thanetephorus cucumeris	VI	Crawford, 1975
Tremella mesenterica	III	Hütter and DeMoss, 1967
Ustilago maydis	III	Hütter and DeMoss, 1967
Cryptococcus laurentii var. *flavescens*	III	Hütter and DeMoss, 1967
Rhodotorula glutinis	III	Hütter and DeMoss, 1967
Sporobolomyces salmonicolor	III	Hütter and DeMoss, 1967

[a] Sedimentation types as in Table 21.

3. The occurrence of the Type II sedimentation pattern is restricted to the Endomycetales, which are thus not likely progenitors of the Basidiomycetes (in agreement with morphological studies).

4. Types VI (in some Homobasidiomycetes) and VII (in *Euglena gracilis* strain G) may be considered as evolutionarily specialized characters.

5. The discovery of Type IV organization in both the Oomycetes and *Anabaena variabilis* is perhaps easy to rationalize but is difficult to explain.

6. In view of the presence of Type I organization in the slime mold *Physarum polycephalum*, it would be of interest to examine other slime molds and protozoa.

Tryptophan is an absolute dietary requirement for some protozoa, including *Peranema trichophorum*, *Crithidia fasciculata* (Kidder, 1967), and many ciliates (Holz, 1964; Hill, 1972). Interestingly, *C. oncopelti*, which contains a bacterial endosymbiont (Gill and Vogel, 1963), does not require tryptophan.

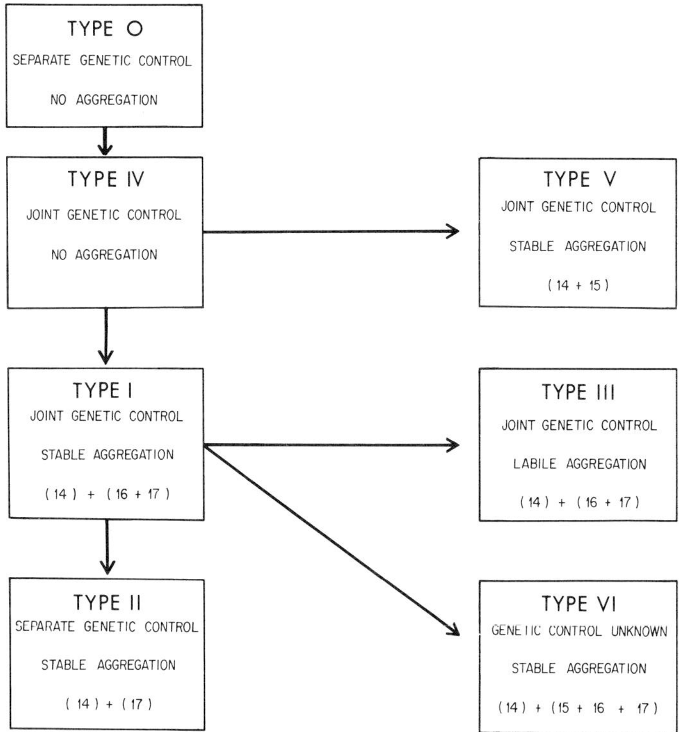

Figure 4. Possible evolution of tryptophan biosynthetic pathways in the Eumycota and Oomycota. After Figure 5 from Hütter and DeMoss (1967). Reproduced by permission of the American Society for Microbiology and the authors. Enzymes: 14: anthranilate synthase; 15: PR transferase; 16: PRA isomerase; 17: InGP synthase.

Tryptophan Synthase

The final enzyme in the tryptophan pathway, tryptophan synthase (EC 4.2.1.20), has been found to catalyze three different reactions *in vitro* (Sakaguchi, 1970; Schmauder *et al.*, 1974):

$$\text{Indole 3-glycerophosphate} + \text{L-serine} \rightarrow \text{L-tryptophan} + \text{D-glyceraldehyde 3-phosphate} + \text{water} \quad \text{(a)}$$

$$\text{Indole} + \text{L-serine} \rightarrow \text{L-tryptophan} + \text{water} \quad \text{(b)}$$

$$\text{Indole 3-glycerophosphate} \rightleftharpoons \text{indole} + \text{D-glyceraldehyde 3-phosphate} \quad \text{(c)}$$

Sites responsible for the "half-reaction" activities (b) and (c) have in several cases been localized on subunits dissociable from the synthase complex.

Two types of subunits prepared from synthases extracted from *Anabaena variabilis*, *Chlorella ellipsoidea*, *Escherichia coli*, and *Salmonella typhimurium* were able to complement each other to different degrees in carrying out reaction (a); no subunit was able to perform this reaction independently (Sakaguchi, 1970). In no case, however, was a subunit from *Neurospora crassa* tryptophan synthase able to complement any of the algal or bacterial subunits, consistent with recent indications that the fungal enzyme is a homodimer (Matchett and DeMoss, 1975). Slight serological cross-reactivities have been observed between green algal and fungal subunits, which could be due either to strongly conserved or to convergent regions.

The greater degree of complementation observed between tryptophan synthase subunits from *Chlorella ellipsoidea* and *Anabaena variabilis* than between either of these and the bacterial enzyme subunits could be rationalized by assuming that the transition from bacteria to blue-green alga brought about, or coincided with, a particularly rapid or marked change in the properties of tryptophan synthase. Just as easily, this result could be due to nonlinearity in the correlation of complementation with amino acid substitution during protein evolution.

Among the gram-negative bacteria, amino acid sequences have been determined for homologous tryptophan synthase α subunits from *Aerobacter aerogenes* (Li and Yanofsky, 1973b), *Escherichia coli* (Guest *et al.*, 1967), and *Salmonella typhimurium* (Li and Yanofsky, 1973a). Partial sequences are known from *Bacillus subtilis*, *Pseudomonas putida*, *Serratia marcescens*, and *Shigella dysenteriae* (Crawford, 1975); tertiary structures are unknown. Subunit complementation techniques (Balbinder, 1964) and immunological tests (Murphy and Mills, 1969; Rocha *et al.*, 1972) reflect the homology observed among the primary structures.

Regulation of Tryptophan Biosynthesis

It has been suggested that Cyanophyceae do not usually rely upon the repression and derepression of genes as a major control mechanism for metabolic processes; instead, these algae apparently rely on modification of previously formed enzymes (Pearce and Carr, 1969). Ingram *et al.* (1972) have, however, provided a counterexample to this generalization. When a mutant of *Agmenellum quadruplicatum* was starved of tryptophan, the activity of its tryptophan synthase B was increased by some twentyfold, and the activities of the four other enzymes in the tryptophan biosynthetic pathway were increased some two- or threefold. The increases were suggested to be due to derepression.

It is of phylogenetic interest, therefore, that one group of bacteria, the pseudomonads, likewise do not tend to use repression control of metabolism. However, the pseudomonads do exhibit the same response to tryptophan starvation observed in *A. quadruplicatum* (Crawford and Gunsalus, 1966; Hegeman and Rosenberg, 1970; Crawford, 1975). Moreover, this "curious coincidence" (Ingram *et al.*, 1972) extends to the differential derepression of the various enzyme activities in the pathway. This "coincidence" is the result either of common ancestry or of convergent evolution. Because it is not known just how intricate and specialized this auxotrophy is at the molecular level, or if it occurs in other bacteria, it is not possible to judge how likely or unlikely convergence might have been.

6.12.3 Chorismate Mutase

Chorismate mutase (EC 5.4.99.5) is the enzyme responsible for committing chorismate to the biosynthesis of tyrosine or phenylalanine (or cinnamate derivatives). It has been studied in bacteria and actinomycetes, Ascomycetes (five species), *Euglena gracilis* strain 1224-5125, *Chlamydomonas reinhardi,* and numerous higher plants. In Ascomycetes and higher plants, three isoenzymes (distinguished by electrophoretic mobilities and inhibition or activation by aromatic amino acids and cinnamate derivatives) have been reported. A much wider range of isoenzymes has been found among bacteria (Woodin and Nishioka, 1973; Zurawski and Brown, 1975). Among filamentous fungi, *Neurospora crassa* contained only isoenzyme CM_1, *Penicillium chrysogenum* possessed CM_1 and CM_3, and *P. duponti* possessed CM_1, CM_2, and CM_3. Further studies using only electrophoretic mobilities, molecular weight data, and inhibitors are unlikely to produce phylogenetically interesting data on this enzyme.

In *Pseudomonas aeruginosa* a bifunctional enzyme complex involved in phenylalanine biosynthesis appears to possess one of the cell's two isoenzymes of chorismate mutase, as well as a molecule of prephenate dehydrogenase (Ahmed and Campbell, 1973).

6.12.4 Tyrosine Biosynthesis

Although tyrosine biosynthesis follows the same general pattern in all protists so far investigated, interesting variations have been observed in the enzymatic details of that pathway, especially among the prokaryotes. Although many enteric bacteria utilize the pathway diagrammed (Figure 3), the blue-green alga *Agmenellum quadruplicatum* converts prephenate first to β-(1-carboxy-4-hydroxy-2,5-cyclohexadien-1-yl)alanine ("pretyrosine"), then synthesizes tyrosine from this intermediate (Stenmark *et al.*, 1974). Whereas *Escherichia coli* utilizes first prephenate dehydrogenase, then 4-hydrophenylpyruvate transaminase, *A. quadruplicatum* possesses, instead, prephenate transaminase and "pretyrosine" dehydrogenase. In effect, the order of dehydrogenation and transamination has been reversed.

Although *A. quadruplicatum* possesses enzymes specific for the pretyrosine pathway, and several bacteria* have enzymes specific for the 4-hydroxyphenylpyruvate pathway, the dehydrogenase from *Pseudomonas aeruginosa* can utilize either prephenate or pretyrosine, and its transaminase either prephenate or 4-hydroxyphenylpyruvate. The regulation of tyrosine biosynthesis in *P. aeruginosa* has characteristics of the tyrosine regulation of both Cyanophyceae and enteric bacteria. These properties could indicate a phylogenetic relationship between blue-green algae and pseudomonads that can furthermore be correlated with similarities in their ecological niches (Jensen and Pierson, 1975).

In the absence of amino acid sequence data and tertiary structures, it is difficult to determine the extent of evolutionary modification needed to alter the substrate specificities of these two enzymes (assuming that they are homologous). As suggested by Stenı..ark *et al.* (1974), it would be interesting to determine which, if either, of these pathways is used by chloroplasts. Kirk and Leech (1972), however, were unable to show significant tyrosine biosynthesis by isolated chloroplasts.

6.12.5 Phenylalanine and Tyrosine Ammonia-Lyases

In many protists, the final products of the shikimate biosynthetic pathway and of the corollary synthetic pathways are simply the three common aromatic amino acids or closely related molecules. Other organisms are able to convert either phenylalanine, or both phenylalanine and tyrosine, into a series of further metabolites based on cinnamic acid. The enzyme responsible for the conversion of phenylalanine into *trans*-cinnamate is L-phenylalanine ammonia-lyase (EC 4.3.1.5). Tyrosine, likewise, is thought to be converted by some organisms into *trans-p*-coumaric acid by an analogous L-tyrosine ammonia-lyase, or "tyrase."

As seen from Table 23, both activities are found only in higher plants (beginning with liverworts) and in certain Basidiomycetes. Tyrosine ammonia-lyase activity is particularly pronounced in members of the family Gramineae (grasses). In other higher plants, tyrase activities are usually an order of magnitude smaller than in the Gramineae. Among the protists, the presence of both activities in some Basidiomycetes merely serves to delineate them from other Basidiomycetes, and (if the absence of these activities in most Basidiomycetes and Ascomycetes is reductive) delineates

* Including members of the genera *Aerobacter, Bacillus, Brevibacterium, Clostridium, Escherichia,* and *Serratia.*

TABLE 23

Distribution of L-Phenylalanine and L-Tyrosine Ammonia-Lyase Activities

Organism	PAL[a]	TAL[a]	Reference
Bacteria and Actinomycetes			
Achromobacter spp. (2 spp.)	–[a]	nd	Ogata *et al.*, 1967
Aerobacter aerogenes	–	nd	Ogata *et al.*, 1967
Alcaligenes faecalis (2 str.)	–	nd	Ogata *et al.*, 1967
Corynebacterium sepedonicum	–	nd	Ogata *et al.*, 1967
Escherichia coli	–	–	Young *et al.*, 1966
Micrococcus spp. (2 spp.)	–	nd	Ogata *et al.*, 1967
Proteus vulgaris	+?	+?	Hirai, 1923
Pseudomonas fluorescens	–	nd	Ogata *et al.*, 1967
Sarcina lutea	–	nd	Ogata *et al.*, 1967
Streptococcus faecalis	–	–	Young *et al.*, 1966
Streptomyces verticillatus	+	–	Bezanson *et al.*, 1970
Ascomycetes[b]			
Cephaloascus fragrans	–	nd	Vance *et al.*, 1975
Ceratocystis fimbriata	–	nd	Vance *et al.*, 1975
Chaetomium globosum	–	nd	Vance *et al.*, 1975
Endomyces hordei	–	nd	Ogata *et al.*, 1967
Hansenula anomala	–	nd	Ogata *et al.*, 1967
Morchella sp.	–	nd	Vance *et al.*, 1975
Nectria cinnabarina	+	+	Vance *et al.*, 1975
Peziza anthracophila	–	nd	Vance *et al.*, 1975
Pichia polymorpha	–	nd	Ogata *et al.*, 1967
Schizosaccharomyces spp. (3 spp.)	–	nd	Ogata *et al.*, 1967
Talaromyces sp.	–	nd	Vance *et al.*, 1975
Zygosaccharomyces spp. (2 spp.)	–	nd	Ogata *et al.*, 1967
Imperfect fungi			
Candida utilis	–	nd	Ogata *et al.*, 1967
Fusarium solani	–	–	Young *et al.*, 1966
Torulopsis candida	–	nd	Ogata *et al.*, 1967
Volucrispora aurantiaca	–	–	Young *et al.*, 1966
Basidiomycetes[b]			
Armillaria mellea	+	nd	Vance *et al.*, 1975
Bovista sp.	+	nd	Vance *et al.*, 1975
Clavaria cristata	+	nd	Vance *et al.*, 1975
Collybia velutipes	+	+	Power *et al.*, 1965
Coprinus domesticus	+	–	Bandoni *et al.*, 1968
Fomes subroseus	+	nd	Vance *et al.*, 1975
Ganoderma lucidum	+	+	Power *et al.*, 1965
G. tsugae	+	+	Power *et al.*, 1965
Lentinus lepideus	+	+	Power *et al.*, 1965
Merulius tremellosus	+	nd	Vance *et al.*, 1975
Platygloea pustulata	+	–	Bandoni *et al.*, 1968
Polyporus adjustus	+	nd	Vance *et al.*, 1975
P. brumalis	+	+	Power *et al.*, 1965

TABLE 23 *(Continued)*

Organism	PAL[a]	TAL[a]	Reference
P. compactus	+	–	Power *et al.*, 1965
P. versicolor	+	+	Power *et al.*, 1965
Pycnoporus sanguineus	+	+	Power *et al.*, 1965
Ramaria secunda	+	nd	Vance *et al.*, 1975
Rhodotorula spp. (25 str.)	+	–	Ogata *et al.*, 1967; Havir and Hanson, 1975
Schizophyllum commune	+	–	Bandoni *et al.*, 1968
Sporidiobolus johnsonii	+	+	Bandoni *et al.*, 1968
Sporobolomyces pararoseus	+	–	Havir and Hanson, 1975
S. roseus	+	+	Bandoni *et al.*, 1968
S. salmoneus	+	+	Bandoni *et al.*, 1968
S. salmonicolor	+	+	Bandoni *et al.*, 1968
Steccherinum adustum	+	–	Bandoni *et al.*, 1968
Stereum hirsutum	+	–	Bandoni *et al.*, 1968
S. pini	–	–	Bandoni *et al.*, 1968
S. sanguinolentum	+	+	Power *et al.*, 1965
Tilletiopsis washingtonensis	+	–	Bandoni *et al.*, 1968
Trametes hispida	+	+	Power *et al.*, 1965
T. suaveolens	+	+	Power *et al.*, 1965
Tricladium splendens	+	nd	Vance *et al.*, 1975
Ustilago bullata	–	–	Bandoni *et al.*, 1968
U. hordei	+	–	Bandoni *et al.*, 1968
U. zeae	+	–	Bandoni *et al.*, 1968
Rhodophyceae			
Polysiphonia lanosa	–	–	Young *et al.*, 1966
Rhodymenia palmata	–	–	Young *et al.*, 1966
Phaeophyceae			
Alaria esculenta	–	–	Young *et al.*, 1966
Haptophyceae			
Isochrysis galbana	–	–	Landymoor, 1976
Bacillariophyceae			
Navicula sp.	–	–	Landymoor, 1976
Chlorophyceae			
Chlorella pyrenoidosa	–	–	Young *et al.*, 1966
Ulva lactuca	–	–	Young *et al.*, 1966
Higher animals (3 spp.)	–	–	Young *et al.*, 1966
Higher plants			
Lichens (2 spp.)	–	–	Young *et al.*, 1966
Mosses (2 spp.)	–	–	Young *et al.*, 1966
Horsetail (1 sp.)	–	–	Young *et al.*, 1966
Liverwort (1 sp.)	+	+	Young *et al.*, 1966
Lycopods (4 spp.)	+	+	Young *et al.*, 1966
Ferns (4 spp.)	+	+	Young *et al.*, 1966
Gymnosperms (8 spp.)	+	+	Young *et al.*, 1966
Monocots (11 spp.)	+	+	Young *et al.*, 1966
Dicots (16 spp.)	+	+	Young *et al.*, 1966

the Basidiomycetes and Ascomycetes from other fungi. Their presence in fungi and higher plants is probably an example of convergent evolution, although mechanisms of gene transfer cannot be ruled out (many Basidiomycetes are "wood-rotting" fungi). In both the Basidiomycetes and the higher plants investigated, light has an effect upon the activity of the ammonia-lyase (Zucker, 1972; Nambudiri *et al.*, 1973; Vance *et al.*, 1975).

The apparent concurrence, in higher plants, of both ammonia-lyases whenever one of them is found (Table 23) is more apparent than real. Several higher plants possess detectable levels of only the phenylalanine ammonia-lyase; the tyrase activity is apparently never found alone. There are, in fact, indications that a single protein may possess both activities at the same catalytic site, and consequently that there may not be a distinct tyrase (Reid *et al.*, 1972). This question remains a topic of current interest in several laboratories.

The distribution of cinnamate-4-hydroxylase, an enzyme in the cinnamate pathway, parallels that of the ammonia-lyases by occurring in the higher plants and in certain specialized Basidiomycetes (Vance *et al.*, 1973).

6.12.6 The 4-Isoprenyltryptophan Pathway

Addition of isoprenyl pyrophosphate or dimethylallylpyrophosphate to tryptophan to form 4-isoprenyltryptophan serves as the starting point for the biosynthesis of a series of secondary metabolites known as ergolines. Common classes of ergolines include the ergot alkaloids and the clavine alkaloids. As the names suggest, they were first isolated from fungi, and have since been found in a range of Zygomycetes, Ascomycetes, Basidiomycetes, and imperfect fungi (Spilsbury and Wilkinson, 1961; Vining, 1973). The discovery of ergolines in several higher plants of the family Convolvulaceae strongly suggests that convergent evolution has occurred in

---

[a] –, activity not detected; +, activity present; nd, not determined.

[b] Both PAL and TAL activities were absent from representatives from the following genera: *Armillaria*, *Bovista*, *Bullera*, *Clavaria*, *Coniophora*, *Dacromyces*, *Daedalea*, *Exobasidium*, *Fomes*, *Ganoderma*, *Itersonilla*, *Lenzites*, *Merulius*, *Montagnea*, *Pleurotus*, *Polyporus*, *Poria*, *Ramaria*, *Sirobasidium*, and *Tremella* (Young *et al.*, 1966; Bandoni *et al.*, 1968). In the Ascomycetes phenylalanine ammonia-lyase could not be detected in *Cephaloascus fragrans*, *Ceratocystis fimbriata*, *Chaetomium globosum*, *Morchella* sp., *Peziza anthracophila*, and *Talaromyces* sp. Although tyrosine ammonia-lyase was not determined for these species, both enzymes were shown to be present in *Nectria cinnabarina* (Vance *et al.*, 1975). These same authors showed PAL was present in *Fomes subroseus*, *Merulius tremellosus*, *Polyporus adustus*, *Ramaria secunda*, and *Tricladium splendens*, all members of the Basidiomycetes. TAL was not determined. These results should be compared with data that are already listed in the table.

these apparently widely separated taxa. Ergolines have not been reported from any algae or lower land plants, nor have they been found in the water molds.

Other alkaloids are also biosynthesized from tryptophan. Nicotinic acid, a precursor to many alkaloids, is biosynthesized from tryptophan in Ascomycetes, Basidiomycetes, and higher animals, whereas in the higher plants, (green?) algae, and bacteria it is biosynthesized, instead, from aspartate and glyceraldehyde via quinolinic acid (Robinson, 1968).

6.13 GLUTAMATE DEHYDROGENASES

Glutamate dehydrogenases are widely distributed enzymes catalyzing the reversible amination of glutamate. Although higher animals possess a glutamate dehydrogenase with both NAD-linked and NADP-linked activities (EC 1.4.1.3), the Ascomycetes (Hypocreales, Sphaeriales, Endomycetales), imperfect fungi, and Basidiomycetes (Agaricales) contain two glutamate dehydrogenase isoenzymes, one linked to NADP (EC 1.4.1.4), the other linked to NAD. NADP-linked forms are also found in *Escherichia coli* (Veronese *et al.*, 1975), ten species of blue-green algae (Carr and Craig, 1970; Neilson and Doudoroff, 1973), *Euglena gracilis* Z (Smillie, 1968), and higher plant chloroplasts (Leech and Kirk, 1968), and *Chlorella pyrenoidosa* may have three distinct glutamate dehydrogenases (Shatilov *et al.*, 1974). Although the bacterial and fungal NADP-linked enzymes possess similar inhibition patterns, molecular weights, subunit sizes, and amino acid compositions, it is not known whether these enzymes are homologous among themselves, nor if they are homologous with NAD-linked glutamate dehydrogenases.

Among the lower fungi and water molds, NAD-linked glutamate dehydrogenases can be classified into four types on the basis of their activators and inhibitors (LéJohn, 1971b, 1974).

Type I: No activators, no inhibitors, no glucose repression of synthesis (GRS). Found in the Chytridiales (Chytridiomycetes) and in most Mucorales (Zygomycetes).

Type II: Activated by Ca^{2+}, Mn^{2+}, and AMP; inhibited by citrate, ATP, FDP, EDTA, and GTP; no GRS. Found in the Blastocladiales (Chytridiomycetes) and in the genus *Absidia* of the Mucorales (Zygomycetes).

Type IIIa: Activated by NADP, NADPH, AMP, acetyl coenzyme A and short-chain derivatives of CoA, and phosphoenolpyruvate; inhibited by ATP, citrate, GTP, long-chain CoA derivatives, Ca^{2+}, and Mg^{2+}; positive GRS. Found in the Hyphochytridiomycetes.

Type IIIb: Same as IIIa except activated (rather than inhibited) by GTP and ATP; not activated by AMP, but instead inhibited. Found in the Oomycetes (see also Wang and LéJohn, 1974a,b).

Using these characters, four groups of lower fungi and water molds can be distinguished. If Types IIIa and IIIb are more similar to each other than to Types I or II (LéJohn, 1974), then the Oomycota (Oomycetes and Hyphochytridiomycetes) can be separated from the lower Eumycota (Chytridiomycetes and Zygomycetes). The Acrasiales (Myxomycota: Plasmodiophoromycetes) apparently possess a Type III dehydrogenase, which remains to be fully resolved.

6.14 LACTATE AND MALATE DEHYDROGENASES

The lactate and malate dehydrogenases have been subjected to tabular treatment (LéJohn, 1974) similar to that for the glutamic dehydrogenases. An enzyme specific for the L(+)-lactate isomer has been found in the Ascomycetes (including the Hemiascomycetes), Basidiomycetes, and Fungi Imperfecti. In the remaining fungi, the D(−)-lactate dehydrogenase (EC 1.1.1.28) exists in two classes. The Chytridiomycetes (both Chytridiales and Blastocladiales) and Zygomycetes (LéJohn, 1971a) possess a class II dehydrogenase characterized by GTP and ATP inhibition. The Hyphochytridiomycetes and Oomycetes possess a class I enzyme (ATP inhibition only). The yeasts (Ascomycetes: Endomycetales) also possess a D(−)-lactate dehydrogenase, but one which is characterized by being cytochrome linked.

All fungal classes possess an NAD-linked malate dehydrogenase, but the Oomycetes and Hyphochytridiomycetes also possess an NADP-dependent malate dehydrogenase. One outstanding feature emerges from the consideration of these three dehydrogenases and isocitrate dehydrogenase (Section 6.3.1). The Oomycetes and Hyphochytridiomycetes (Oomycota) are quite distinct from all other fungi. This and other information indicates only a very distant and tenuous phylogenetic relationship of the Oomycota to the Eumycota and reinforces the taxonomic separation into two divisions.

6.15 NITRATE REDUCTASES

Nitrate reductase (EC 1.6.6.1), the enzyme responsible for the assimilatory reduction of nitrate to nitrite, has been reported in bacteria, fungi, algae, and higher plants. In at least some higher plants, it consists of two enzyme activities, an NAD(P)H dehydrogenase and an FMNH nitrate reductase (Rigano, 1971). In *Neurospora crassa,* where it has been extensively studied, the enzyme requires a molybdenum cofactor; it is in such instances that mineral requirements (e.g., molybdenum) might occasionally be of interest in biochemical phylogenetics. The cofactor requirement is one of the few biochemical properties of nitrate reductases that have been examined in a range of different organisms; even this character may prove difficult to examine, as nitrate reductases are sometimes bound to cytochromes and other enzymes in cell extracts.

In the diatom *Thalassiosira pseudonana* (Amy and Garrett, 1974), most Chlorophyceae, and the higher plants, NADH is the required electron donor for nitrate reductase; in *Dunaliella parva,* however, NADH, NADPH, and FMNH are effective electron donors (Heimer, 1975), as in *Cyanidium caldarium* (Rigano, 1971). *Neurospora crassa* and *Aspergillus nidulans* (McDonald and Coddington, 1974) require NADPH for nitrate reductase activity, whereas *Hansenula anomala* has a nitrate reductase capable of utilizing either NADH or NADPH (Pichinoty and Méténier, 1967). In the bacteria that possess an active nitrate reductase activity, a wide range of cofactor requirements has been found, including FMNH, FADH, NADH, and ferredoxin; some assimilatory nitrate reductases require NADPH (Stanier *et al.*, 1970; Forget, 1974; Van't Riet and Planta, 1975) or NADH (Herrera and Nicholas, 1974). Other bacteria apparently require neither NADH nor NADPH (Pichinoty, 1966; Forget, 1974). *Escherichia coli* has a respiratory nitrate reductase utilizing NADH (Nason, 1962).

It is therefore of interest that *Anabaena cylindrica* possesses an inducible nitrate reductase capable of using electrons derived from NADH (via a diaphorase), NADPH (in the presence of ferredoxin), or, indirectly, from Fd itself (Hattori and Myers, 1967; Hattori and Uesugi, 1968). FMN and FAD are less suitable electron donors for the *A. cylindrica* enzyme (Naylor, 1970). A comprehensive examination of various orders of fungi (Bresinsky and Schneider, 1975) has shown that nitrate reductase (assayed by production of nitrite in the medium) is present in most of the Ascomycetes (Eurotiales, Pezizales, Helotiales, Hypocreales, and Sphaeriales), except the Endomycetales. It appears to be absent in the Oomycetes, Chytridiomycetes, Zygomycetes, and Basidiomycetes (except the Tricholomataceae in the order Agaricales). However, the nature of the test, and the inconsistency in distributions within orders, would preclude assigning phylogenetic significance at this time.

7

Metabolites: I

7.1 MONOSACCHARIDES AND OLIGOSACCHARIDES

For a group of compounds to be useful in biochemical phylogenetics, it must consist of molecules both structurally varied and of sufficiently widespread distribution. Although the monosaccharides meet the former condition moderately well, with few exceptions they are so widespread and uniformly distributed among the protists that few phylogenetic relationships can be elucidated by their distribution (see Table 24).

This cosmopolitan distribution can be explained by their fundamental involvement in the central energy-yielding metabolic pathways. As vital metabolites, they are present in all or virtually all organisms. Other simple sugars are biosynthesized from these fundamental ones by single enzymatic steps. Because many of the monosaccharides listed in Table 24 have probably been present in organisms from early stages of evolution, enzymes further elaborating these sugars into closely related molecules would have had ample opportunity to be selected for, perhaps more than once.

Although some of these sugars have been reported in relatively few groups of organisms, this by no means precludes their presence in many other taxa. Similarly, some sugars listed in Table 24 for a given taxon may have been reported in only one or two members of that taxon. Some of these sugars are more commonly found as phosphorylated derivatives, in glycosides, or in polymers. Others usually may not be accumulated in the cell during normal conditions of growth. Moreover, many of these compounds are often regarded as "uninteresting," and are not often looked for or reported in the modern literature.

TABLE 24

Distribution of Selected Monosaccharides and Disaccharides[a]

Saccharide	Bacteria	Actinomycetes	Cyanophyceae	Oomycetes	Zygomycetes	Ascomycetes-imperfect fungi	Basidiomycetes	Chytridiomycetes	Protozoa	Euglenophyceae	Rhodophyceae	Dinophyceae	Cryptophyceae	Bacillariophyceae	Phaeophyceae	Chrysophyceae	Haptophyceae	Xanthophyceae	Chlorophyceae	Prasinophyceae	Higher plants
Arabinose	+	+	+			+			+		+	+		+			+		+	+	+
Deoxyribose	+	+	+	+	+	+	+	+	+	+	+	+	+	+	+	+	+	+	+	+	+
Fructose	[b]		+							+	[c]	+				+	+		+		+
Fucose	+		+		+	+	+		+		[c]	+		+	+				+		+
Galactose	+	+	+	+	+	+	+	+	+		+	+		+	+	+	+		+	+	+
Glucose	+	+	+	+		+	+	+	+	+	+	+		+	+	+	+	+	+		+
Maltose										+	+	+					+		+	+	+
Mannose	+	+	+	+	+	+	+	+	+		+	+		+	+	+	+	+	+	+	+
Rhamnose	+	+	+	+		+			+			+		+		+	+		+		+
Ribose	+	+	+	+	+	+	+	+	+	+	+	+	+	+	+	+	+	+	+	+	+
Sucrose			+							+	+				?	+		+	+		+
Trehalose	[d]		+			+	+		+	+	+										+
Xylose	+		+			+	[e]		+		+	+		+	+	+	+	+	+		+

[a] Free, phosphorylated, in polymers (including DNA or RNA), or as glycosides. Sugars of limited distribution have been omitted. Based on data presented by Abdel-Fattah *et al.* (1974), Allen and Northcote (1975), Allen *et al.* (1974), Aspinall (1967), Auriol (1974), Barashkov (1956), Bartnicki-Garcia (1970), Burlakova *et al.* (1971), Craigie (1974), Craigie *et al.* (1968), Crook and Johnston (1962), Cummins and Harris (1956), Elbein (1974), Fleming *et al.* (1967), Gunnison and Alexander (1975), Haug (1974), Lewin (1958), Lewis and Smith (1967), Manners *et al.* (1967), Norris *et al.* (1955), Novaes-Ledieu *et al.* (1967), Percival and McDowell (1967), Philippe *et al.* (1975), Rönnerstrand (1968), Rogers and Perkins (1968), Ryley (1967), Siegel and Siegel (1973), Sutherland and Smith (1973), Thiemann *et al.* (1969), Työrinoja *et al.* (1974), Vining and Taber (1964), Weise *et al.* (1970), Wolk (1973), Wright (1964).

[b] D-Fructose has been reported in hydrolysates of several soil bacteria (Wurst *et al.*, 1974).

[c] Based on old references (see Davis, 1915).

[d] In actinomycetes and corynebacteria (Elbein, 1974).

[e] Apparently absent from *Rhodotorula glutinis* (Zalashko and Pidoplichko, 1974).

7.1.1 Sugar Derivatives

Just as the distribution of simple sugars is all but useless as an indicator of protistan phylogeny, the distribution of simple derivatives of these sugars proves to have little phylogenetic use. Among the more widespread derivatives are amino sugars; the amino group is attached by transamination of the corresponding sugar. The transaminating enzymes are probably ubiquitous in protists, and most are of unknown specificity. *N*-Acetylated derivatives of amino sugars are particularly common in prokaryotes, but have been reported widely throughout eukaryotes as well.

Many amino sugars occur primarily as polymers; chitin, found in Hyphochytridiomycetes and Eumycota, is a polymer of variable proportions of glucosamine and *N*-acetylglucosamine, linked by β-1,4 bonds. Chitin, hyaluronic acid (an *N*-acetylglucosamine-containing polymer), *N*-acetylgalactosamine, and *N*-acetylmannosamine are widespread throughout higher animals. Uronic acids are not common in bacteria (Sutherland and Smith, 1973), but are present in Cyanophyceae (Wolk, 1973) and in many eukaryotes, especially Rhodophyceae and Phaeophyceae (Percival and McDowell, 1967). Sulfated polyuronides are major constituents of Phaeophyceae, Bacillariophyceae, Rhodophyceae, and Chlorophyceae.

7.1.2 Polyols and Cyclitols

Straight-chain and cyclic polyalcohols (polyols and cyclitols) are present in most investigated organisms. They usually are biosynthesized by enzymatic reduction of the corresponding saccharide: glycerol is formed from glyceraldehyde 3-phosphate, inositols from glucose, and mannitol from fructose. These polyalcohols possess a variety of functions in living systems, serving as carbon reserves (mannitol), in lipid biosynthesis (glycerol, inositol), and in regulation of osmotic balance in phytoplankton.

The distribution of these metabolites (Table 25) demonstrates that they are not likely to be of use in phylogenetic reconstruction. The ubiquity of glycerol and mannitol is, as in the case of simple saccharides, due to their basic role in cellular metabolism. Moreover, very little is known of the appropriate biosynthetic enzymes. It is therefore not certain that the formation of ribitol from ribose in Chlorophyceae is homologous with ribitol formation in bacteria. Among the higher plants, however, there are cases where the distribution of polyols and cyclitols may be of taxonomic or even of phylogenetic significance at the lower phyletic levels (Plouvier, 1963). Kremer (1976a) has suggested, however, that mannitol may not be a natural metabolite, at least in common occurrence in the Rhodophyceae.

TABLE 25

Distribution of Polyols and Cyclitols[a]

Polyols and cyclitols	Bacteria	Cyanophyceae	Ascomycetes	Basidiomycetes	Protozoa	Euglenophyceae	Rhodophyceae	Dinophyceae	Bacillariophyceae	Phaeophyceae	Chrysophyceae	Haptophyceae	Xanthophyceae	Chlorophyceae	Prasinophyceae	Higher plants
Polyols																
Arabitol			+	+												
Dulcitol			+	+			+									+
Erythritol			+	+				?	?			+		+		+
Glycero-ido-heptitol			*b*	*b*												
Glycerol	+	+	*b*	*b*	*c*	+	+	+	+	+	+	+		+		+
Heptitol		+														
Mannitol		+	+	+		+		+	+	+	+	+	+	+	+	+
Sorbitol	+	+	*b*	+			+	+						+		+
Ribitol	+													+		+
Threitol			*b*	*b*							+					+
Volemitol			*b*	*b*			+			+						+
Xylitol			*b*	+												
Cyclitols																
Cyclohexane-tetrols							+				+					+
Cyclohexane-pentols														+		
myo-Inositol	+		+	+	+	+	+		+	+	+	+		+		+
L-Inositol																+
D-Inositol														*d*		+
scyllo-Inositol							+		+							+
Mytilitol							+							+		
Laminitol							+		+	+				+		

[a] Data based on reports by Allen and Northcote (1975), Bender (1975), Burlakova *et al.* (1971), Craigie (1974), Hellebust (1965), Ikawa *et al.* (1968), Kempner and Miller (1972), Laycock and Craigie (1970), Lewis and Smith (1967), Percival and McDowell (1967), Plouvier (1963), Ramanathan *et al.* (1966), and Työrinoja *et al.* (1974).

[b] Present in "fungi" (Lewis and Smith, 1967).

[c] Present in *Dictyostelium discoideum* (Dewey, 1967).

[d] D-*chiro*-Inositol found in some green algae (Craigie, 1974).

The site of the biosynthesis of the most widespread polyol, mannitol, has recently been discovered to be the chloroplast, at least in the brown algae (Willenbrink and Kremer, 1973). It is nonetheless present in chloroplast-free organisms such as the higher fungi (Lewis and Smith, 1967). Two biosynthetic pathways to mannitol have been discovered in fungi (Lewis and Smith, 1967), but the phylogenetic implications of their distribution are not obvious. Two pathways of mannitol catabolism are known in the bacteria (Doelle, 1969).

Teichoic acids, originally regarded as present only in gram-positive prokaryotes (Weise *et al.*, 1970; Wicken and Knox, 1975), are long-chain, phosphodiester-linked polymers of ribitol phosphate, choline phosphate, *N*-acetylgalactosamine, *N*-acetyldiaminotrideoxyhexoses, and glucose (Watson and Baddiley, 1974). There are indications that they are involved in antigenic reactions to these prokaryotes. There are now also indications (Hewitt *et al.*, 1976) that lipoteichoic acids are also present in gram-negative bacteria.

7.1.3 Floridoside and Isofloridoside

One of the most common low molecular weight carbohydrates in the Rhodophyceae is floridoside, 2-*O*-glycerol α-D-galactopyranoside. It is biosynthesized from UDPgalactose and α-glycerol phosphate (Bean and Hassid, 1955). A structural isomer (isofloridoside) is also known, and a lack of specificity in one of the biosynthetic enzymes leading to floridoside could result in isofloridoside biosynthesis (Craigie *et al.*, 1968).

Floridoside itself is often thought to be characteristic of the Rhodophyceae, but has been found in Cyanophyceae and Cryptophyceae (Craigie, 1974). Isofloridoside occurs in all red algae investigated by Craigie and coworkers (1968), but may be more predominant in the Bangiophycidae and in certain Florideophycidae. Isofloridoside has also been found in members of the Chrysophyceae (Craigie, 1974).

The distribution of these relatively simple molecules is not easy to explain by algal phylogenies based on the majority of the other criteria. Given the simplicity of their molecular structures and their apparently straightforward biosyntheses, convergent evolution would not be an unreasonable explanation. It would also be possible to entertain hypotheses suggesting multiple losses. In either case, little information is gained that is useful to biochemical phylogenetics. In taxonomic studies, however, some useful information might be forthcoming. Kremer and Vogl (1975) have shown that the floridoside labeling pattern from HCO_3^- in the Ceramiales is markedly different from other rhodophycean orders.

7.2 POLYSACCHARIDES

Polysaccharides constitute a large portion of the biomass of many protists. Because they are key food reserves, and because some are of economic importance, polysaccharides have been investigated in considerable detail. Gross physical properties such as solubility and gel formation have historically served as the basis for classification of polysaccharides, with further subdivision often based on cytochemical staining properties. Stoloff and Silva (1957) have utilized these gross properties in a taxonomic investigation of twenty-two species of Rhodophyceae, concluding that there is a rough correlation between type of polysaccharide and taxonomic position (based on morphology and reproduction). Such classifications alone cannot indicate phylogeny, however, because it is not possible to assess from gross physical properties alone whether a "gelan" or a carrageenan is the more advanced evolutionarily, or which may have been derived from the other. Indeed, many of the available carbohydrate data have very limited phylogenetic use. They do have more use for systematics, but again it is limited, and applicable only in broad generalizations.

A more biochemical treatment of the primary structure of a polysaccharide recognizes two basic properties of its biosynthesis: the nature of its subunits and the manner in which they are linked. Individually, neither the nature of subunits (Table 26) nor the linkages between these subunits (Table 27) provides interesting phylogenetic information. Consideration of both these properties simultaneously (Table 28) is more enlightening, if more complicated. The major protistan polysaccharides fall into several structural (and probably biosynthetic) classes, of which the α-(1,4)-linked glucans (starches), β-(1,4)-linked glucans (celluloses), and β-(1,3)-linked glucans will be discussed in greatest detail.

7.2.1 α-(1,4)-Glucans

Starches are distinguished by the presence of α-(1,4)-linked glucose residues. Linear arrangements of these units comprise *amylose*, whereas *amylopectin* is further branched through α-(1,6) bonds. Different protists produce starches differing in chain length, degree of branching, and length of branches (Percival and McDowell, 1967; Craigie, 1974; Haug, 1974); thus, the literature describes "floridean starch," "myxophycean starch," "chlorophycean starch," etc., names that have in many cases been handed down since their original descriptions in the late 1800's. Animal starch is more highly branched than most others, and is called glycogen. Although the biosynthesis of starch has been thoroughly studied in a few organisms,

TABLE 26

Distribution of Polysaccharides[a,b]

Polysaccharide	Bacteria-Actinomycetes	Cyanophyceae	Oomycota	Chytridiomycetes	Zygomycetes	Ascomycetes-imperfect fungi	Basidiomycetes	Protozoa	Higher animals	Euglenophyceae	Rhodophyceae	Dinophyceae	Cryptophyceae	Bacillariophyceae	Phaeophyceae	Chrysophyceae	Xanthophyceae	Chlorophyceae	Higher plants
Arabans																		+	+
Fructans	+	+																+	+
Galactans	+										+							+	
Glucans	+	+	+	+	[c]	+	+	[d]	+	+	+	+	+	+	+	+	+	+	+
Mannans	+					+	+	+			+							+	+
Polyuronides	[e]				+									+	+				
Xylans								+			+							+	+

[a] Grouped by component monosaccharides.

[b] Data from Aaronson (1970), Bartnicki-Garcia (1970), Craigie (1974), Dodd and McCracken (1972), Falk *et al.* (1966), Lewin (1955), Manners *et al.* (1967), Novaes-Ledieu *et al.* (1967), Percival and McDowell (1967), Rogers and Perkins (1968), Rosell and Birkhed (1974), Sokatch (1968), Sutherland and Smith (1973), Wolk (1973).

[c] In spore wall of *Mucor rouxii* (Bartinicki-Garcia, 1968).

[d] In cellular slime molds (Bartnicki-Garcia, 1970) and protozoa (Manners *et al.*, 1967; Ryley, 1967).

[e] Present only under extraordinary conditions (Percival and McDowell, 1967).

there are insufficient data on a sufficiently wide range of protists for starch biosynthesis to be used in biochemical phylogenies (Augier, 1965).

7.2.2 β-(1,4)-Glucans

Cellulose is often considered to be virtually ubiquitous among algae (e.g., Scagel *et al.*, 1966), although the chemical and biophysical data required to support its presence have been accumulated for relatively few of these organisms. Current data support its presence in some (but certainly not all) Rhodophyceae, Phaeophyceae, Xanthophyceae, and Chlorophyceae. It is widespread among the higher plants, and appears to be present in at least

TABLE 27

Distribution of Linkage Types in Polysaccharides[a]

Linkage type	Bacteria-Actinomycetes	Cyanophyceae	Oomycota	Chytridiomycetes	Zygomycetes	Ascomycetes-imperfect fungi	Basidiomycetes	Protozoa	Higher animals	Euglenophyceae	Rhodophyceae	Dinophyceae	Cryptophyceae	Bacillariophyceae	Phaeophyceae	Chrysophyceae	Xanthophyceae	Chlorophyceae	Higher plants
α-(1,1)[b]	+	+				+	+	+		+	+								+
α-(1,2)	+					+	+												+
α-(1,3)	+					+	+				+								+
α-(1,4)	+	+	+			+	+	+	+	+	+	+	+		+			+	+
α-(1,6)	+	+				+	+	+			+	+	+					+	
β-(1,2)	+																	+	
β-(1,3)			+	+		+	+			+	+	+		+	+	+	+	+	+
β-(1,4)	+	+	+	+	+	+	+		+		+	+		+			+	+	+
β-(1,6)	+		+	+		+	+							+	+	+	+		+
β-(2,1)	+																		+
β-(2,6)	+																		

[a] From data of Aspinall (1967), Bartnicki-Garcia (1970), Craigie (1974), Dodd and McCracken (1972), Falk *et al.* (1966), Haug (1974), Jacobi (1962), Lewin (1974), Mackie and Preston (1974), Manners *et al.* (1967), Novaes-Ledieu *et al.* (1967), Percival and McDowell (1967), Rogers and Perkins (1968), Sokatch (1968), Sturgeon (1974).

[b] In trehalose.

some amoebae, cellular slime molds, tunicates, and mammals (Aaroson and Hutner, 1966; Klein and Cronquist, 1967). It is important to recognize a basic morphological distinction that may have a major bearing on the "presence" of cellulose. On the one hand, there are nonflagellated vegetative cells or thalli, which almost invariably have a "rigid" cell wall and hence often reveal the presence of cellulose, a material well suited for rigidity with flexibility. On the other hand, there are the flagellated vegetative cells. These usually lack a "rigid" material like cellulose, possessing instead substances that provide pliability. The Rhodophyceae, for example, show no flagellated cells, and could be expected always to show cellulose, although there is question about the presence of a distinct cellulose type wall in such mucilage-secreting coccoids as *Porphyridium cruentum*.* Vegetative Phaeophyceae are always nonflagellated. Do the flagellated gametes or zoospores, however, possess cellulose? Likewise, in such classes as the Xanthophyceae, Dinophyceae, and Chrysophyceae one must distinguish flagellated and nonflagellated vegetative cells. Significantly, then, the presence or absence of cellulose can in part be a reflection of the morphological habit.

A very few bacteria biosynthesize a cellulose-like glucan, although not all of these retain it in their cells (see note *e*, Table 28). The presence of cellulose in Oomycota, and its apparent absence from virtually all Eumycota (cf. LéJohn, 1974), supports the basic phylogenetic separation of these two groups. LéJohn (1974) has pointed out that many Oomycetes are plant parasites, and that those fungi synthesizing cellulose have a history of association with higher plants. This raises the possibility that other parasites, such as rusts and downy mildews, might synthesize cellulose.

The apparent absence of cellulose from the Euglenophyceae is of considerable interest in view of its seeming absence from most fungi, which also utilize the AAA pathway of lysine biosynthesis (Section 6.9). However, two caveats must be stated: First, this is an "absence" character, and it may be considerably more likely for two groups to have independently lost a biochemical character than for two groups to have developed one; and second, very few euglenoids other than *Euglena* spp. have been thoroughly examined for the presence of cellulose. In view of our earlier comments about the presence of cellulose and the relationship to the morphological habit, it is important that nonflagellated (vegetative phase) Euglenophyceae, such as *Colacium*, be examined.

7.2.3 β-(1,3)-Glucans

Storage materials in many protistan taxa consist of β-(1,3)-linked glucans. Laminarin and laminarin-like polymers have been reported in Phaeophyceae, fungi, and water molds. Among these polymers, variability exists with respect to the degree of polymerization, characteristics of branching, and the number of terminal mannitol residues. Chrysolaminarin, mycolaminarin, leucosin, and paramylon—present in a range of algae, euglenoids, and fungi (Table 28)—are merely somewhat more extreme variants on a fundamentally laminarin-like theme. Many fungal glucans

* Kieras and Chapman (1977) and Kieras *et al.* (1977) have recently shown that this extracellular mucilage is a polypeptide-containing sulfated polysaccharide in which the amino acid–sugar linkages involve serine–xylose and possibly threonine.

TABLE 28

Distribution of Selected Polysaccharides[a]

Polysaccharide	Bacteria-Actinomycetes	Cyanophyceae	Hyphochytridiomycetes	Oomycetes	Chytridiomycetes	Zygomycetes	Ascomycetes	Basidiomycetes	Protozoa	Higher animals	Euglenophyceae	Rhodophyceae	Dinophyceae	Cryptophyceae	Bacillariophyceae	Phaeophyceae	Chrysophyccac	Haptophyceae	Xanthophyceae	Chlorophyceae	Prasinophyceae	Charophyceae	Higher plants
α-(1,4)-Linked																							
Starches		+					+	+	+			+	+	+						+	+	+	+
Amyloses							+	+					+	+						+	+	+	+
Amylopectins		+							+			+	+	+						+	+	+	+
Glycogens	[b]	+			[c]		+		+	+										[d]			
β-(1,4)-Linked																							
Cellulose	[e]	+	+	+			[f]	[f]	[f]	+		+	?	?	?	+	+		+	+		+	+
β-(1,3)-Linked																							
Laminarin				[g]			[g]									+							
Chrysolaminarin				[g]									+		+		[h]	+	+	[g]			
Paramylon											+												

Other																			
Agarose + agaropectin											+								
Alginic acid	[i]														+				
Fucoidan															+				
Carrageenans											+								
Chitin			+	[j]	+	+	+	+	[k]	+		[l]							[l]
Chitosan						+		+											
Peptidoglycans	+	+																	

[a] Data from Aaronson and Hutner (1966), Bartnicki-Garcia (1968, 1970), Craigie (1974), Craigie *et al.* (1971), Dodd and McCracken (1972), Evans (1974), Green and Jennings (1967), Harrison (1963), LéJohn (1971a), Mackie and Preston (1974), Mautner (1954), Meeuse and Hall (1973), Novaes-Ledieu *et al.* (1967), Percival and McDowell (1967), Rogers and Perkins (1968), Ryley (1967), Siegel and Siegel (1973), Soma (1960), Steward *et al.* (1970), Sturgeon (1974), Sutherland and Smith (1973), Suzuki (1974), Wolk (1973).

[b] "Purple bacteria can store . . . a glycogen-like polysaccharide" (Kondrat'eva, 1965; see also Wolk, 1973).

[c] "Glycogen-like" material in *Blastocladiella emersonii* (Pannbacker and Wright, 1967).

[d] In *Prototheca zopfii* (Manners *et al.*, 1967; Ryley, 1967).

[e] Many bacteria secrete cellulose into their environment, but no bacteria are known to use it in their cell walls (Doyle, 1970). Cellulose has been reported to accumulate in *Acetobacter xylinum*, *Sarcina ventriculi* (Aaronson and Hutner, 1966), and *A. acetigenum* (Anderson, 1963).

[f] Cellulose is present in cellular slime molds (Klein and Cronquist, 1967), amoebae (Aaronson and Hutner, 1966), and perhaps in a very few Ascomycetes and Basidiomycetes (Bartnicki-Garcia, 1968), but is apparently absent from true slime molds (Klein and Cronquist, 1967).

[g] "Laminarin-like" polysaccharides have been reported in *Neurospora crassa* (Lewis and Smith, 1967) and in *Phytophthora* spp. (Wang and Bartnicki-Garcia, 1974), and *Caulerpa simpliciuscula* (Howard *et al.*, 1976).

[h] A paramylon-like material, presumably chrysolaminarin, has been reported in some Chrysophyceae (see Craigie, 1974).

[i] Alginate is an extracellular product of a few bacteria (Haug and Larsen, 1974).

[j] Chitin has been demonstrated in the leptomitalean oomycete *Apodachlya* sp. (Lin and Aronson, 1970), but has not been detected in the majority of Oomycetes investigated.

[k] Chitin is apparently absent from true slime molds.

[l] Chitin has occasionally been reported, on the basis of cytochemical staining reactions, from various algae (Bartnicki-Garcia, 1970). "Chitan" has been identified in two species of diatoms (see text). Most, although not necessarily all, reports of true chitin in the algae have proved to have been erroneous (Percival and McDowell, 1967; Mackie and Preston, 1974).

(Bartnicki-Garcia, 1970) and glucans of protozoa probably also resemble laminarin. Further characterization of glucans from fungi and water molds, particularly with attention to the Chytridiomycetes and Hyphochytridiomycetes, may provide some insights into the phylogenetic relationships of these organisms.

Examination of the storage reserve pattern reveals a general phylogenetic dichotomy in the algae. One line (e.g., Bacillariophyceae, Dinophyceae, Phaeophyceae, Chrysophyceae, Haptophyceae, Xanthophyceae) is characterized by the β-1,3-linked polysaccharides. The remaining algal classes (plus Dinophyceae) emphasize the α-1,4 linkage. This basic phylogenetic dichotomy is borne out by other criteria (Sections 9.4, 10.2).

7.2.4 Other Polysaccharides

The dichotomy between prokaryotes and eukaryotes is reflected in the distribution of lipopolysaccharides, teichoic acids, and mucopeptides, which are widely distributed among the prokaryotes but have no obvious counterparts among the eukaryotes (Weise *et al.*, 1970). These characters group the actinomycetes and Cyanophyceae with the bacteria (Aaronson and Hutner, 1966).

Many other polysaccharides have been used to characterize certain taxa of organisms. Although the distribution of these molecules (e.g., agarose, agaropectin, carrageenans, alginic acid, chitosan) may be of taxonomic importance, phylogenetic information can be gathered only by studying their structure, hence their biosynthesis, in detail. In many cases this will involve the enzymology of polysaccharide biosynthesis as well as the genetic and physiological aspects of polysaccharide formation.

Chitin (see also Section 7.1.1), present in Eumycota, Hyphochytridiomycetes, and animals, may be analogous to some cell wall carbohydrates in the algae and higher plants. Its absence from investigated members of the Oomycetes (except *Apodachlya*) makes it unlikely that Hyphochytridiomycetes are derivatives of the Oomycetes. The distribution of chitin among the protozoa could be of interest in reconstructing the early phylogenetic history of fungi and animals as well as the modern groups of protozoa. Although chitin has occasionally been reported in algae, most (although not necessarily all) of these reports are open to serious question (Percival and McDowell, 1967; Mackie and Preston, 1974; Siegel and Siegel, 1973). A structurally related β-(1,4)-linked polymer of 2-acetamido-2-deoxy-D-glucose has been reported in *Thalassiosira fluviatilis* (Falk *et al.*, 1966) and in *Cyclotella cryptica* (McLachlan and Craigie, 1966).

7.2.5 The Gram Reaction

The gram reaction (Gram, 1884)* conveniently divides the bacteria into two groups: those that retain the stain (gram-positive) and those that do not (gram-negative). Thus, the latter condition must be regarded as are other "absence" characters. Although the chemical basis of the reaction is not entirely understood, it involves biochemical as well as structural properties of the cell wall (Stanier *et al.*, 1970; De Ley and Kersters, 1975).

Such a dichotomous division of the bacteria is indeed convenient; however, convenience may be characteristic of taxonomies but never of phylogenies. Phylogenetically, it seems reasonable to suspect that the gram-negative condition has evolved several times (Bisset, 1962). Although Aaronson and Hutner (1966) have suggested that (primary) gram-negative organisms were ancestral to all bacteria, the opposite conclusion of Hall (1971) better rationalizes why such catalase-negative obligate anaerobes as some clostridia are gram-positive. Gram-negative organisms encompass divergent groups, including many aerobic bacteria, sulfate reducers, carbonate reducers, photosynthetic bacteria, and Cyanophyceae (Echlin and Morris, 1965; Hall, 1971).

Several biochemical characters appear to be well correlated with the presence or absence of the gram reaction (Bishop *et al.*, 1962; Aaronson and Hutner, 1966), which tends to strengthen its use in bacterial phylogenies. Immunological investigation of the bacteria has also reinforced the gram-positive vs. gram-negative dichotomy, and has suggested *Diplococcus* spp. as a possible "cross-link" between these two groups (Kwapinski and Kwapinski, 1975). Bacterial phylogeny is a large and complex matter, and is further complicated by the presence of plasmids and other gene transfer mechanisms serving as randomizing factors. Further information on bacterial phylogeny is available in *Bergey's Manual* (Buchanan and Gibbons, 1974), volume twelve of the *Symposia of the Society for General Microbiology* (1962), Sagan (1967), De Ley (1968), Broda (1971a,b) Margulis (1972, 1974), Horvath (1974) and De Ley and Kersters (1975).

7.3 AMINO ACID DISTRIBUTION

It might be expected, *a priori*, that any compounds as basic to life as are the common amino acids would be strongly conserved throughout evolu-

* Gram-positive organisms retain the color caused by successive staining with crystal violet, iodine, and safranine.

tion. Organisms losing the ability to biosynthesize a certain amino acid would either die, or become dependent upon a dietary source of that amino acid. Any novel amino acids biosynthesized would only rarely be incorporated into polypeptides.

This is indeed what has been found. Although different biosynthetic pathways have occasionally been found from taxon to taxon (e.g., Sections 6.9 and 6.11), the distribution of common amino acids is quite uniform. Extensive surveys of their distribution in the protists have been published (Lewis and Gonzalves, 1960, 1962; Chuecas and Riley, 1969a; Madgwick and Ralph, 1972), and it appears infrequent that either qualitative or quantitative data can suffice to distinguish different species of protists, let alone to provide any phylogenetic information.

Uncommon amino acids—of which there are far too many to list here—often characterize a single genus or species of protist, but their phylogenetic usefulness is severely limited by the restricted distributions and by lack of knowledge about their biosynthesis. Miyazawa (1971) has presented a review of the uncommon amino acids in algae and Impellizzeri *et al.* (1975) have recently examined a number of red algae. Fungi also contain many unusual amino acids. Other examples are common in the higher plants: e.g., lathyrine is found only in eleven species of the genus *Lathyrus* (Turner, 1967).

One possible exception to the above generalities is the occurrence of 4-hydroxyproline in the protists (Table 29). This amino acid is formed by the hydroxylation of proline, often after it has been incorporated into a polypeptide (there is no hydroxyproline codon). The distribution of hydroxyproline thus parallels the distribution of an active proline 4-hydroxylase. Hydroxyproline is thought to be involved in the extensibility of some plant cell walls (Lamport, 1965), although it is not concentrated in cell walls of Phaeophyceae, *Chlorella pyrenoidosa,* or *Nitella* sp. (Gotelli and Cleland, 1968). Miller *et al.* (1972) have studied the specificity of the hydroxyproline to glycosyl residue bond in various cell wall fraction proteins in higher plants and Chlorophyceae. These results demonstrate that this linkage is formed more specifically (to fewer different sugars) in the higher plants than in the investigated Chlorophyceae.

The only groups apparently lacking hydroxyproline, and consequently an active proline hydroxylase, are the bacteria, the red algae, some protozoa, and Eumycota. There is no obvious relationship among all of the hydroxyproline-containing organisms, but the often-proposed immediate relationship between Rhodophyceae and Cyanophyceae is weakened by the

TABLE 29

Distribution of 4-Hydroxyproline (Free or in Polypeptides)

Organism	Present	Reference
Bacteria	nd[a]	
Cyanophyceae		
Eucapsis sp.	Traces	Gotelli and Cleland, 1968
Lyngbya sp.	Traces	Gotelli and Cleland, 1968
Merismopedia sp.	Present	Gotelli and Cleland, 1968
Nostoc sp.	Present	Punnett and Derrenbacker, 1966
Phormidium faveolarum	Traces	Gotelli and Cleland, 1968
Plectonema boryanum	nd	Gotelli and Cleland, 1968
Rhodophyceae (14 spp.)	nd[b]	Gotelli and Cleland, 1968
Phaeophyceae (11 spp.)	Present	Gotelli and Cleland, 1968
Bacillariophyceae (2 spp.)	Present[c]	Gotelli and Cleland, 1968
Xanthophyceae (1 sp.)	Present	Gotelli and Cleland, 1968
Chrysophyceae (1 sp.)	Present	Aaronson, 1970
Chlorophyceae (16 spp.)	Present	Gotelli and Cleland, 1968; Aaronson, 1970
Charophyceae (1 sp.)	Traces	Gotelli and Cleland, 1968
Oomycetes (4 spp.)	Present	Crook and Johnston, 1962, Aaronson, 1970
Zygomycetes (1 sp.)	nd	Crook and Johnston, 1962
Ascomycetes and imperfect fungi		
Hemiascomycetidae (5 spp.)	nd	Crook and Johnston, 1962
Filamentous fungi (5 spp.)	nd	Crook and Johnston, 1962
Candida albicans	Present[d]	Chattaway *et al.*, 1968
Basidiomycetes (3 spp.)	nd	Crook and Johnston, 1962
Protozoa		
Gromia oviformis	Present	Aaronson, 1970
Paramecium spp. (3 spp.)	nd	Kidder, 1967; Aaronson, 1970
Tetrahymena pyriformis	[e]	Kidder, 1967
Higher animals	Present	Aaronson, 1970
Higher plants	Present	Miller *et al.*, 1972

[a] nd, not detected.

[b] Reports to the contrary (Lewis and Gonzalves, 1960, 1962) are discussed by Gotelli and Cleland (1968).

[c] A 3,4-dihydroxyproline has been reported in eight species of marine diatoms (Nakajima and Volcani, 1969).

[d] Both 3-hydroxyproline and 4-hydroxyproline (Chattaway *et al.*, 1968).

[e] Present in one of the eight strains tested (Kidder, 1967).

presence of 4-hydroxyproline in the latter but not in the former organisms. However, this distribution could merely indicate that the appearance of hydroxyproline in some blue-green algae is convergent. An examination of the respective proline hydroxylases may help resolve this question. There have been no reports suggesting that chloroplasts are involved with hydroxyproline biosynthesis.

8

Metabolites: II

8.1 ACETATE-DERIVED BIOSYNTHETIC PATHWAYS

In attempting to organize the huge numbers of primary and secondary metabolites found in the protists, it is necessary to employ a comprehensive and powerful scheme. The biosynthetic approach has much to recommend it; it tends to group homologous compounds, stresses the importance of biosynthetic enzymes (which in turn are more closely related to molecular genetic processes in the cell), and provides a natural means for identifying characters as being simple or complex. Major or significant deviations from or modifications upon a basic biosynthetic pathway would represent both taxonomically and phylogenetically significant events (Birch, 1973a,b).

Among the most fundamental biosynthetic pathways are the acetate-derived pathways (Figure 5), responsible for the production of fatty acids, many phenolic compounds, part of all flavonoids, and other specialized metabolites to be described below (Packter, 1973). Most acetate-derived compounds are known from only one or a few organisms, and their distribution is consequently of little phylogenetic interest; many of these compounds have been listed by Birkinshaw (1965), Geissman and Crout (1969), and Turner (1971). Other groups of acetate-derived metabolites, which may eventually be of interest in protistan phylogeny, include flavonoids, naphthaquinones, phloroglucinols, alkanes, and alkenes. Although currently believed to occur in only a few taxa of protists, further research may uncover more widespread distribution; examples of this may be seen in the recent discoveries of a flavonoid in *Aspergillus candidus* (Marchelli and Vining, 1973) and of phloroglucinol derivatives in bacteria (Reddi and Borovkov, 1969) and Phaeophyceae (Glombitza *et al.*, 1973; Glombitza and

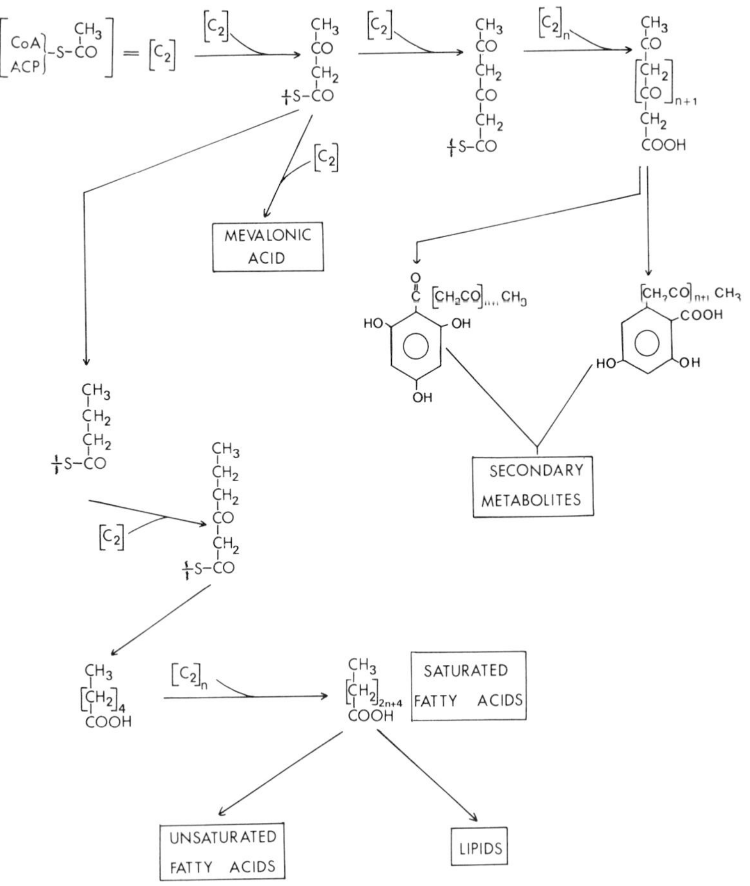

Figure 5. Acetate-derived biosynthetic pathways.

Sattler, 1973; Glombitza and Rösener, 1974; Glombitza *et al.*, 1975a,b; Ragan and Craigie, 1976).

Fatty acids and lipids are the acetate-derived compounds most widely used in phylogenetics, and will be discussed in detail in the subsequent pages.

8.2 FATTY ACIDS

Fatty acids are found in all organisms except the viruses, although a few protists (e.g., *Leptospira* spp. and *Treponema* spp.) cannot synthesize their own *de novo*, and require a dietary source (Livermore and Johnson, 1974).

A great number of papers describing fatty acid compositions of various protists may be found in the literature. In the more recent of these, identification of fatty acids is usually made on the basis of retention times during gas chromatography, usually coupled with chemical modification techniques and occasionally with mass spectrometry. Although most identifications made in this manner are trustworthy, some margin for error persists, particularly in the identification of double-bond isomers and of minor constituents. Indeed, some reviewers systematically ignore fatty acids reported present in amounts not exceeding some limit, usually 1% of the total fatty acids. Although such reports are admittedly less certain than might be desirable, it will be assumed for the purposes of this chapter that even "trace" amounts are valid components of the organisms being examined. The usual convention of identifying fatty acids by their carbon number and positions of desaturation* will be followed here in lieu of relying on trivial names (Table 30).

Because fatty acids are biosynthesized from units of acetyl coenzyme A (or ACP), many of the common fatty acids possess even numbers of carbon atoms; the distribution of these even-carbon, fully saturated fatty acids yields little phylogenetic information (Table 31). Although variations in the concentrations of individual fatty acids have often been reported from organism to organism, these quantitative differences are often as much the result of metabolic states of the organisms, culture conditions, light regimes, and a host of other poorly defined environmental factors as they are of the genetic factors more germane to biochemical phylogeny. There have been distressingly few experiments examining how much variation, both qualitative and quantitative, can be induced by changes in these environmental conditions. Consequently, use of quantitative variations is usually difficult, although occasionally it may reinforce other taxonomic data (e.g., Ackman *et al.*, 1970).

It is possible that evolutionary modifications may result in unexpected distributional patterns for individual fatty acids. The apparent absence of 20:0 from the Cyanophyceae, for instance, would tend to set these organisms off from other protists, but would not aid in reconstruction of their phylogeny. Similarly, the presence of relatively large quantities of 10:0 in some Cyanophyceae might be useful in taxonomy but not in phylogeny. Branched-chain fatty acids are often considered to be characteristic of bacteria and actinomycetes (Ballio and Barcellona, 1968; Pommier and Michel, 1973) and higher animals (Ackman and Sipos, 1965).

* For instance, an unbranched fatty acid with twelve carbons and no double bonds is identified as 12:0. Double bonds are specified in the unsaturated acids: the fatty acid 18:2 (9,12) has eighteen carbons and two double bonds (between carbons nine and ten, and between carbons twelve and thirteen.)

TABLE 30

Trivial Names of Common Fatty Acids

Arachidic acid	20:0
Arachidonic acid	20:4(5,8,11,14)
Behenic acid	22:0
Capric acid	10:0
Cerotic acid	26:0
Eicosenoic acid	20:1
Erucic acid	22:1(13)
Gladoleic acid	20:1(9)
Lactobacillic acid	11,12-methylenestearic acid
Lignoceric acid	24:0
Lauric acid	12:0
Linoleic acid	18:2(9,12)
α-Linolenic acid	18:3(9,12,15)
γ-Linolenic acid	18:3(6,9,12)
Margaric acid	17:0
Myristic acid	14:0
Nervonic acid	24:1(15)
Oleic acid	18:1(9)
Palmitic acid	16:0
Palmitoleic acid	16:1(7)
Stearic acid	18:0
Tuberculostearic acid	10-methylstearic acid
Vaccenic acid	*cis*-18:1(11)

Iso: $CH_3CH(CH_3)(CH_2)_nCOOH$ (with CH_3 branch on the CH)

Anteiso: $CH_3CH_2CH(CH_3)(CH_2)_nCOOH$ (with CH_3 branch on the CH)

Branched-chain acids are widely distributed in protists, however, albeit in somewhat smaller amounts (Table 31). Cyclopropane fatty acids are found in bacteria and in some zooflagellates (Erwin, 1973). In general, however, fatty acid distributions have very little phylogenetic use and only a little more use for systematic or taxonomic purposes.

8.2.1 Biosynthesis of Monounsaturated Fatty Acids

Desaturation of the basic fatty acid complement of a protist could obviously give rise to an almost limitless variety of monounsaturated fatty acids. Interestingly, only a moderate number of these compounds have been found (Table 31), suggesting that the enzymes are relatively specific for certain types of desaturations. The biosynthesis of monounsaturated fatty acids has been described by Erwin and Bloch (1964), Erwin (1973), and Weete (1974). Three major pathways are apparent:

1. The anaerobic pathway, limited to but widespread among the bacteria, including photosynthetic bacteria but excluding Cyanophyceae.
2. The direct oxidative desaturation pathway, or Desaturase System I. This pathway is found in some gram-positive bacteria and actinomycetes, in Cyanophyceae, *Penicillium chrysogenum, Saccharomyces cerevisiae, Candida utilis, Dictyostelium discoideum, Acanthamoeba* sp., *Hartmannella rhysoides,* zooflagellates, ciliates, higher animals, *Porphyridium cruentum,* a cryptomonad (?), Chrysophyceae (*Ochromonas malhamensis, Poteriochromonas stipitata*), autotrophically grown *Chlorella vulgaris, Astasia longa,* and heterotrophically grown *Euglena gracilis* Z (Erwin *et al.*, 1964; Erwin, 1973). The enzymology of this pathway has been reviewed by Erwin (1973).
3. The plant-type pathway, or Desaturase System II. Desaturation occurs before the final elongation step. The enzymes of this pathway are probably present in *Penicillium chrysogenum*, Chlorophyceae, light-grown *Euglena gracilis* Z, and higher plants (Erwin, 1973; Richards and Quackenbush, 1974).

Both distributional data and biochemical investigations have suggested that the Desaturase System II enzymes may be localized in the chloroplast, where their action is somehow coupled to photosynthesis (Erwin, 1973). Fatty acid biosynthesis by isolated chloroplasts has recently been demonstrated (Hawke *et al.*, 1974), but mitochondrial and microsomal fractions were needed for significant production of monounsaturated fatty acids. These data are apparently consistent with the endosymbiotic theory of chloroplast origin, but are perhaps more easily explained by the gradual evolution theory. These findings do not elucidate the origin of the Euglenophyceae, but suggest that there is a phyletic relationship between the chloroplasts of the Chlorophyceae and those of the Euglenophyceae.

Erwin (1973) has suggested that Desaturase II enzymes could have evolved from Desaturase I enzymes, which it should be possible to test once the appropriate amino acid sequences and tertiary structures are known. Other explanations could be consistent with the few available data.

8.2.2 Biosynthesis of Polyunsaturated Fatty Acids

The distribution of polyunsaturated fatty acids has been well studied in many classes of protists, although there are still significant gaps in our

TABLE 31

Distribution of Saturated and Monosaturated Fatty Acids[a]

Fatty acid	Bacteria	Cyanophyceae	Hyphochytridiomycetes	Oomycetes	Chytridiomycetes	Zygomycetes	Ascomycetes	Imperfect fungi	Basidiomycetes	Rhodophyceae	Dinophyceae	Cryptophyceae	Bacillariophyceae	Phaeophyceae	Chrysophyceae	Haptophyceae	Xanthophyceae	Chlorophyceae	Prasinophyceae	Charophyceae	Higher plants	Euglenophyceae	Ciliatea	Mastigophorea	Sarcodina	Myxomycota
12:0	+	+			+	+	+	+	+	+	+	+	+	+	+	+	+	+	+		+	+	+	+	+	
12:unsat.	+					+		+		+	+	+	+	+		+	+					+				
14:0	+	+	+	+	+	+	+	+	+	+	+	+	+	+	+	+	+	+	+		+	+	+	+	+	+
14:unsat.	+	+		+		+	+	+	+	+	+	+	+	+	+		+	+				+	+	+		
15:0	+	+	+			+	+	+	+	+	+	+	+	+		+	+	+	+		+	+	+	+		
16:0	+	+	+	+	+	+	+	+	+	+	+	+	+	+	+	+	+	+	+		+	+	+	+	+	+
16:1	+	+	+	+	+	+	+	+	+	+	+	+	+	+	+	+	+	*b*	+	+	+	*b*	+	+	+	+
17:0	+	+	+		+	+	+	+	+	+	+	+	+	+	+	+	+	+	+		+	+	+	+		
17:unsat.		+	+		+	+	+	+	+	+		+	+		+		+	+	+			+				+
18:0	+	+	+	+	+	+	+	+	+	+	+	+	+	+	+	+	+	+	+		+	+	+	+	+	+
18:1	*c*	+	+	+	+	+	+	+	+	+	+	+	+	+	+	+	+	*c*	+	+	+	*c*	+	+	+	
19:0	+		+		+		+		+	+	+	+		+	?	+	+	+	+			+	+			
19:unsat.	+		+		+							+	+					+				+				+
20:0	+			+	+	+	+	+	+	+	+	+	+	+	+	+	+	+	+		+	+	+			

20:1	+			+	+	+	+	+	+	+	+	+	+	+	+	+	+	+	+			+			+	
22:0	+	+		+		+	+	+	+	+		+	+	+	+	+		+			+	+			+	
22:1	+					+		+	+			+	+	+	+	+	+	+				+				
24:0	+					+		+	+	+	+	+	+	+	+	+	+	+								
Hydroxy	+			+			+	+	[d]															+		
Branched	+					+	+	+	+	?	+	+	+	?	+	+	+	+	+				+	+	+	+

[a] Data for bacteria: Bean *et al.* (1972), Drucker (1974), Goldfine and Bloch (1961), Ikawa (1967), Nichols *et al.* (1965), Parker *et al.* (1967), Radunz (1969), Sokatch (1968), Uchida (1974), Wood *et al.* (1965); Cyanophyceae: Beach *et al.* (1970), Das and Smith (1968), Nichols (1973), Nichols *et al.* (1965), Parker *et al.* (1967), Radunz (1969), Shaw (1966), Weise *et al.* (1970); Oomycota: Bean *et al.* (1972), Erwin (1973), Shaw (1965, 1966), Weete (1974); Eumycota: Bean *et al.* (1972), Byrne and Brennan (1975), Dedyukhina and Bekhtereva (1970), Erwin (1973), Shaw (1966), Weete (1974); all eukaryotic algae; Ackman *et al.* (1968), Chuecas and Riley (1966, 1969a), Shaw (1966); Rhodophyceae: Allen *et al.* (1970), Beach *et al.* (1970), Erwin and Bloch (1964), Erwin *et al.* (1964), Ikan and Seckbach (1972), Klenk *et al.* (1963), Nichols and Appleby (1969), Pohl *et al.* (1968), Radunz (1968, 1969); Dinophyceae: Erwin (1973), Harrington *et al.* (1970), Lee and Loeblich (1971); Cryptophyceae: Antia *et al.* (1974), Beach *et al.* (1970), Erwin (1973), Glasl and Pohl (1974), Lee and Loeblich (1971); Bacillariophyceae: Beach *et al.* (1970), Erwin (1973), Lee and Loeblich (1971), Tornabene *et al.* (1974); Phaeophyceae: Klenk *et al.* (1963), Lee and Loeblich (1971), Pohl *et al.* (1968), Radunz (1969), Youngblood and Blumer (1973); Chrysophyceae and Haptophyceae: Beach *et al.* (1970), Erwin (1973), Erwin *et al.* (1964), Lee and Loeblich (1971), Nichols and Appleby (1969); Xanthophyceae: Mercer *et al.* (1974), Nichols and Appleby (1969); Chlorophyceae: Bean *et al.* (1972), Brush and Percival (1972), Klenk *et al.* (1963), Pohl *et al.* (1968); Prasinophyceae: Ackman *et al.* (1970); Charophyceae: Erwin (1973); higher plants: Beach *et al.* (1970), Nichols (1970), Radunz (1969), Schwertner and Biale (1973); Euglenophyceae: Dewey (1967), Lee and Loeblich (1971), Nichols and Appleby (1969), Shaw (1966); Protozoa and slime molds: Dearborn and Korn (1974), Erwin (1973), Erwin *et al.* (1964), Glasl and Pohl (1974), Shaw (1966).

[b] Fatty acids 16:1(9), 16:1(7), and 16:1(3) have been reported in *Euglena gracilis* strain 1225/53 and in several algae (Harwood and James, 1975, Klenk *et al.*, 1963, Wood *et al.*, 1965, Nichols and Appleby, 1969).

[c] Fatty acid 18:1(9) is found in most investigated organisms. 18:1(11) (*cis*-vaccenic acid) is more characteristic of bacteria, although it may occur in small amounts in eukaryotic algae. 18:1(11) may comprise up to 35% of the total 18:1 of some Cyanophyceae (Nichols, 1973).

[d] Especially in spores of the rust fungi (Erwin, 1973).

knowledge (e.g., amoeboflagellates, Chloromonadophyceae, Hyphochytridiomycetes). The biosynthesis of polyunsaturated fatty acids utilizes the common tactic of serial desaturation (Erwin, 1973), which reflects similarities (and perhaps homologies) in the enzyme systems involved. The biosynthesis of the most commonly encountered polyunsaturated fatty acids is given in Figure 6. Further interconversions between ω3 and ω6 fatty acids* have been reported, especially in *Euglena gracilis* strain Z (Shaw, 1966), so it must be borne in mind that the mere presence of an ω6 fatty acid, for example, does not constitute adequate proof of the operation of the entire ω6 fatty acid biosynthetic pathway.

Modifications of the biosynthetic pathways shown in Figure 6 are believed to occur in *Chlorella vulgaris* and *Euglena gracilis* Z, in which fatty acids may also be desaturated after incorporation into lipids. This is not likely to represent a major evolutionary divergence from all other protists, although no final assessment can be made until the biochemical basis of this modification is understood.

The presence of 18:3 isomers has been used by several authors to establish two major lines of phylogenetic divergence among the eukaryotic algae: an ω3 line (Chlorophyceae and Euglenophyceae) and an ω6 line (other investigated eukaryotic algae). This approach is at best a severe oversimplification of a complex matter. As examples, Cryptophyceae and Haptophyceae would easily qualify as ω3 algae, whereas several green algae possess 18:3 (6,9,12) in amounts up to 6% of the total fatty acid level (Wood, 1974).

Indeed, both ω3 and ω6 fatty acids have been reported in all algal classes, although Erwin (1973) has indicated otherwise for the Xanthophyceae. However, the higher fungi possess few polyunsaturated fatty acids, with only 18:3 (9,12,15) found in most. Greater biosynthetic capacity is seen in the lower fungi and water molds. The condition in the higher fungi must have been reductive unless a direct fungal derivation from bacteria is postulated (and thereby two origins of the eukaryotic condition).

Within the ω6 biosynthetic pathway, a route from 18:2 (9,12) to 20:2 (11,14) has been established in *Acanthamoeba* sp., *Hartmannella rhysoides*, *Physarum polycephalum*, *Euglena gracilis* Z, *Astasia longa*, and rat liver (Erwin, 1973). The intermediate 20:2 (11,14) itself has been extracted from *Porphyridium cruentum*, *Ochromonas danica*, *Leishmania tarentolae*, slime molds, and possibly from Chytridiomycetes and Oomycetes. Nichols and Appleby (1969) report, however, that the pathway of 20:3 biosynthesis is through 18:3 (6,9,12) in *P. cruentum* and *O. danica*, as in *Polytoma uvella*,

* An ω*n* fatty acid possesses one or more double bonds, with the double bond closest to the carboxyl end of the molecule being between the *n*th and $n + 1$st carbons from that end.

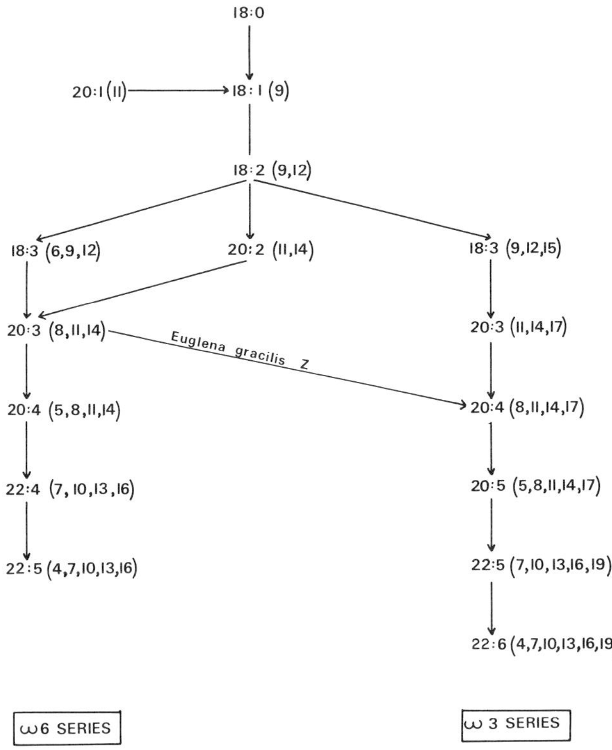

Figure 6. Biosynthesis of polyunsaturated fatty acids. After Figure 1 from Shaw (1966). Reproduced by permission of Academic Press and the author.

zooflagellates, ciliates, Zygomycetes, and higher animals (Erwin, 1973). The utilization or nonutilization of the pathway through 20:2 (11,14) does not appear to be related to the site of fatty acid biosynthesis.

The absence of polyunsaturated fatty acids from most bacteria* suggests that, in this respect, they are a primitive group: desaturation requires one or more specialized enzymes. It is, of course, possible that the capacity for desaturation has been lost by the bacteria (as presumably by fungi), but this assumption is unnecessary and unfounded. The photosynthetic bacteria have fatty acid compositions similar to those of the Eubacteriales, although one strain of *Rhodopseudomonas spheroides* may possess linoleic acid (Hands and Bartley, 1962). The Cyanophyceae are able to biosynthesize linolenic acid (ω3 or ω6, depending on the species: see Watanabe and Yamamoto, 1972).

* Fulco (1970) has reported 16:2 (5,10) and 16:2 (7,10) in *Bacillus licheniformis*.

TABLE 32

Distribution of Selected Polyunsaturated Fatty Acids[a]

Fatty acid	Bacteria	Cyanophyceae	Hyphochytridiomycetes	Oomycetes	Chytridiomycetes	Zygomycetes	Ascomycetes	Imperfect fungi	Basidiomycetes	Rhodophyceae	Dinophyceae	Cryptophyceae	Bacillariophyceae	Phaeophyceae	Chrysophyceae	Haptophyceae	Xanthophyceae	Chlorophyceae	Prasinophyceae	Charophyceae	Higher plants	Euglenophyceae	Ciliatea	Mastigophorea	Sarcodina	Myxomycota
ω3																										
16:4										+	+	+	+	*b*	+	+	+	*b*	*b*			+	+			
18:3	*c*	*d*	*e*	+	+	+	*f*	*f*		+	+	+	+	+	+	+	+	+	+	+	+	+		*g*		
18:4					+					+	+	+	+	+	+	+	+	+	+		+	+				
20:3			*e*	*h*	*i*	*i*								*j*	+		+			*k*		+		+		
20:4	*c*				+						+	+	+	+	+	+	+	+	+					+		
																						+				
20:5				+	+					+	+	+	+	+	+	+	+	+	+		*l*	+		+		
22:5										*m*	+	+	+	+	+	+	+	+	+			+		+		
22:6				+	+					+	+	+	+	+	+	+	+	+	+		*l*	+		+		
ω6																										
18:2	*c*	+	+	+	+	+	+	+	+	+	+	+	+	+	+	+	+	+	+	+	+	+	+	+	+	*n*
18:3	*c*	*d*	*e*	+	+	+	*f*	*f*		+	+	+	+	+	+	+	+	+	+		*o*	+	+	+		
20:2				*h*	*h*	*i*			*p*	+				*p*	+				+			+		+	+	+
20:3			*e*	*h*	*i*	*i*				+		+	+	*j*	+	+	+	+		*k*		+		+	+	+
20:4	*c*			+	+	+				+	+	+	+	+	+	+	+	+	+	+	*o*	+,		+	+	+
22:4											+	+	+		+	+		+				+	+	+		
22:5										*m*	+		+		+	+		+	+			+	+	+		
ω9																										
20:2				*h*	*h*	*i*			*p*		+	+	+	*p*	+	+	+	+								

[a] Data for bacteria: Bean *et al.* (1972), Drucker (1974), Goldfine and Bloch (1961), Ikawa (1967), Nichols *et al.* (1965), Parker *et al.* (1967), Radunz (1969), Sokatch (1968), Uchida (1974), Wood *et al.* (1965); Cyanophyceae: Beach *et al.* (1970), Das and Smith (1968), Nichols (1973), Nichols *et al.* (1965), Parker *et al.* (1967), Radunz (1969), Shaw (1966), Weise *et al.* (1970); Oomycota: Bean *et al.* (1972), Erwin (1973), Shaw (1965, 1966), Weete (1974); Eumycota: Bean *et al.* (1972), Byrne and Brennan (1975), Dedyukhina and Bekhtereva (1970), Erwin (1973), Shaw (1966), Weete (1974); all eukaryotic algae; Ackman *et al.* (1968), Chuecas and Riley (1966, 1969a,b), Shaw (1966); Rhodophyceae: Allen *et al.* (1970), Beach *et al.* (1970), Erwin and Bloch (1964), Erwin *et al.* (1964), Ikan and Seckbach (1972), Klenk *et al.* (1963), Nichols and Appleby (1969), Pohl *et al.* (1968), Radunz (1968, 1969); Dinophyceae: Erwin (1973), Harrington *et al.* (1970), Lee and Loeblich (1971); Cryptophyceae: Antia *et al.* (1974), Beach *et al.* (1970), Erwin (1973), Glasl and Pohl (1974), Lee and Loeblich (1971); Bacillariophyceae: Beach *et al.* (1970), Erwin (1973), Lee and Loeblich (1971), Tornabene *et al.* (1974); Phaeophyceae: Klenk *et al.* (1963), Lee and Loeblich (1971), Pohl *et al.* (1968), Radunz (1969), Youngblood and Blumer (1973); Chrysophyceae and Haptophyceae: Beach *et al.* (1970), Erwin (1973), Erwin *et al.* (1964), Lee and Loeblich (1971), Nichols and Appleby (1969); Xanthophyceae: Mercer *et al.* (1974), Nichols and Appleby (1969), Jamieson and Reid (1976); Chlorophyceae: Bean *et al.* (1972), Brush and Percival (1972), Klenk *et al.* (1963), Pohl *et al.* (1968); Prasinophyceae: Ackman *et al.* (1970); Charophyceae: Erwin (1973); higher plants: Beach *et al.* (1970), Nichols (1970), Radunz (1969), Schwertner and Biale (1973); Euglenophyceae: Dewey (1967), Lee and Loeblich (1971), Nichols and Appleby (1969), Shaw (1966); Protozoa and slime molds: Dearborn and Korn (1974), Erwin (1973), Erwin *et al.* (1964), Glasl and Pohl (1974), Shaw (1966).

[b] Unidentified 16:4 isomer in Phaeophyceae (Radunz, 1968). Chlorophyceae and Prasinophyceae contain both the $\omega 1$ and $\omega 3$ isomers of 16:4 (Chuecas and Riley, 1969a, Klenk *et al.* 1963).

[c] Fatty acid 18:2(9,12) has been reported from *Rhodopseudomonas spheroides* (Hands and Bartley, 1962). Shaw (1966) notes that there is an unconfirmed report of an unidentified 20:4 in another bacterium. 18:3 is present in some biotypes of *Treponema refringens*, but is not biosynthesized *de novo* by this organism (Livermore and Johnson, 1974). In general, bacteria do not accumulate polyunsaturated fatty acids.

[d] For distribution in individual species of Cyanophyceae, see Beach *et al.* (1970), Hitchcock and Nichols (1971), Kenyon (1972), Levin *et al.* (1964), Nichols (1970, 1973), Shaw (1966), Watanabe and Yamamoto (1972), Wood (1974).

[e] Unidentified 18:3 and 20:3 in *Rhizidiomyces apophysatus* (Bean *et al.*, 1972).

[f] In small quantities in *Chaetomium cochliodes* and *Dactylaria ampulliforme* (Safe and Brewer, 1973; Erwin, 1973, respectively).

[g] Constitutes 18% of the total fatty acids of *Leishmania tarentolae* (Erwin, 1973).

[h] Unidentified 20:2 or 20:3 (Erwin, 1973).

[i] Fatty acid 20:2 detected in only one of fifty Zygomycetes surveyed, and also present in *Blakeslea trispora*; unidentified 20:5(5,8,11,14,17) in one of fifty Zygomycetes surveyed, but present at 27% of total fatty acid concentration (Erwin, 1973; Dedyukhina and Bekhtereva, 1970); 20:3 in one of fifty Zygomycetes and in *Blakeslea trispora*.

[j] Unidentified 20:3 in Phaeophyceae (Klenk *et al.*, 1963; Shaw, 1966).

[k] Unidentified isomer (Hitchcock and Nichols, 1971).

[l] Rare in higher plants (Ackman *et al.*, 1968).

[m] Unidentified 22:5 in *Ceramium rubrum* (Klenk *et al.*, 1963) but not usually detected in Rhodophyceae.

[n] Fatty acid 18:2(9,12) is present in *Physarum polycephalum* but is not accumulated in *Dictyostelium discoideum*, which instead accumulates 18:2(5,9) and 18:2(5,11) (Shaw, 1966).

[o] Not detected in most plants, but present in certain seed oils (Shaw, 1966).

[p] Unidentified 20:2 isomer in *Agaricus bisporus* (Byrne and Brennan, 1975), *Cyanidium caldarium* (Allen *et al.*, 1970; Ikan and Seckbach, 1972), and in many Phaeophyceae (Lee and Loeblich, 1971; Hitchcock and Nichols, 1971).

Differences in fatty acid composition between the Ciliatea and the Mastigophora are of interest. On the one hand, the former group has a restricted biosynthetic capacity for polyunsaturated fatty acids, accumulating no $\omega 3$ fatty acids at all. On the other hand, zooflagellates accumulate most of the polyunsaturated fatty acids listed in Table 32. This could represent a significant evolutionary divergence between the two groups, or could have resulted from numerous secondary losses. It is also of interest, as noted by Erwin (1973), that euglenoids biosynthesize a very wide range of polyunsaturated fatty acids.

8.2.3 Location of Fatty Acid Biosynthesis

The subcellular site of fatty acid biosynthesis apparently varies from fatty acid to fatty acid, and from one organism to the next. This is probably to be expected, in view of the diverse functions of fatty acids and of the complexity of lipid biosynthesis and of membrane assembly. Some fatty acid biosynthesis is chloroplastic in many algae and higher plants, confirming the observations of the early light microscopists (e.g., Schmitz, 1883). Although pea chloroplasts can biosynthesize 18:3 from 18:2, conversion of 18:1 to 18:2 is largely microsomal (Tremolières and Mazliak, 1974). Hawke *et al.* (1974) have also observed participation of microsomal and mitochondrial fractions in unsaturated fatty acid biosynthesis by chloroplasts isolated from corn.

It is likely that chloroplasts play a significant role in fatty acid biosynthesis. Because there is no evidence, however, supporting the localization of genes for any of the relevant enzymes in the chloroplast genome, it cannot be affirmed that any of the fatty acid biosynthetic patterns observed in eukaryotes are due to chloroplasts per se. However, it might be countered that the significantly more developed biosynthetic capabilities of photosynthetic eukaryotes could be best rationalized by a fairly radical all-or-nothing step such as the ingestion of a protochloroplast or a protomitochondrion.

Erwin (1973) has explored this question in some detail. He has concluded that the fatty acid distributional data are inconsistent with the evolutionary scheme proposed by Margulis (1970), and has pointed out parallels between these distributional data and the phylogenies proposed by Klein and Cronquist (1967) and by other advocates of a gradual origin of chloroplasts. This does not mean that the fatty acid data are inconsistent with an endosymbiotic origin of the chloroplast, however. Using the comprehensive data from Tables 31 and 32, a phylogeny involving an endosymbiotic origin of the chloroplast can be accommodated as easily as can the gradualistic hypotheses. Innovations in fatty acid biosynthetic capability would be required at four points (Cyanophyceae, Chlorophyceae, Rhodophyceae, higher fungi) in Erwin's proposed phylogeny (Section 12.1), whereas a biochemical phylogeny (Section 12.2) could require only two innovations (Cyanophyceae, and the line to nonphotosynthetic eukaryotes).

8.3 LIPIDS

Lipids* constitute a rather heterogeneous group of compounds involved in a moderately wide range of biological functions in protists. One of the major functions of lipids is their participation in the structure and function of biomembranes (Section 8.3.4), but some lipids serve as cellular carbon reserves. The biosynthesis of protistan lipids has recently been reviewed in a volume edited by Erwin (1973). Unfortunately, very little is known about lipid-biosynthesizing enzymes.

All organisms except most viruses contain lipids. In recent years, lipid identification has been aided greatly by the introduction of thin-layer chromatography and gas chromatography, although conventional chemical methods are still in use. In spite of the advances in methodology, several phylogenetically interesting groups of organisms have received comparatively little attention recently: these include the Oomycota, Cryptophyceae, Charophyceae, and Prasinophyceae.

Lipids are classified into structural groups for ease of discussion. Commonly studied groups include phospholipids, sulfolipids, and acylglycerides; other groups have also been described. Although there is a moderate degree of variation in the distribution of these classes of lipids (Table 33), phylogenetically more interesting information will probably come from a detailed analysis of individual lipids (i.e., including differences in the fatty acids esterified to the lipid "backbone"). Minimal information on individual lipid structures is available for protists.

8.3.1 Phospholipids

Phospholipids have recently been reviewed by Mangnall and Getz (1973), who describe details of structure and biosynthesis, especially for the protists *Acanthamoeba castellani, Dictyostelium discoideum, Saccharomyces* spp., *Tetrahymena pyriformis*, and *Trypanosoma* spp.

Phosphatidylcholine is biosynthesized by a phosphocholinetransferase mechanism in *Neurospora crassa, Saccharomyces cerevisiae, Crithidia fasciculata, Entodinium caudatum, Plasmodium knowlesi, Tetrahymena*

* Excluding steroids (see Section 9.5).

pyriformis, and higher animals (Rock, 1971; Palmer, 1974; Broad and Dawson, 1975). Its biosynthesis proceeds instead by serial methylation of phosphatidylethanolamine in investigated bacteria, *Euglena gracilis* Z, *Ochromonas malhamensis,* some diatoms, and (in addition to the phosphocholinetransferase mechanism) in *T. pyriformis* and *C. fasciculata* (Tipton and Swords, 1966; Opute, 1974b).

The distribution of phospholipids in the Actinomycetales has been used as a significant part of their taxonomic classification (Pommier and Michel, 1973). By their classification scheme, *Nocardia mediterranei* and *N. turbata* should be reassigned to other genera because they contain no phosphatidylinositol and no phosphatidylethanolamine, respectively. Although "absence" of a biochemical character is a dangerous datum when wielded by phylogeneticists, in this instance the authors showed a good correlation between these phospholipid data and differences in accumulation of fatty acids and of *meso*-diaminopimelic acid (for the use of other biochemical characters in taxonomy of Actinomycetales, see Lechevalier *et al.*, 1971).

8.3.2 Sulfolipids

Sulfolipids have been characterized in several protists, but distributional and biosynthetic data are quite incomplete. One of the more intriguing sulfolipids known is 6-sulfoquinovosyl diglyceride, which has been reported in *Rhodomicrobium vannielii, Rhodopseudomonas spheroides, Rhodospirillum rubrum,* Cyanophyceae, and all eukaryotic photosynthesizers investigated. In the Chlorophyceae and higher plants, it is believed to be involved in the structure and/or function of the chloroplast membrane, and may somehow be involved in the photosynthetic act itself (Haines, 1973), suggesting a common ancestry for the photosynthetic apparatus of all photosynthesizers.

8.3.3 Acylglycerides

Glycolipids (galactosyl diglycerides) of the chloroplast may be biosynthesized at the chloroplast outer membrane (Douce, 1974), possibly reinforcing the view that this membrane is derived from the cytoplasmic endoplasmic reticulum. The function of lipids in chloroplast activity is also suggested by the characteristic association of certain lipids with chloroplast fractions, and by the correlation between lipid concentrations and photosynthetic state in eukaryotic algae. As discussed by Rosenberg (1973), galactosyl diglycerides and sulfoquinovose appear to be important in energy transformation phenomena in chloroplasts, and seem to be involved in the function of the Hill reaction. It is also of note that although the phospholipid composition of the various photosynthetic prokaryotes is quite variable, phosphatidylglycerol is always present.

8.3.4 Biomembranes

Although bacteria and Cyanophyceae are morphologically, and in some respects biochemically, simple organisms, it appears that their lipid compositions are not extraordinarily primitive (Table 33). It may be that the most recent common ancestor to all protists was far from "primitive" with respect to lipid composition.

It may not be entirely coincidental that an important feature of this common protistan ancestor was the biomembrane delimiting this organism from the surrounding environment. Early organisms were most likely under strong selective pressures regarding the composition of this biomembrane. All membrane models, from the early picture of Danielli and Davson (1935) to the fluid membrane model of Singer and Nicolson (1972), stress the importance of lipids in membrane construction and function.

The ultrastructure of different biomembranes appears remarkably uniform at the current limits of electron microscopic resolution, although this may be due in part to fixation procedures. Differences in gross composition among different membranes have occasionally been reported, some of which may be directly due to mutational events as opposed to variations in environmental conditions (Keith *et al.*, 1973). Nonetheless, gross compositional differences do not appear promising as a field for further phylogenetic study. Phylogenetic investigations will have to deal with amino acid sequences of biomembrane polypeptides, individual lipids present in biomembranes (including individual fatty acids esterified in lipids), membrane steroids, and the complex process of membrane biogenesis.

Biogenesis of organelle membranes is a field of active investigation at the present time. Chloroplasts are capable of biosynthesizing most of their own membrane components (Getz, 1972), but the location of the relevant genes is unknown. More work has been done on mitochondrial membrane biogenesis, and an ever more complex picture is emerging. The primary data have been summarized by Getz (1972) and will not be repeated here. Any insight into the origin of organelles gained through studies of biomembranes is likely to come not through overt comparisons between biomembranes of prokaryotes and of chloroplasts or mitochondria, but rather in the intricacies of membrane biogenesis and its control.

Membranes have recently been implicated in the "biological clock," which regulates many periodic phenomena in eukaryotes (Njus *et al.*, 1974).

TABLE 33

Distribution of Lipids[a]

Lipid	Gram +ve bacteria	Gram −ve bacteria	Actinomycetales	Thiorhodaceae	Athiorhodaceae	Chlorobacteriaceae	Cyanophyceae	Euglenophyceae	Rhodophyceae	Dinophyceae	Cryptophyceae	Bacillariophyceae	Phaeophyceae	Chrysophyceae	Xanthophyceae	Chlorophyceae	Higher plants	Chytridiomycetes	Zygomycetes	Ascomycetes-imperfect fungi	Basidiomycetes	Protozoa
Triglycerides			+					+	+	+	+	+	+	+	+		+	+	+	+	+	+
Diglycerides																						
Monogalactosyl	b			+	?	+	+	+	+	+	+	+	+	+	+	+	+	+				
Digalactosyl	b			+			+	+	+		+	+	+	+	+	+	+	+				
Trigalactosyl				+												+	+					
Other glycolipids							c									+						+
Sulfoquinovosyl diglyceride					+		+	+	+			+	+	+	+	+	+					d
Other sulfolipids	+	+					+				+	+	+	+		+	+			+	+	+
Cardiolipin	+	+	+	+	+			+	+			+	+	+		+	+			+		+
Phosphatidylglycerol	+	+		+	+		+	+	+		+	+		+	+	+	+			+		+
Phosphatidylcholine	+	+			+			+	+		+	+	+	+	+	+	+	+	+	+	+	+
Phosphatidylethanolamine	+	+	+	+	+			+	+		+	+	+	+	+	+	+	+	+	+	+	+
Phosphatidylserine	+	+						+		+			+			+	+	+	+	+	+	e
Phosphatidyl-*myo*-inositol	+		f		+		f	+	+		+	+	+	+	+	+	+	+	+	+	+	g
Phosphatidic acid	+	+	+		+							+	+				h			+		+
Sphingolipids	+	+															+		+	i	+	i
Other peptidolipids	+		+		+																	
Halolipids														+								
Lipopolysaccharides		+			+		+															
Phosphonolipids										+				+								+

[a] Data for bacteria: Benson *et al.* (1959), Constantopoulos and Bloch (1967), Cruden and Stanier (1970), Goldfine (1972), Gupta *et al.* (1974), Haines (1973), Ikawa (1967), Livermore and Johnson (1974), Nichols (1970), Pommier and Michel (1973), Short and White (1970), Steiner *et al.* (1969), Stoffel *et al.* (1975), Sutherland and Smith (1973), Weise *et al.* (1970), Wood *et al.* (1965); Cyanophyceae: Beach *et al.* (1970), Cruden and Stanier (1970), Kates and Volcani (1966), Nichols *et al.* (1965), Wolk (1973); Euglenophyceae: Nichols and Appleby (1969); Rhodophyceae: Allen *et al.* (1970), Beach *et al.* (1970), Craigie *et al.* (1968), Nichols (1970), Nichols and Appleby (1969), Shibuya (1960); Dinophyceae: Harrington *et al.* (1970), Kittredge and Roberts (1969); Cryptophyceae: Antia *et al.* (1974), Beach *et al.* (1970), Glasl and Pohl (1974): Bacillariophyceae: Kates and Volcani (1966), Opute (1974a,b); Phaeophyceae: Alam *et al.* (1971), Beach *et al.* (1970), Kennedy and Collier (1963); Pham Quang and Laur (1974, 1975), Tulloch *et al.* (1973); Chrysophyceae and Xanthophyceae: Alam *et al.* (1971), Baldwin and Braven (1968), Beach *et al.* (1970), Haines (1973), Mooney *et al.* (1972), Nichols (1970), Nichols and Appleby (1969); Chlorophyceae: Benson and Maruo (1958), Benson and Strickland (1960), Benson *et al.* (1959), Brush and Percival (1972), Cruden and Stanier (1970), Kennedy and Collier (1963), Nichols (1970), Nichols *et al.* (1965), Shibuya (1960), Wolfersberger and Pieringer (1974); higher plants: Benson and Strickland (1960), Benson *et al.* (1959), Brush and Percival (1972), Harwood and James (1975), Ikawa (1967), Kennedy and Collier (1963), Nichols (1970), Nichols *et al.* (1965), Tulloch *et al.* (1973); Chytridiomycetes: Mills *et al.* (1974); Zygomycetes: Jack (1966), Weete (1974); Ascomycetes, Basidiomycetes and imperfect fungi: Gerasimova *et al.* (1974), Haines (1973), Jack (1966), Mangnall and Getz (1973), Weete (1974); Protozoa: Benson *et al.* (1959), Comes and Kleinig (1973), Ellingson (1974), Haines (1973), Hill (1972, pp. 52–53), Horiguchi and Kandatsu (1959), Ikawa (1967), Kittredge and Roberts (1969), Mangnall and Getz (1973), Metzner (1973), Palmer (1974), Smith *et al.* (1970).

[b] Glucosyl diglycerides present in *Staphylococcus aureus* (Beining *et al.* (1975).

[c] Glycolipids encompass a wide range of compounds, and are probably of wider distribution than these data indicate. Unusual glycolipids of Cyanophyceae are discussed by Bryce *et al.* (1972).

[d] In *Crithidia fasciculata* (Haines, 1973).

[e] Phosphatidylserine (PS) has been found in *Acanthamoeba castellani, Dictyostelium discoideum* (Mangnall and Getz, 1973), *Tetrahymena pyriformis* (Hill, 1972; Palmer, 1974), and higher animals. It has not been detected in *Crithidia fasciculata* (Palmer, 1974) or *Physarum polycephalum* (Comes and Kleinig, 1973). Phosphatidyl-*N*-(2-hydroxyethyl)alanine may occur instead of PS in rumen protozoa (Mangnall and Getz, 1973).

[f] Some phosphatidylinositol may be present in *Spirulina maxima* (Hudson and Karis, 1974); PI-mannosides in actinomycetes (Ikawa, 1967).

[g] Phosphatidyl-*myo*-inositol is present in *Dictyostelium discoideum* (Ellingson, 1974), *Physarum polycephalum* (Cruden and Stanier, 1970), rumen protozoa (Mangnall and Getz, 1973), and higher animals (Benson *et al.* (1959); a slightly modified form may exist in *Acanthamoeba castellani* (Mangnall and Getz, 1973). Polyphosphoinositides have been reported in yeasts, protozoa, and animals (Palmer, 1973, 1974).

[h] Localized in the chloroplast (Schwertner and Biale, 1973).

[i] Sphingomyelin present in four species of trypanosomes (Mangnall and Getz, 1973), *Tetrahymena pyriformis* (Taketomi, 1961), and higher animals. Sphingolipids have been found in several fungi (Weiss *et al.*, 1973; Kish and Jack, 1974, Smith and Lester, 1974, Työrinoja *et al.*, 1974; Weete, 1974).

8.4 ACETYLENIC COMPOUNDS

Acetylenes are distinguished from all other compounds by the presence of one or more carbon to carbon triple bonds. Although this classification is not purely biosynthetic, most acetylenes are derived from the acetate pathway, or from acetate and mevalonate. Acetylenes are particularly widely distributed among the higher plants, where they may be of interest in phylogenetic studies (Sørensen, 1968). A monograph listing most known acetylenes has recently appeared (Bohlmann *et al.*, 1973). Acetylenic carotenoids are discussed with the other carotenoids (Section 9.4).

Identification of individual acetylenic compounds is usually made on spectroscopic evidence, often only on ultraviolet absorption spectroscopy sometimes backed up by gas chromatography or thin-layer chromatography. All of these methods are relatively sensitive, but are usually insufficient to give the exact molecular structure.

Among the protists, the Basidiomycetes have been most thoroughly examined for acetylenic compounds, due in part to the antibiotic properties of fungal acetylenes such as mycomycin (Table 34). The few acetylenes discovered so far in algae appear to bear no biogenetic resemblance to those in higher plants (Bohlmann *et al.*, 1973). Although reports of acetylenic antibiotics in the Actinomycetales have appeared, it is likely that these "actinomycetes" were erroneously identified eukaryotic fungi (Sørensen, 1968, p. 200). As pointed out by both Sørensen (1968) and Bohlmann *et al.* (1973), acetylenes of Basidiomycetes are almost always detected in the culture medium—rarely in the fungus itself. It is possible that other acetylenes are present in the mycelia of these and other fungi.

Some phylogenetic information may be found in the distribution of acetylenes in the Basidiomycetes if certain assumptions are made. The most critical assumption necessary is that the enzymes responsible for acetylene biosynthesis are more closely related to each other than any is to, for instance, an enzyme generating a double bond (i.e., that the acetylenes are biosynthetically homologous, not convergent). Testing this assumption would require purification and characterization of the enzymes, which has not yet been accomplished.

One phylogenetic result is that no acetylenes have been reported among fungi normally classified as Heterobasidiomycetidae. Further distributional patterns within the Homobasidiomycetidae are complicated by the different views on the taxonomy of these organisms (Sørensen, 1968). Adopting for the moment the classification of Ainsworth and Bisby (Figure 9 of Sørensen, 1968), it is seen that acetylenes have been found in only two of the eight orders (Polyporales and Agaricales). Within the Polyporales, they have been found in three of the six families (Hydnaceae, Polyporaceae, and

TABLE 34

Distribution of Acetylenic Compounds in the Protists[a]

Bacteria
Bacterium coli
Actinomycetes[b]
Streptomyces chibaensis
S. griseus
S. reticuli var. *aquamyceticus*
Imperfect fungi
Helminthosporium siccans g-207
Basidiomycetes
Homobasidiomycetidae[c]
Agrocybe dura
Aleurodiscus roseus
Clitocybe spp. (14 spp.)
Coprinus quadrifidus
Cortinellus berkeleyanus
Daedalea juniperina
Drosophila (=Psathyrella) spp. (3 spp.)
Fistulina spp. (2 spp.)
Flammula sapinea
Fomes annosus
Lepista diemii
Leptoporus kymantodes
Marasmius ramealis
Merulius lacrymans
Odontia bicolor
Omphalia umbilicata
Papulospora polyspora
Pleurotus ulmarius
Polyporus spp. (3 spp.)
Poria spp. (7 spp.)
Psilocybe sarcocephala
Rhodotorula glutinus var. *lusitanica*[c]
Tricholoma spp. (4 spp.)

[a] Data from Sørensen (1968) and Bohlmann *et al.* (1973).

[b] See text.

[c] There is no general agreement on the taxonomic position of *Rhodotorula*. Bohlmann *et al.* (1973) separate it from the Basidiomycetes; Casperson and Purz (1974) consider it to be a basidiomycete; Storck (see Section 4.2.2) has suggested that it may best be classified with the Heterobasidiomycetidae. On the basis of acetylene production it might be transferred to the Homobasidiomycetidae. However, all morphological and biochemical data will have to be considered before formal reclassification is justified.

Thelephoraceae); among the Agaricales, they have been found in only one (Agaricaceae) of the five families.

The classification of Nannfeldt (Figure 8 of Sørensen, 1968) allows a taxonomically broader occurrence of acetylenes. Nannfeldt divides the subclass into two orders, the Gasteromycetes and the Hymenomycetes; acetylenes are found only in the latter taxon, but are distributed throughout both of its major divisions (Agaricales and Aphyllophorales). The distribution of polyacetylenes does not allow a choice between these two classifications, but (with the exception of acetylenes in *Rhodotorula* spp.) reinforces the dichotomy between the Heterobasidiomycetidae and the Homobasidiomycetidae proposed on the basis of gross morphology and cell wall chemistry (Bartnicki-Garcia, 1968, 1970).

Many other factors suggest only a distant phylogenetic relationship between higher plants and fungi. Therefore the presence of acetylenic compounds in both groups may be ascribed to convergent evolution or to gene transfer. Hydrocarbon acetylenes, common in the higher plants and especially in the Compositae, are rare in the Basidiomycetes (Sørensen, 1968).

8.5 COMPOUNDS DERIVED FROM SHIKIMIC ACID

The shikimic acid pathway (Figure 3, Section 6.12) is the biosynthetic pathway leading to phenylalanine, tyrosine, and cinnamate derivatives (Haslam, 1974). Many compounds derived from this pathway are of very limited distribution, and are of interest in biochemical phylogenies only when their biosynthesis has been established. Some other shikimate-derived compounds of interest are as follows:

1. Lignin, a heteropolymer of cinnamic acid derivatives, is usually considered to be restricted to the higher plants (possibly including the bryophytes: Erickson and Miksche, 1974). Variations in the composition of lignin have been used in the taxonomy of families of higher plants. Reports of incompletely characterized "lignin-like" compounds in Phaeophyceae (Ripa and Borquez, 1967; Gol'man *et al.*, 1973) are almost certainly mistaken, and may be due to "type 2" polymeric polyphloroglucinols (Ragan and Craigie, 1976). There are unconfirmed reports of an uncharacterized lignin-like substance in Chlorophyceae (Gunnison and Alexander, 1975) and of cinnamate derivatives in seven species of Cyanophyceae (Kozitskaya, 1974).
2. Tannins derived from gallic acid, found in the higher plants, are also known from *Spirogyra arcta* (Nakabayashi, 1955). Gallic acid itself

has been found in other protists, including *Phycomyces blakesleeanus* (Birkinshaw, 1965).

3. Lunularic acid, a dihydrostilbene first characterized as a growth inhibitor in liverworts (Pryce, 1971), has been identified in various algae on the basis of thin-layer and gas chromatography. Because the range of algae thought to possess lunularic acid is so wide (members of the Cyanophyceae, Rhodophyceae, Bacillariophyceae, Phaeophyceae, Xanthophyceae, and Chlorophyceae: Pryce, 1972), it is difficult to use this character in algal phylogeny. It was not accumulated by the one fungus examined, a basidiomycete.

4. Brominated phenols have been reported in the Rhodophyceae and more recently in the Cyanophyceae and Phaeophyceae (Craigie and Gruenig, 1967; Pedersén and DaSilva, 1973; Pedersén and Fries, 1975). The biosynthesis of these compounds has not been studied, but their hydroxyl substitution pattern suggests a derivation from shikimic acid. These compounds may eventually be of limited use in red algal taxonomy (Fenical, 1975).

8.6 OTHER ROUTES TO AROMATIC COMPOUNDS

Although the polyketide condensation and shikimic acid pathways are the major routes to aromatic compounds in the protists, other pathways are known (Peterson, 1963; Neish, 1964):

1. The isoprenoid route, used in biosynthesis of aromatic carotenoids in the bacteria (Moshier and Chapman, 1973), is also utilized for gossypol biosynthesis (in cotton plants), estradiol biosynthesis (in mammals), thymol biosynthesis, and biosynthesis of phenylterpenoids (in the Cupressaceae).
2. *Pseudomonas beijerinckii* can form tetrahydroxybenzoquinone from inositol.
3. *P. aeruginosa* converts glycerol or dihydroxyacetone to pyocyanine.
4. *P. testosteroni* and *Septomyxa affinis* aromatize sterols.
5. *Eremothecium ashbyii* and *Candida flareri* biosynthesize riboflavin from acetoin.
6. Mammals can dehydrogenate hexahydrobenzoate to hippuric acid (not yet reported from protists).
7. Mammals can convert *trans*-cyclohexadiene-1,2-diol to catechol.
8. *Xanthomonas* synthesizes bromopolyenes (Andrewes *et al.*, 1973). Although they are formally similar to the aromatic carotenoids, there is no evidence, one way or the other, for an isoprenoid origin.

9. *Aspergillus niger* may produce under certain nutritional conditions an aromatic polyene, asperrubrol (Rabache *et al.*, 1974). The biosynthetic pathway is unknown. As in *Xanthomonas*, the compound is formally similar to an aromatic carotenoid. The temptation, however, to draw conclusions from "paper chemistry" must be avoided. Recently, Bu'Lock's group (De Rosa *et al.*, 1972, 1974) have shown that a cyclohexane fatty acid is derived, at least partially, from shikimic acid.

The evidence supporting some of these proposed pathways is stronger than for others. Each of these pathways, with the possible exception of the first one listed, is likely to seem a rather specialized evolutionary modification without general phylogenetic significance, at least until the mutational events responsible are more fully elucidated. The discovery of three "minor" pathways in the genus *Pseudomonas* is suggestive of the diversity of this group.

9

Metabolites: III

9.1 THE MEVALONIC ACID PATHWAY: ISOPRENOIDS

The mevalonic acid pathway and the biosynthesis of isoprenoids begin with the appearance of acetyl coenzyme A. The mevalonate pathway diverges from the acetate pathway per se, however, in the formation of the six-carbon compound β-hydroxy-β-methylglutarate. Individual steps in the mevalonate pathway have been examined in great detail, and the absolute stereochemistry of many of these steps is known (Clayton, 1965a; Geissman and Crout, 1969; Goodwin, 1971a,b). Five-carbon units are formed by decarboxylation of β-hydroxy-β-methylglutarate to yield mevalonic acid, which in turn is decarboxylated at the C-2 position to produce isopentenyl pyrophosphate (Figure 7). A structural isomer, dimethylallyl pyrophosphate, condenses with isopentenyl pyrophosphate to yield geranyl pyrophosphate. Subsequent condensations with isopentenyl pyrophosphate (Figure 8) yield the family of chemical compounds the isoprenoids or terpenoids. Monoterpenes (C_{10} compounds) have been found in Chlorophyceae, Rhodophyceae, and Phaeophyceae (Katayama, 1964). Sesquiterpenes (C_{15} compounds), diterpenes (C_{20} compounds), and triterpenes other than sterols (C_{30} compounds) have occasionally been reported to accumulate in protists, and certainly are involved in the biosynthesis of carotenoids and sterols. Squalene, formed by the tail to tail condensation of two C_{15} farnesyl pyrophosphate units, has been discovered in bacteria, higher animals, many benthic algae, and phytoplankton (Blumer *et al.*, 1971; Halsall and Hills, 1971), and is probably present in small amounts in all protists capable of sterol biosynthesis, since presqualene pyrophosphate (Altman *et al.*, 1971)

Figure 7. Isopentenyl pyrophosphate (IPP) and dimethylallyl pyrophosphate (DMAPP).

and squalene 2,3-epoxide appear to be the parent C_{30} compounds of the sterols.

Because carotenoids and sterols occur in all protistan groups, it is likely that most protists possess a functional mevalonic acid pathway. Exceptions would include protists displaying an absolute dietary requirement for mevalonate-derived compounds. The accumulation of classes of terpenoids (e.g., monoterpenes) is not likely to be of great phylogenetic interest for this reason. The distribution of individual terpenes may be of more interest, especially when biosynthetic steps are known. As will be seen in the sterols, alternate pathways exist for some isoprenoids, further complicating phylogenetics.

The distribution of some isoprenoids appears to be of taxonomic interest. Heptacyclic terpenes may be a characteristic of members of the Dictyo-

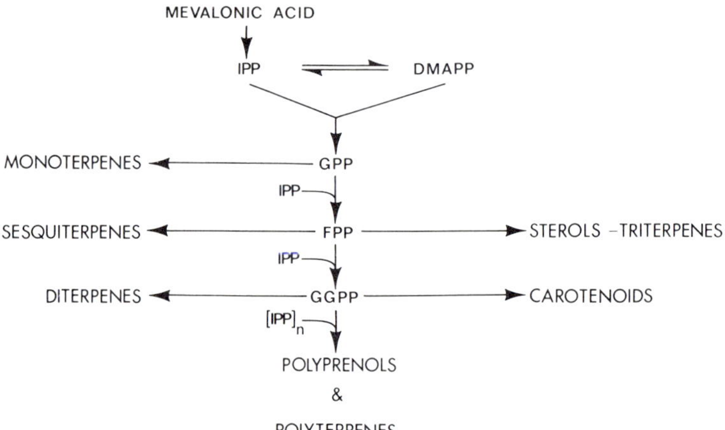

Figure 8. Major pathways of terpenoid biosynthesis. IPP, isopentenyl pyrophosphate; DMAPP, dimethylallyl pyrophosphate; GPP, geranyl pyrophosphate; FPP, farnesyl pyrophosphate; GGPP, geranylgeranyl pyrophosphate. The diagram does not include all known terpenoid groups. It is drawn to illustrate the basic relationships between the central, or common pathway, and the various biosynthetic classes derived from this central pathway.

taceae (Phaeophyceae) (Hirschfeld *et al.*, 1973; Moore and Yost, 1973). Halogenated sesquiterpene alcohols have been found in certain groups of Rhodophyceae (Fenical, 1975). Even quite specialized chemical structures, for instance the (–)-kaurene nucleus, may be of unexpectedly wide distribution; kaurene forms the basis of gibberellic acids, which have been identified in fungi and higher plants. Gibberellin-like activities have been found in extracts from Rhodophyceae, Phaeophyceae, Chlorophyceae, and Prasinophyceae (Provasoli and Carlucci, 1974). The taxonomic utilization of terpenoid data has been further discussed by Ponsinet *et al.* (1968).

Among the compounds traditionally used in biochemical phylogenies are, of course, the carotenoids, and sterols have more recently been examined in this regard. These two classes of mevalonate-derived compounds will be examined in detail, and mention will be made of other mevalonate-based molecules that may be of some interest in more restricted phylogenetic investigations.

9.2 PHYTOL, GERANYLGERANIOL, AND FARNESOL

Phytol is the C_{20} terpenoid alcohol esterified to the cyclic tetrapyrrole moiety of chlorophylls *a* and *b* and some bacterial chlorophylls (Section 10.2). Chemically, phytol is the tetrahydro derivative of geranylgeraniol (the alcohol parent of geranylgeranyl pyrophosphate). In certain Athiorhodaceae (Section 10.2), geranylgeraniol may also serve as the esterifying alcohol of some bacteriochlorophylls, along with or in place of phytol. In the bacteriochlorophyll *c* and *d* series, farnesol, the C_{15} alcohol analog of geraniol, is the esterifying alcohol. An esterifying C_{15} analog to farnesol has not been identified. In *Rhodopseudomonas spheroides*, phytol biosynthesis is tightly coupled to biosynthesis of the pyrroles used in bacteriochlorophyll formation, which is of added interest because these two compounds are derived from different pathways (Brown and Lascelles, 1972). Localization of phytol biosynthesis in the chloroplasts of eukaryotic cells would be of interest in light of the endosymbiotic theory of chloroplast origin. However, technical difficulties have hampered this work; enzymes responsible for the biosynthesis of phytoene and of its diterpene precursors are usually lost during the isolation of chloroplasts in aqueous media. Later-stage enzymes, which elaborate phytoene into carotenoids, xanthophylls, and other derivatives, are inhibited by isolation in nonaqueous media (Buggy *et al.*, 1974). Enzymes for phytol biosynthesis, presumably from geranylgeranyl pyrophosphate, have not been characterized. Indeed, one may question whether phytol is in fact biosynthesized in the free form. Rüdiger *et al.*

(1977) have shown that geranylgeranyl pyrophosphate is the esterifying moiety for chlorophyllide *a*.

There is a single report of phytol biosynthesis by isolated chloroplasts (Charlton *et al.*, 1967). Chloroplasts isolated from *Phaseolus vulgaris* by nonaqueous techniques were, however, incapable of phytol biosynthesis, although capable of biosynthesizing phytoene (Buggy *et al.*, 1974). Until the experimental problems are resolved, and until the genes responsible for phytol biosynthetic enzymes are localized, evolutionary relationships between chloroplasts and photosynthetic prokaryotes cannot be argued for or against on the basis of phytol biosynthesis.

9.3 QUINONES

The study of quinones in biochemical phylogenies may be of interest because quinones have been detected in fossil sediments (Thomson, 1971). Biosynthesis of these molecules proceeds from acetate via shikimate (the quinone nucleus) and from isopentenyl pyrophosphate (the terpenoid side chain) (Thomson, 1971).

Most quinones can be classified into four structural, and probably biosynthetic, classes: benzoquinones, naphthaquinones, anthraquinones, and anthracyclinones. Benzoquinones are particularly abundant in higher plants, higher animals, and fungi; however, ubiquinone, plastoquinone, and α-tocopherolquinone are benzoquinones that are extremely widely distributed in protists. Another benzoquinone derivative, rhodoquinone, is found in *Rhodospirillum rubrum, Euglena gracilis* var. *bacillaris,* and (curiously) in some nematodes.

Naphthaquinones are common in higher plants, fungi, bacteria, and starfish. Phylloquinones, including vitamin K_1, are probably present in all photosynthetic organisms; one of these naphthaquinones, 5′-monohydroxyphylloquinone, has been reported in *Anacystis nidulans, Chlorella pyrenoidosa,* and in photoautotrophically grown *Euglena gracilis* Z. It is absent in streptomycin-bleached *E. gracilis* Z (Law *et al.*, 1973). Chlorobiumquinone, another naphthaquinone, is found only in green photosynthetic bacteria. Anthraquinones are found in bacteria, higher plants, Ascomycetes, and lichens; they are less commonly reported from Basidiomycetes. Anthracyclinones have been reported only in *Streptomyces* spp.

Although a few taxonomically interesting patterns are seen in the above-described distribution of classes of quinones, individual quinones may prove to be of greater phylogenetic interest. Yamada and Kondô (1973) have reported that fungi of the genera *Cryptococcus, Rhodosporidium, Rhodo-*

torula, and *Sporobolomyces* have differences in their coenzymes Q (ubiquinones) that correlate well with variations observed in DNA base compositions and serological cross-reactivities. These coenzymes Q were differentiated on the basis of the number of isoprenyl units in their side chains. Although most of the coenzymes Q in the basidiomycete-like fungi had ten isoprenyl units (Q_{10}: fifty carbon atoms), several strains of *Rhodotorula* possessed Q_9, and *R. infirmo-miniata* had a Q_8. The strains of *R. infirmo-miniata* differed from all other fungi examined in requiring biotin for growth, assimilating inositol, and biosynthesizing starchlike polymers. Coenzyme Q is apparently absent from most gram-positive bacteria, actinomycetes, and Cyanophyceae (Aaronson and Hutner, 1966; Denis *et al.*, 1975).

The naphthaquinone derivative vitamin K_2 (menaquinone) is not found in eukaryotic algae and higher plants (Nichols, 1973). Among prokaryotes, vitamin K_2 is found in most bacteria. Gram-negative bacteria may have ubiquinone alone or in combination with vitamin K_2, but vitamin K_2 is present alone in gram-positive bacteria (Bishop *et al.*, 1962). Among Cyanophyceae, *Anacystis nidulans* is set apart from most other investigated blue-green algae by its ability to biosynthesize 5′-monohydroxyphylloquinone and by its apparent lack of α-tocopherolquinone (Powls and Redfearn, 1967). Early reports claiming that the quinones of Cyanophyceae were fundamentally different from those of higher plants were misled by their reliance entirely on the quinones of *A. nidulans*. Ubiquinone has not been identified in any cyanophycean examined so far (Wolk, 1973), but is found in gram-negative bacteria (Bishop *et al.*, 1962), including photosynthetic bacteria (Kondrat'eva, 1965; Halsey and Parson, 1974), and in many eukaryotic algae (Nichols, 1973). Interestingly, *Tetrahymena pyriformis* may biosynthesize ubiquinone entirely from shikimate and five-carbon units (Hill, 1972).

Plastoquinone is found in photosynthetic bacteria, Cyanophyceae, and probably universally throughout photosynthetic organisms (Aaronson and Hutner, 1966). The side chains of vitamin K_1, tocopherolquinones, and plastoquinones are thought to be biosynthesized within the chloroplast in higher plants, although biosynthesis of ubiquinone is probably extrachloroplastic in higher plants (Goodwin, 1967).

9.4 CAROTENOIDS

Carotenoids and their oxidized derivatives, the xanthophylls, have been extensively used in phylogenetic and taxonomic studies of photosynthetic organisms because they are widely distributed, structurally variable, and

Basic skeleton

[End group]—[In chain]—[End group]

In chains

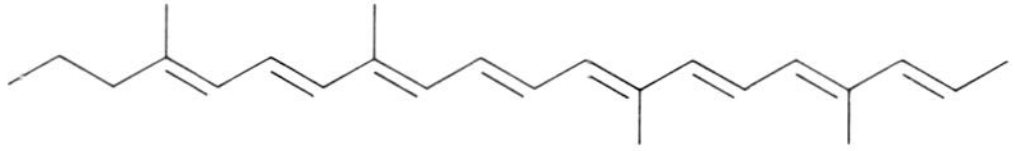

R_1

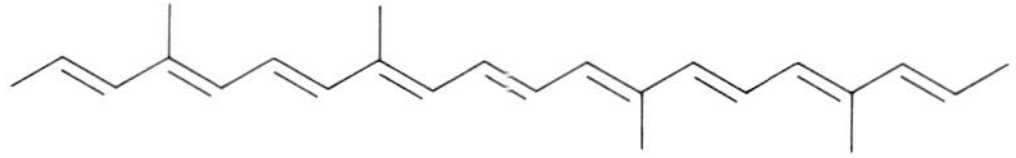

R_2

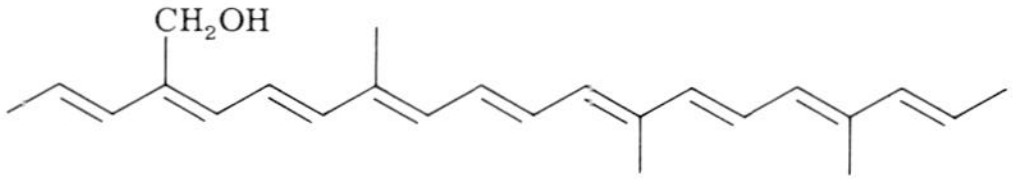

CH_2OH

R_3

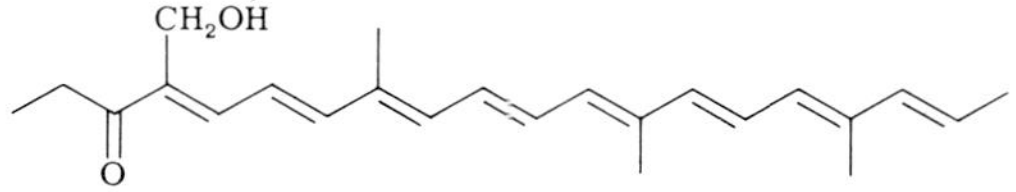

CH_2OH

O

R_4

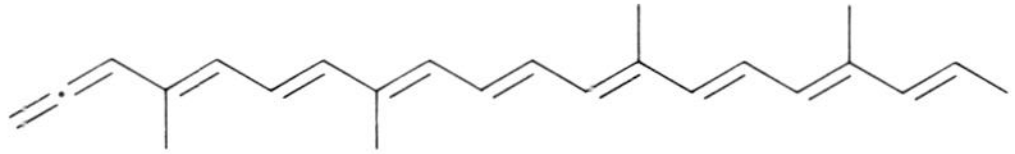

R_5

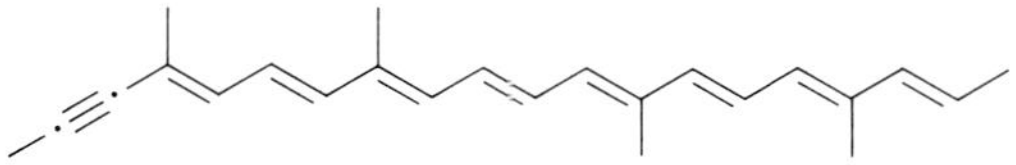

R_6

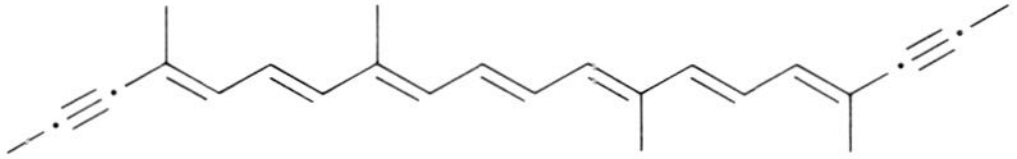

R_7

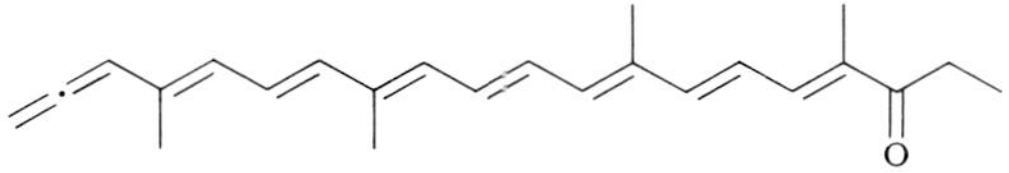

O

R_8

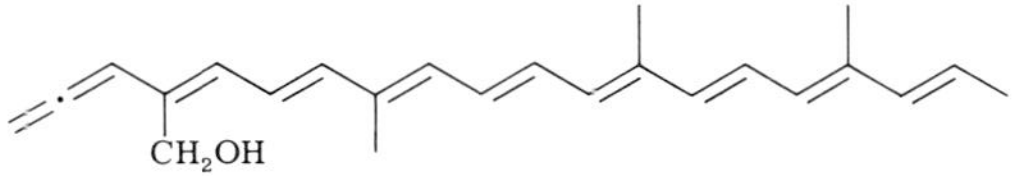

CH_2OH

R_9

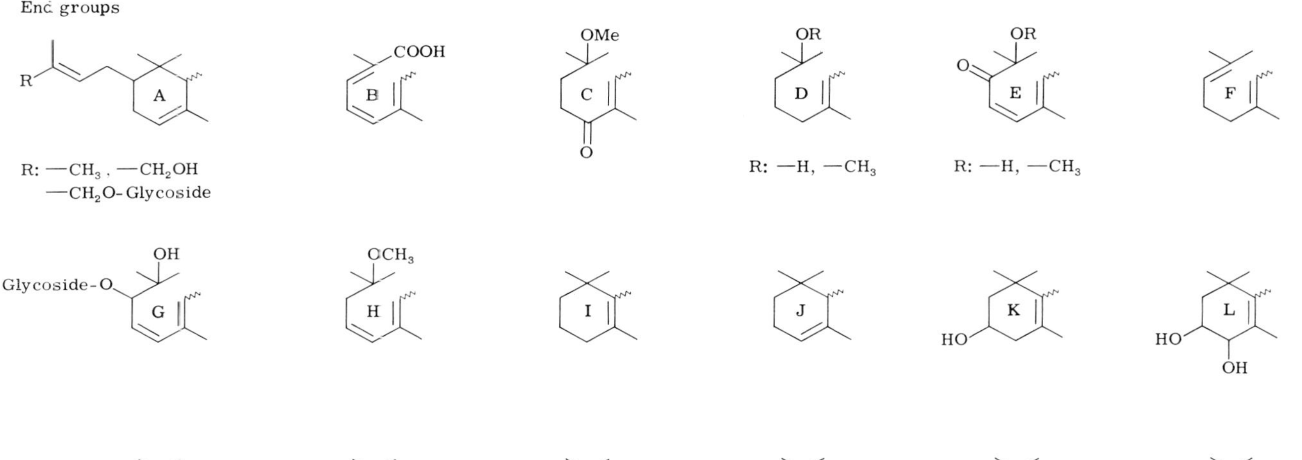

Figure 9. Representative carotenoid structures illustrating principal biosynthetic modifications. This diagram does not show all the known in-chains or end groups, but only those that are in the carotenoids that are represented in the majority of protists. The actual carotenoid structures are given in key form in Table 35. A complete structural compendium is provided by Straub (1971).

relatively easily identified. Chromatographic techniques have been applied to their identification since the work of Cohn (1867) and Sorby (1873). In recent years chromatography has been combined with ultraviolet spectroscopy, mass spectrometry, chemical reactivity, and to a lesser extent nuclear magnetic resonance spectrometry to give unambiguous identifications of many individual carotenoids. In contrast to recent advances in chemical identification of carotenoids, biological factors affecting carotenoid biosynthesis and accumulation have received less attention. Culture conditions (light regimes, nitrogen availability, minerals) and the metabolic state of the cultured organisms affect carotenoid compositions, and intermediates in carotenoid biosynthesis may be observed only if large quantities of nonmutant cells, or favorable mutants, are extracted (Chapman, 1965).

Structures and trivial names of the hundreds of known carotenoids are outside the scope of this treatment; for full details see the compendium by Straub (1971).

Acyclic carotenoids (Figure 9) other than basic biosynthetic intermediates are found almost exclusively among the prokaryotes, where they are often oxygenated at C-1 or C-2 and are sometimes covalently linked to glucose, mannose, rhamnose, or methyl groups. Cyclic carotenoids are, however, synthesized by Cyanophyceae (Wolk, 1973) and flexibacteria (Weedon, 1971), green bacteria, and *Rhodomicrobium vannielii* (Conti and Benedict, 1962). Aromatic carotenoids, unknown among eukaryotes,* have been reported from members of the genera *Phaeobium, Chlorobium, Chromatium, Thiothece, Chloropseudomonas, Mycobacterium, Streptomyces,* and *Corynebacterium* (Weedon, 1971; Moshier and Chapman, 1973). Unusual C_{45} and C_{50} carotenoids have been characterized from gram-positive bacteria (Goodwin, 1971a–c).

A number of approaches have been utilized in the construction of phylogenies and taxonomies based wholly or partially on carotenoid distributions. These have usually followed the pattern of sorting into groups with similar carotenoid distributions. This approach has served quite well for strictly taxonomic aspects, but has produced little in the way of phylogenetic input. Despite the large number of carotenoids that have been positively identified in protists, only a few (e.g., fucoxanthin, peridinin, alloxanthin, chlorobactene, spirilloxanthin, myxoxanthophyll) are individually significant in terms of phylogenies or taxonomies. As with the fatty

* Eukaryote "animals" are excluded. Animals do not synthesize carotenoids *de novo* from mevalonic acid, but instead derive them, with or without subsequent metabolic modification, from dietary sources. For this reason the aromatic carotenoids of the poriferan *Reniera* are not included in this statement of distribution.

acids (Section 8.2), the biosynthesis, and hence the presence or absence, of many carotenoids is influenced by such parameters as mineral deficiency (e.g., ketocarotenoids of Chlorococcales and other Chlorophyceae), light–dark cycles (e.g., epoxide cycle involving such carotenoids as violaxanthin, diadinoxanthin), and aerobiosis or anaerobiosis (e.g., spheroidene–spirilloxanthin series of Athiorhodaceae). Many other carotenoids have such limited and/or anomalous distributions (e.g., ϵ-carotene) or have such a widespread distribution (e.g., zeaxanthin, β-carotene) that they have virtually no taxonomic or phylogenetic usefulness.

The carotenoids represent a classic example of the limitations, strengths, and problems of employing classes of compounds, especially episemantic molecules, in biochemical taxonomies and phylogenies. Obviously "presence or absence" can be used only on a selective basis—but on what basis does one make the selection? The authors have considered this problem (Chapman and Ragan, 1977) and suggested that an alternative approach lies in considering biosynthetic pathways, and major deviations from these (Birch, 1973a,b). One is not attempting, then, to sort out individual carotenoids or assign relative "weights" to their significance, nor to distinguish between inability to synthesize (i.e., lack of appropriate DNA coding) and the capacity to synthesize under certain circumstances (i.e., possession of appropriate DNA coding, but with synthesis and appearance under strict environmental control). Admittedly, the decision as to what is a "major" biosynthetic deviation is subjective. However, we believe there is less subjectivity involved than in the more classic approach of considering individual carotenoids and trying to reach decisions on such problems as "relative importance." If one considers a biosynthetic sequence

$$A \rightarrow B \rightarrow C \rightarrow D \rightarrow E$$

that is derived from a central metabolic pathway, one can make a strong argument for the absence of this pathway, and hence a key enzyme (and presumably appropriate gene), if none of the five compounds is detected. Individually, the five compounds need not be considered. If, however, a situation arises (as it frequently does) in which one group of organisms synthesizes all five compounds A, B, C, D, E and another group appears to synthesize only A, B, C, D, one is in a quandary. Is the absence of E due to the lack of an enzyme (i.e., to a genetic difference) or is it a problem of biosynthetic control or methodological inability to detect? Which of these two alternatives in the case is vitally important for the taxonomist and phylogeneticist. The carotenoids represent an ideal case study.

Based on fragmentary biosynthetic studies and comparison of known molecular structures, a number of major pathways of carotenoid biosynthesis may be recognized (Figure 10). The distribution of these pathways and

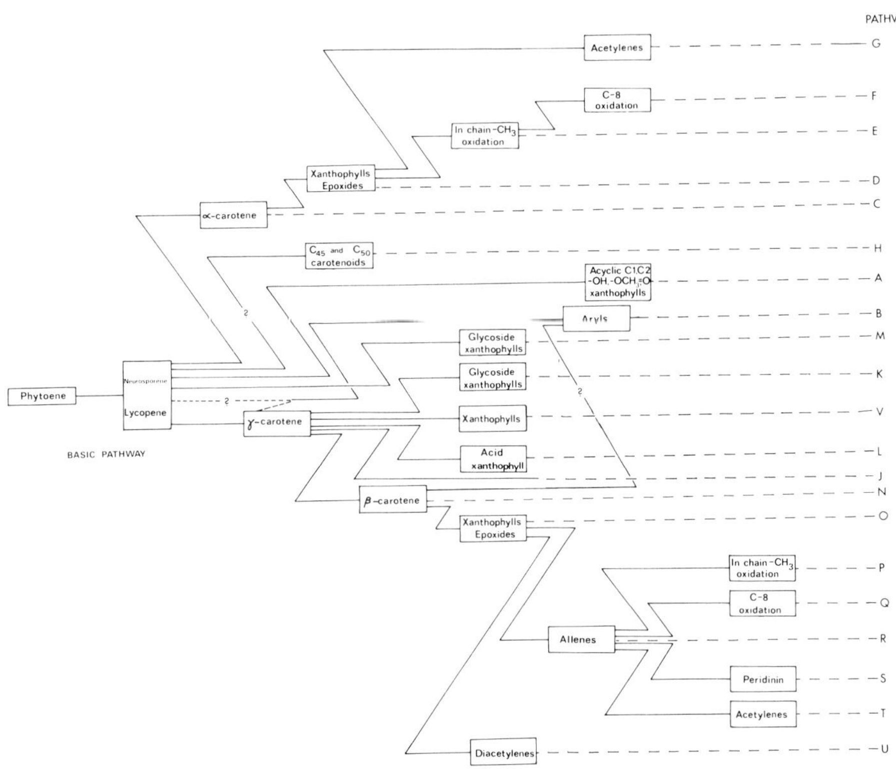

Figure 10. Proposed major pathways of carotenoid biosynthesis. The diagram of pathways does not include all known biosynthetic modifications, but only the principal types. Likewise the appearance of a pathway in a group of organisms does not mean that all intermediates of that pathway have been identified in each organism. With the exceptions of pathways A, C, D, J, N, V, and O, definitive experimental proof is lacking, and the proposals are based primarily on comparative chemistry and known chemical conversions. The taxonomic distribution of the pathways is given in Table 36.

of modifications thereof (Tables 35, 36) is likely to prove more significant for biochemical phylogenetics than would distributions of individual carotenoids. Although subjectivity is still a factor, we believe no one would seriously question that the biosynthetic deviations involved are significant and undoubtedly occur because of the presence of key enzymes. The "deviations" may be briefly summarized as follows: formation of alicyclic xanthophylls; formation of monocyclic xanthophylls; aromatization; cyclization to an α-ionone ring type; formation of an acetylene bond; formation of an allene bond; oxidation of the "in-chain" methyl groups;

TABLE 35

Representative Carotenoids for Each of the Basic Pathways Outlined in Figure 10

Pathway	Carotenoid	Chemical structure[a]	
A	Spheroidene	F-R_1-D	(—OCH_3)
	Spheroidenone	F-R_1-E	
	Rhodovibrin	D-R_2-H	(—OH)
	Spirilloxanthin	H-R_2-H	
	Phillipsiaxanthin	E-R_2-E	(—OH)
B	Chlorobactene	U-R_2-F	
	Okenone	V-R_2-C	
	Isorenieratene	U-R_2-U	
C	α-Carotene	I-R_2-J	
D	Lutein	K-R_2-N	
E	Loroxanthin	K-R_3-N	
	Pyrenoxanthin	[b]	
F	Siphonaxanthin	K-R_4-N	
G	Monadoxanthin	K-R_6-N	
H	Sarcinene	A-R_2-A	(—CH_3)
	Bacterioruberin	S-R_2-S	
	Corynexanthin	A-R_2-A	(—Glycoside)
	Decaprenoxanthin	A-R_2-A	(—CH_2OH)
J	γ-Carotene	I-R_2-F	
K	Myxoxanthophyll	K-R_2-G	
	Aphanizophyll	L-R_2-G	
L	Torularhodin	I-R_2-B	
M	Oscillaxanthin	G-R_2-G	
N	β-Carotene	I-R_2-I	
O	Zeaxanthin	K-R_2-K	
	Echinenone	M-R_2-I	
	Cryptoxanthin	K-R_2-I	
	Canthaxanthin	M-R_2-M	
	Nostoxanthin	R-R_2-R	
	Violaxanthin	O-R_2-O	
P	Vaucheriaxanthin	P-R_9-O	
Q	Fucoxanthin	P-R_8-O	(P:C-3 = Acetate ester)
R	Neoxanthin	P-R_5-O	
S	Peridinin	C_{39} carotenoid	
T	Heteroxanthin	K-R_6-Q	
	Diatoxanthin	K-R_6-K	
	Diadinoxanthin	K-R_6-O	
	Crocoxanthin	K-R_6-J	
U	Alloxanthin	K-R_7-K	

TABLE 35 (*Continued*)

Pathway	Carotenoid	Chemical structure[a]	
V	Flexixanthin	T-R_2-H	(—OH, not —OCH_3)
	Saproxanthin	K-R_2-H	(—OH, not —OCH_3)
	Rubixanthin	K-R_2-F	
	Plectaniaxanthin	I-R_2-G	(Nonglycoside)
K	Phleixanthophyll	I-R_2-G	(O—glycoside, not —OH, —OH on C-2, not O—glycoside)
	Myxobactone	M-R_2-H	(—O—glycoside, not —OCH_3)

[a] See Fig. 9.
[b] Methyl group on C-13 oxidized to alcohol, not C-9 as in loroxanthin.

isoprenyl substitution on the C-2 position; oxidation at the C-8; carbonyl formation. These are not mutually exclusive and neither are they the only modifications. Peridinin has an unusual structure, with an abbreviated "in-chain." Obviously, peridinin is biosynthesized by a pathway unique to this carotenoid and its analogs.

The phytoene → lycopene sequence is designated as the basic pathway, common to all carotenogenic organisms. This does not imply that an identical enzyme complex is operating in all organisms. Indeed, the evidence to date suggests that such a system may be a feature only of the fungi (de la Guardia *et al.*, 1971; Garber *et al.*, 1975; Subden and Turian, 1970). The remaining pathways are formulated on the basis of either definitive experimental evidence (Goodwin, 1971b,c; Davies, 1973, 1975; Britton, 1976) or comparative biochemistry and a consideration of analogous chemical structures (Chapman and Antia, 1977). This latter approach, especially in the case of the eukaryote xanthophylls, produces hypothetical but plausible pathways, and is an extension of earlier proposals (Chapman and Ragan, 1977; Chapman and Antia, 1977) with regard to algal xanthophylls. The distribution of these pathways is shown in Tables 35 and 36. It is obvious that considerable taxonomic importance can be attached to these distributions. Nevertheless, one can also observe certain phylogenetic trends.

The Athiorhodaceae and Cyanophyceae appear to represent a phylogenetic sequence through the acyclic xanthophylls.

The Chlorobiaceae and Thiorhodaceae may be phylogenetic offshoots from a main line, based on the presence of the aryl pathway.

The Cyanophyceae are essentially synthesizers through the basic carotenoid pathways (viz., they lack complex modifications). The similarity with the Rhodophyceae suggests a phylogenetic connection.

The Chlorophyceae, Charophyceae, and Prasinophyceae show a high degree of uniformity, together with the higher plants, suggesting a common evolutionary line. Although the carotenoids of these groups are chemically more complex, they lack the complexity of other algae. The Chrysophyceae, Bacillariophyceae, Haptophyceae, and Phaeophyceae have, on the basis of the carotenoids, a common phylogenetic heritage, and may well have a common origin with the Xanthophyceae and Chloromonadophyceae. One can propose, using these data, a phylogeny in which the Xanthophyceae, Eustigmatophyceae, and Chloromonadophyceae form, or are related to, the ancestral group of the Chrysophyceae, Bacillariophyceae, Haptophyceae, and Phaeophyceae, and are in turn most closely affiliated with the divergence of all these classes from the "green algae–higher plant" line.

The Dinophyceae are anomalous but are probably related, distantly, to the former evolutionary line. Carotenoid data provide little in the way of a solution for the Cryptophyceae. Apart from the acetylene function the carotenoids are chemically uncomplicated, and this feature, together with the lack of epoxides (see later), suggests that the Cryptophyceae may well be a primitive group.

The Euglenophyceae pose a problem. The nature of the carotenoid distribution suggests an affinity with the "brown algal line." This is in contrast to deductions based on other chorophyll evidence. A possible explanation may be found in proposals suggesting that the euglenophycean chloroplast is of recent symbiotic (green algal?) origin. Such symbiosis with the probable modification and reduction would also explain the absence, for example, of an α-carotene xanthophyll pathway.

In surveying the entire range of photosynthetic protists, some generalizations can be promulgated.

The progression of evolution has been paralleled by increasing cyclization of carotenoids and by increasing oxidation as represented by epoxides and carbon-chain oxidations (to yield allenes and acetylenes). It is interesting to speculate on the oxidative increase and the appearance of aerobiosis and aerobic photosynthesis.

The distribution of carotenoid biosynthetic pathways reinforces the establishment of the class Eustigmatophyceae with certain algae formerly classified as Xanthophyceae (Hibberd and Leedale, 1970). Assuming correct identification of the carotenoids of these two classes (Norgård *et al.*, 1974a; Whittle and Casselton, 1975a,b), both groups posses Pathways N, O, and P; some Eustigmatophyceae lack Pathway R and virtually all Xanthophyceae possess Pathway T (diadinoxanthin and diatoxanthin), which is lacking in the Eustigmatophyceae.

Fungal carotenoids have received considerably less attention than have

TABLE 36

Distribution of Carotenoid Biosynthetic Pathways in Protists[a]

Group	Pathways known																	
Gram +ve bacteria						H		K			N	O[b]						
Gram −ve bacteria						H		K			N							
Actinomycetales		B				H		K										V
Thiorhodaceae	A[c]	B																
Chlorobacteriaceae		B																
Athiorhodaceae	A[c]									M								
Flexibacteria										M								V[d]
Cyanophyceae								K		M	N	O[b]						
Myxomycetes							J		L		N							
Zygomycetes							J				N							
Ascomycetes + Fungi Imperfecti	A[e]						J		L		N							V
Basidiomycetes							J		L		N	O						
Oomycetes																		
Rhodophyceae			C	D							N	O	Q[f]	R				
Cryptophyceae			C		G						N						U	
Dinophyceae							J				N	O	Q[f]	R	S	T		
Bacillariophyceae			C								N	O	Q	R		T		
Chrysophyceae											N	O	Q	R		T		
Haptophyceae			C				J				N	O	Q	R		T		

Phaeophyceae	C										N	O		Q	R		
Xanthophyceae											N	O	P				T
Chloromonadophyceae											N	O	P				T
Eustigmatophyceae											N	O	P		R		
Euglenophyceae	C	D									N	O			R		T
Prasinophyceae	C	D	E	F			J				N	O			R		
Charophyceae	C	D	E	F			J				N	O			R		
Chlorophyceae	C	D	E	F			J				N	O			R		
Higher plants (chloroplasts)	C	D									N	O			R		

[a] Data for bacteria: Conti and Benedict (1962), David (1974), Goodwin (1952, 1971); Cyanophyceae: Chapman (1966), Goodwin (1966), Hager and Stransky (1970b), Ogawa *et al.* (1970), Parsons (1961), Stransky and Hager (1970b), Wolk (1973); eukaryotic algae: Goodwin (1966), Hager and Stransky (1970a,b), Riley and Wilson (1967), Stransky and Hager (1970a,b,c), Thomas and Goodwin (1965); Rhodophyceae: Chapman (1966), Bjornland and Aguilar-Martinez (1976); Dinophyceae: Jeffrey (1968a), Johansen *et al.* (1974), Parsons (1961); Cryptophyceae: Chapman (1965), Chapman and Haxo (1963); Bacillariophyceae: Jeffrey (1968a); Phaeophyceae: Jeffrey (1968a), Nitsche (1974), Seely *et al.* (1972); Chrysophyceae: Craigie *et al.* (1971), Dales (1960), Parsons (1961); Haptophyceae: Craigie *et al.* (1971), Jeffrey (1968a); Xanthophyceae: Egger *et al.* (1969), Riley and Segar (1969), Strain *et al.* (1970), Whittle and Casselton (1975a,b); Eustigmatophyceae: Bisalputra (1974), Norgård *et al.* (1974a), Whittle and Casselton (1975a,b); Chloromonadophyceae: Chapman (1966), Chapman and Haxo (1966); Euglenophyceae: Chapman (1966), Gross *et al.* (1975); Prasinophyceae: Ricketts (1967, 1971b), Riley and Segar (1969); Chlorophyceae: Czygan (1966, 1968), Jeffrey (1968a), Parsons (1961), Ricketts (1971a); higher plants: Goodwin (1952, 1966, 1971a), Jeffrey (1968a), Weedon (1971). General: Chapman and Antia (1977); Liaaen-Jensen and Andrewes (1972); Goodwin (1971a, 1974); Fungi, Fiasson (1968).

[b] Epoxides absent in prokaryotes.

[c] Rhodopinal series. In-chain oxidation to aldehyde (Thiorhodaceae). Rhodopinal and spirilloxanthin series in Athiorhodaceae.

[d] Flexibacteria (flexixanthin, saproxanthin) differ from Ascomycetes (plectaniaxanthin).

[e] Different (phillipsiaxanthin) from rhodopinal or spirilloxanthin series.

[f] See text concerning Dinophyceae with fucoxanthin, and Bjørnland and Aguilar-Martinez (1976) for Rhodophyceae.

algal and bacterial ones (Goodwin, 1952; Haxo, 1955; Arpin, 1968). Interestingly, xanthophylls have apparently not been detected among the Oomycetes, Hyphochytridiomycetes, Chytridiomycetes, and Zygomycetes. In those fungi in which they are present, they are often acidic (e.g., torularhodin, neurosporaxanthin), a feature distinctive of the fungi. The other noteworthy feature is the chemical simplicity of the carotenoids of fungi. They lack the oxidative modifications (epoxides*, allenes, acetylenes) characteristic of algal carotenoids. This probably should not be interpreted to mean that the fungi are necessarily primitive, nor that they were derived from a primitive algal ancestor (e.g., Rhodophyceae). The absence of the highly oxidized carotenoids is more likely the result of the nonphotosynthetic nature of the fungi.

9.4.1. Xanthophyll Cycles

It has been recognized since the work of Sapozhnikov (Sapozhnikov *et al.*, 1957; Sapozhnikov, 1973) that illumination can bring about changes in the pigment compositions of plants. These changes often involve the interconversion of hydroxy and epoxy groups of xanthophylls. Investigation of most algal classes has demonstrated two patterns in these "light-induced xanthophyll cycles."

The Euglenophyceae, Xanthophyceae, Chloromonadophyceae, Chrysophyceae, Dinophyceae, and Bacillariophyceae have a xanthophyll cycle involving the conversion of diadinoxanthin to diatoxanthin in the presence of light. A violaxanthin–antheraxanthin–zeaxanthin cycle was reported in Phaeophyceae, Chlorophyceae, and one member of the Eustigmatophyceae (Stransky and Hager, 1970c). The report of the latter cycle in *Ochromonas stipitata* may be artifactual; the diadinoxanthin–diatoxanthin cycle is more likely to occur naturally. It is possible that algae of the classes Xanthophyceae and Eustigmatophyceae can be distinguished on the basis of their differing xanthophyll cycles (see Whittle and Casselton, 1975b).

Unfortunately, the biochemical basis of this character is not well understood: Is it enzyme-mediated? Is the enzyme specific, or does it use "whatever xanthophyll is handy?" Consequently, the phylogenetic significance of xanthophyll cycles, as apart from being another manifestation of the presence of certain xanthophylls and not others, is unclear. A lutein–lutein monoepoxide cycle has recently been reported in tomatoes (Rabinowitch *et al.*, 1975).

* Valadon and Mummery (1976) have now claimed the presence of an epoxide in a cantharelloid fungus.

9.4.2 Sporopollenin

It is common to associate carotenoids exclusively with photosynthetic (algal, higher plant) and other (bacterial, fungal) pigments. Yet there are carotenoids that apparently serve as structural compounds. Sporopollenin is a polymer composed of oxidized carotenoids and carotenoid esters, and is of unusually resilient nature. First discovered by John in 1814 (Shaw, 1970), it has more recently been demonstrated in the cell walls of fossil and present-day Bacillariophyceae, Chlorophyceae, and Charophyceae, and in fossil (but not recent) Prasinophyceae (Atkinson *et al.*, 1972). Sporopollenin-like material has also been recovered from carbonaceous chondrites (Shaw, 1970). A sporopollenin-like material has also been observed in the cell wall of zygomycetous zygospores (Gooday *et al.*, 1973).

The distribution of sporopollenin in biological materials parallels the presence of trilaminar cell wall construction and the presence of "secondary carotenoids." Phylogenetically, it is possible that this material, likely heterogeneous, is a primitive feature utilized in past ages by the common ancestor of most modern algae, and has been retained by some of these organisms. It would be unjustified, however, to consider this character as strongly supporting or opposing a given phylogeny, at least until its structure and biosynthesis are better understood, and until it can be ascertained if the sporopollenin fractions isolated from different protists are biosynthetically homologous.

9.5 STEROLS

Sterols* have been reported from almost all organisms closely investigated, including many bacteria and Cyanophyceae. Organisms that do not biosynthesize sterols usually either have a dietary requirement and metabolic capacity for exogenous sterols, or, more rarely, produce other analogous triterpenoids. Anaerobic organisms apparently do not synthesize sterols, due to the obligate requirement for oxygen in the cyclization of squalene. In this regard, it would be very interesting to determine if photosynthetic bacteria synthesize sterols, and if so, whether this is dependent upon maintenance of the bacteria (e.g., Athiorhodaceae) in an aerobic condition. Because the widespread distribution of sterols coincides with considerable variation in their molecular structures, it might be

* Including 4α-methylated and 4,4-dimethylated molecules.

expected that sterol distributional data would provide a deep insight into protistan phylogeny. Unfortunately, such is not yet the case.

A major factor preventing such application is the complexity of sterol biosynthesis. Although the earlier stages are well known and need not be reiterated here (Clayton, 1965a,b; Goad and Goodwin, 1972; Goodwin, 1973a,b; Weete, 1974), later steps in the biosynthesis of individual sterols are poorly understood. Moreover, as detailed for ergosterol biosynthesis (Figure 11), phytosterol biosynthesis (Figure 12), stigmasterol biosynthesis (Grunwald, 1975), and ergosta-5,7,22-trien-3-β-ol biosynthesis (Weete, 1973), numerous interlocking pathways apparently exist for many sterols, even within a single organism. It follows that the presence of a sterol in two different protists cannot directly imply that homologous biosynthetic pathways or homologous enzymes exist. What will be phylogenetically interesting are not the mere distributional patterns of sterols, but rather, variations in their biosynthetic pathways, differences in sterol biosynthetic enzymes, and possibly mechanisms for control of sterol accumulation.

Although a sterol may be synthesized by a protist, it may not accumulate sufficiently to be detected. Sterol accumulation may be influenced by environmental as well as biochemical factors (Safe, 1973).

A more practical problem is the identification of sterols. This is particularly difficult when dealing with pairs of epimeric sterols such as poriferasterol and stigmasterol, which differ only in their geometry at C-24. The methods in widest use, including gas chromatography and thin-layer chromatography, will not usually resolve such pairs of sterols, and only minor differences are often seen in their mass spectra. Only recently has use of high-resolution proton and carbon-13 nuclear magnetic resonance spectrometry allowed unambiguous determination of stereochemistry. It is of great importance to keep this difficulty in mind when examining the sterol literature for phylogenetically relevant information.

Moreover, many earlier reports have overlooked the presence of water-soluble sterols and sterol derivatives. Steryl esters and steryl glycosides have now been discovered in Zygomycetes (Mercer and Bartlett, 1974), other fungi (Jack, 1966; Mills *et al.*, 1974), higher animals, Rhodophyceae (Feige, 1975), Cryptophyceae (Glasl and Pohl, 1974), Chrysophyceae (Beastall *et al.*, 1972), Chlorophyceae and Charophyceae (Goodwin, 1973a), and higher plants (Bush and Grunwald, 1974; Grunwald, 1975). Steryl esterases have been found in prokaryotes as well as in eukaryotes (Sih *et al.*, 1963).

If it is currently difficult or impossible to arrange sterols in a biosynthetic grouping for phylogenetic purposes, is there any useful approach to be followed other than numerical taxonomy? Some of the approaches currently under examination are as follows.

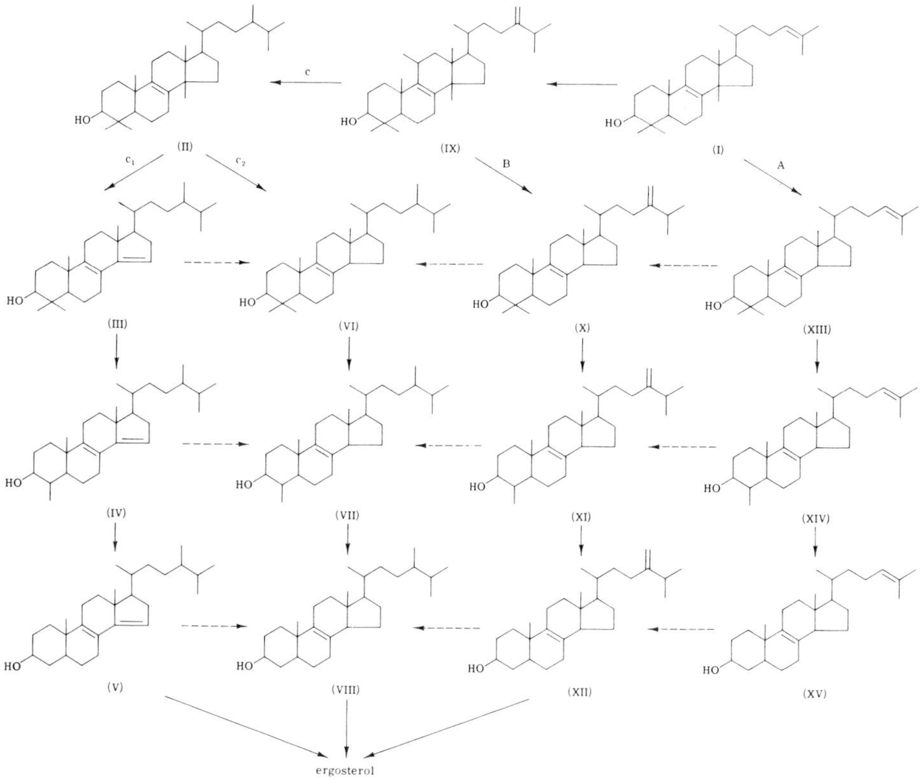

Figure 11. Potential biochemical pathways for the conversion of lanosterol to Δ^{8}-4-demethyl sterols as they may occur in the Eumycota. Principal pathways (→); interconversions between principal pathways (----→). The systematic names of each sterol component in this figure are listed below and can be identified by the corresponding Roman numeral: (I) lanosta-$\Delta^{8,24}$-dienol (lanosterol)*†; (II) 24-methyl-lanosta-Δ^{8}-enol (24-methyl-dihydrolanosterol)*†; (III) 4,4-dimethyl-ergosta-$\Delta^{8,14}$-dienol; (IV) 4α-methyl-ergosta-$\Delta^{8,14}$-dienol; (V) ergosta-$\Delta^{8,14}$-dienol†; (VI) 4,4-dimethyl-ergost-Δ^{8}-enol; (VII) 4α-methyl-ergost-Δ^{8}-enol; (VIII) ergost-Δ^{8}-enol; (IX) 24-methylene-lanosta-Δ^{8}-enol (24-methylene-dihydrolanosterol)*†; (X) 4,4-dimethyl-ergosta-$\Delta^{8,24(28)}$-dienol*; (XI) 4α-methyl-ergosta-$\Delta^{8,24(28)}$-dienol*†; (XII) ergosta-$\Delta^{8,24(28)}$-dienol*; (XIII) 4,4-dimethyl-cholesta-$\Delta^{8,24(25)}$-dienol*; (XIV) 4α-methyl-cholesta-$\Delta^{8,24(25)}$-dienol*†; (XV) cholesta-$\Delta^{8,24(25)}$-dienol.* Asterisk indicates a fungal product. Dagger indicates product can be incorporated into ergosterol by a fungal system. Reproduced from Scheme 2 from Weete (1973) by permission of Pergamon Press, Inc., and the author.

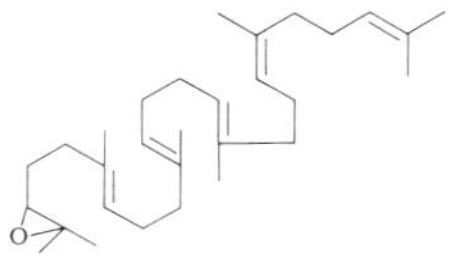

SqO 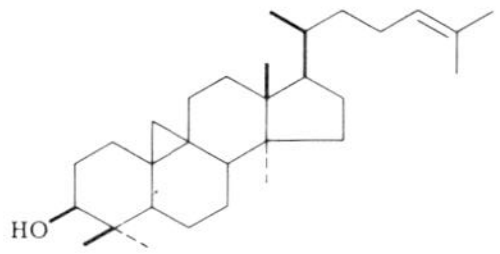 Cy 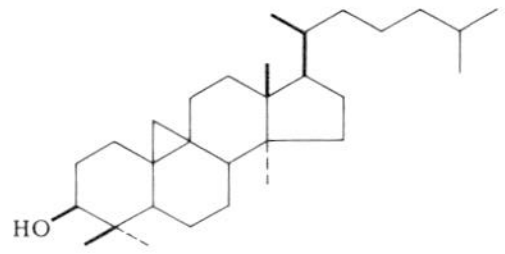 H-Cy 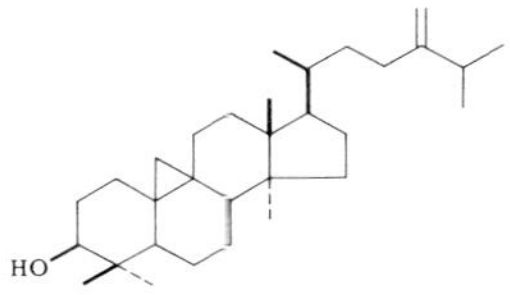 24-Cy

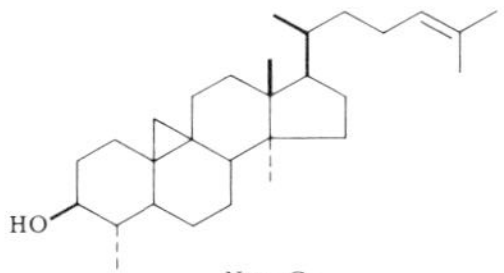

Nor-Cy 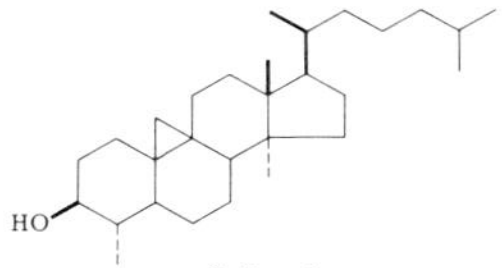 H-Nor-Cy 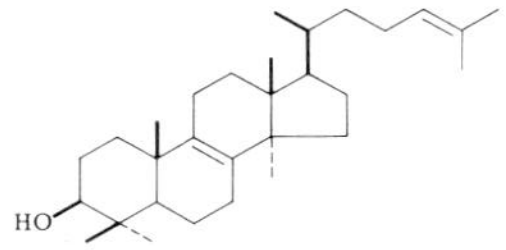 La 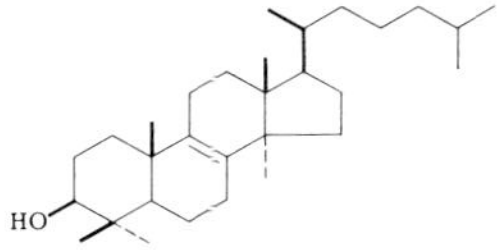 H-La

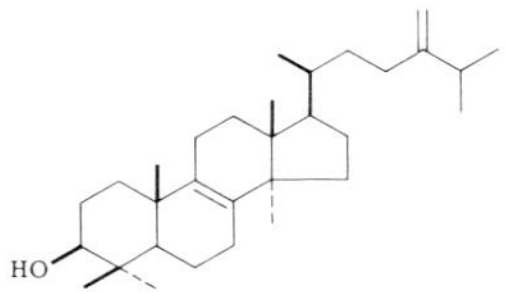

24-La 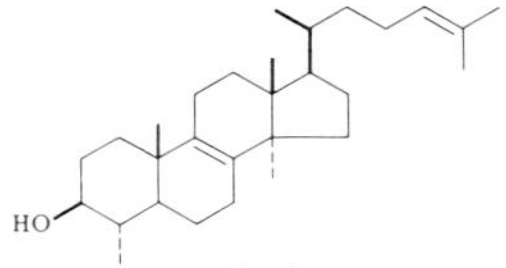 Nor-La 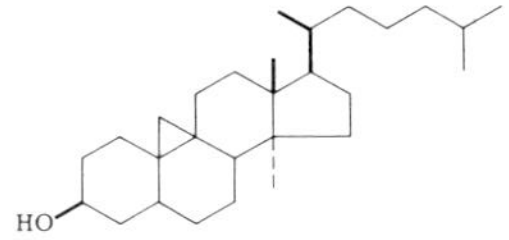 Pol 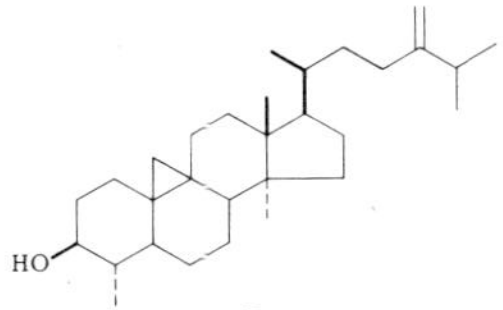 Cyeu

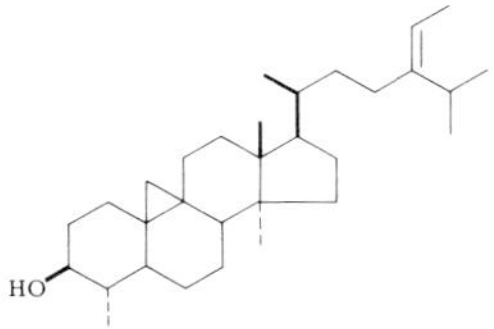

Et-Cyeu 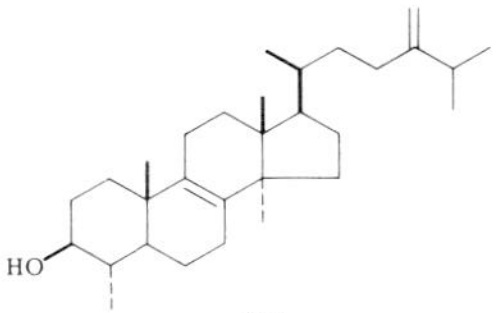 Obt 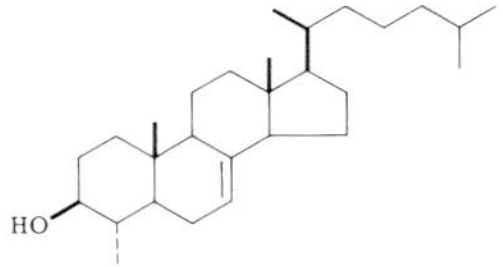 Lo 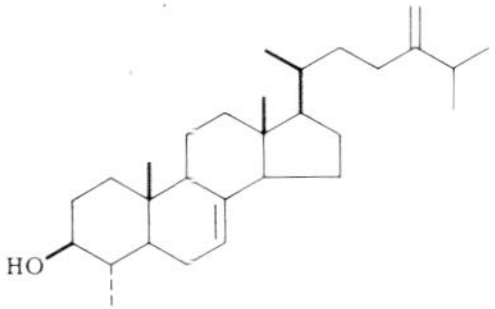 Me-Lo

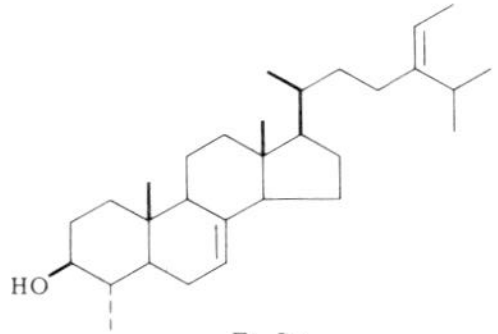

Et-Lo 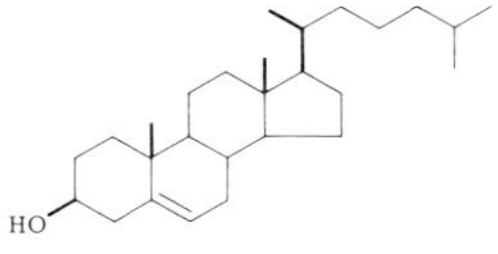 Chol 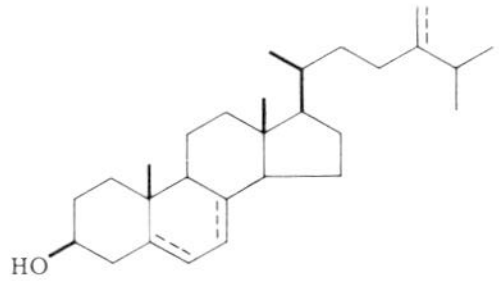 Me-St 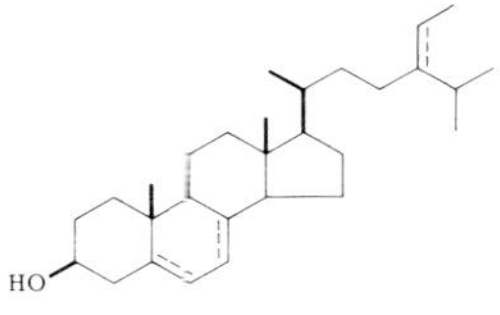 Et-St

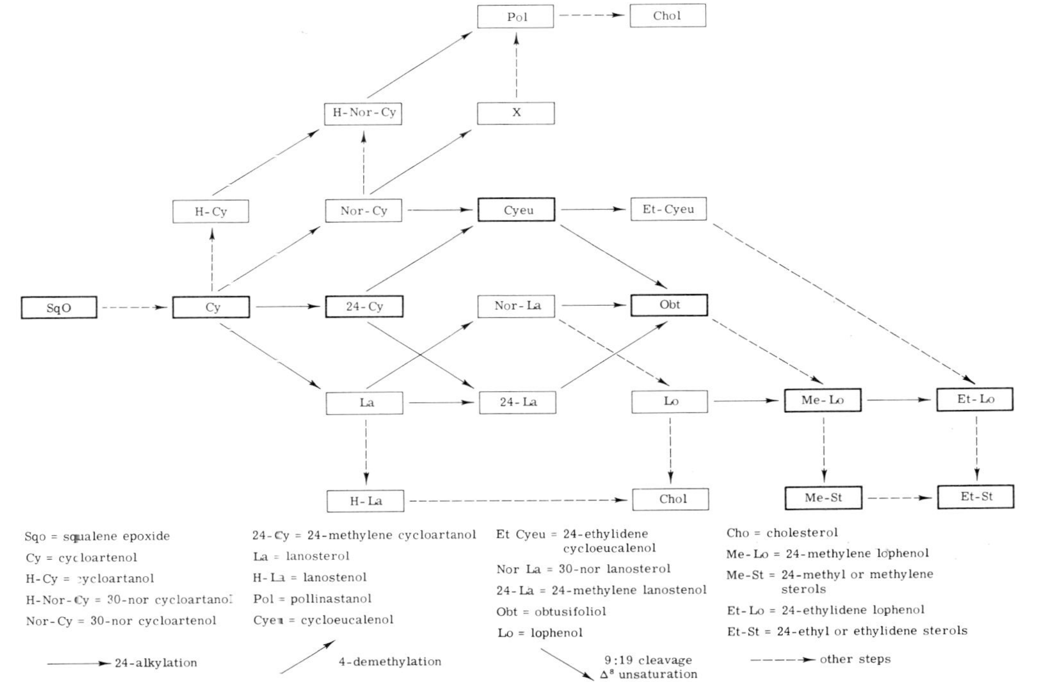

Figure 12. Phytosterol biosynthetic pathways. Reproduced from Chart 3 from Anding *et al.* (1972) by permission of the Royal Society of London and the authors.

9.5.1 General Biosynthetic Patterns

Perhaps the most basic dichotomy that can be drawn among the known sterols is based on the bifurcation in sterol biosynthesis after squalene 2,3-epoxide; closure of the sterol molecule into the familiar polycyclic structure gives either lanosterol or cycloartenol, depending on the mechanism of closure. This step is doubtless under genetic control, since cell preparations have been found that selectively form either lanosterol or cycloartenol from squalene 2,3-epoxide (Beastall *et al.*, 1974).

In general, phytosterols (plant and algal sterols) arise from cycloartenol, and fungal and animal sterols are formed from lanosterol (Goodwin, 1973a,b; Goad and Goodwin, 1972). Sterols of the Oomycetes are very unlike those from Zygomycetes, and probably do not arise from lanosterol (McCorkindale *et al.*, 1969; Bu'Lock, 1973; Bu'Lock and Osagie, 1976). However, the conversion of cycloartenol to lanosterol (Anding *et al.*, 1972) or to dihydrolanosterol (Beastall *et al.*, 1974) has been demonstrated, and interconversions may occur even beyond this point. Although the enzymology of sterol biosynthesis is very poorly known, observations such as the conversion of at least forty-two different sterols to cholesterol by animal tissues (Schroepfer *et al.*, 1972) strongly suggest the presence of nonspecific or mixed function enzymes. If so, the arrangement of sterol biosynthetic enzymes in multienzyme complexes or on membranes may determine many aspects of sterol biosynthetic routes followed in individual protists.

9.5.2 Individual Sterols

Comprehensive coverage of all known protistan sterols is beyond the scope of this discussion; the distribution of the more widely occurring sterols is given in Table 37. Perhaps the most striking feature of this distribution is the presence of cholesterol in virtually all organisms studied, exceptions currently being the Hyphochytridiomycetes and the Bacillariophyceae (Kanazawa *et al.*, 1971). Although usually more predominant in the animals, cholesterol is found in plants, fungi, animals, and prokaryotes, in heterotrophs and autotrophs. It is the major sterol of *Euglena gracilis* Z (Anding *et al.*, 1972). Although its precise biosynthetic pathway may be of phylogenetic interest, clearly its distribution alone will be of minimal significance.

The old animal–plant dichotomy is more compatible with the distribution of cycloartenol and lanosterol. The former sterol has not been reported from any animal, and the latter has been found in only one group of plants (the Euphorbiaceae). Interestingly, *Euglena gracilis* Z contains both lanosterol and cycloartenol (Anding *et al.*, 1972); if this strain is representative of

TABLE 37

Distribution of Selected Sterols[a]

Sterol[b]	Bacteria	Cyanophyceae	Rhodophyceae	Cryptophyceae	Bacillariophyceae	Phaeophyceae	Chrysophyceae	Xanthophyceae	Euglenophyceae	Charophyceae	Chlorophyceae	Higher plants	Hyphochytridiomycetes	Oomycetes[c]	Chytridiomycetes	Zygomycetes	Ascomycetes	Imperfect fungi	Basidiomycetes	Protozoa
C-24																				
A																				
Campesterol	+	+	+									+	+		+					*d*
Schottenol									?			+							?	?
Sitosterol	+	+	+									+	+		+					*d*
α-Spinasterol											?	+								?
Stigmasterol	+	+	+	+			?		?			+	+			?	?	+		*d*
B																				
Brassicasterol		+	+		*e*	+	+			+	+	*f*						+		*g*
Chondrillasterol					?				+		+									
Clionasterol					+	+	+	+	+	+	+									?
5-Dihydroergosterol	+		+						+							+	+		+	
22,23-Dihydroergosterol																+	+	+	+	+
Δ^5-Ergostenol			+				+		+	+	+									
Δ^7-Ergostenol	+								+		+					+	+	+	+	
Ergosterol	+		?	+			+		?		+	*f*				+	+	+	*h*	+
Poriferasterol						+	+		+		+								+	+

TABLE 37 (*Continued*)

Sterol[b]	Bacteria	Cyanophyceae	Rhodophyceae	Cryptophyceae	Bacillariophyceae	Phaeophyceae	Chrysophyceae	Xanthophyceae	Euglenophyceae	Charophyceae	Chlorophyceae	Higher plants	Hyphochytridiomycetes	Oomycetes[c]	Chytridiomycetes	Zygomycetes	Ascomycetes	Imperfect fungi	Basidiomycetes	Protozoa
C																				
Cholesterol	+	+	+			+	+	+	+	+	+	+		+	+	+	+	+	+	+
Cycloartenol			+			+	+	+	+		+	+								
22-Dehydrocholesterol		+	+			+									+					
Desmosterol			+			?						+		+			+			
Lanosterol	+								+			[f]				+	+	+	+	+
Zymosterol											+						+			
D																				
Eburicol																+			+	
Episterol									+							+	+		+	
Fucosterol			+		+	+	+			+	+	+	+	+		+				
28-Isofucosterol										+	+	+								
24-Methylenecholesterol			+			+		+	+	+	+	+		+						+
24-Methylenecycloartanol						+	+	+	+		+	+								
24-Methylenelophenol									+		+	+					+	+		?
Obtusifoliol			+						+			+							+	
Δ^4-3-Ketosteroids	+		+																	+

[a] Data for all groups: Goad and Goodwin (1972); Goodwin (1973a,1974); bacteria: Bean (1973), Iizuka *et al.* (1969), Lefebvre *et al.* (1974), Schubert *et al.* (1964, 1967, 1968); Cyanophyceae: deSouza and Nes (1968), Forin *et al.* (1975), Nichols (1973), Patterson (1971), Reitz and Hamilton (1968), Teshima and Kanazawa (1972); Rhodophyceae: Alcaide *et al.* (1969), Bean (1973), Beastall *et al.* (1974), Ikan and Seckbach (1972), Kanazawa and Yoshioka (1971, 1972), Patterson (1971); Cryptophyceae: Avivi *et al.* (1967); Bacillariophyceae: Kanazawa *et al.* (1971), Low (1955), Rubinstein and Goad (1974b), Tornabene *et al.* (1974); Phaeophyceae: Goad and Goodwin (1969), Ikan and Seckbach (1972), Ikekawa *et al.* (1972), Knights (1970), Patterson (1968, 1971), Smith *et al.* (1973); Chrysophyceae: Beastall *et al.* (1972), Collins and Kalnis (1969), Gershengorn *et al.* (1968), Kanazawa and Yoshioka (1972), Patterson (1971); Xanthophyceae: Mercer and Harries (1975), Mercer *et al.* (1974); Euglenophyceae: Anding *et al.* (1972), Avivi *et al.* (1967), Bean (1973), Patterson (1971); Charophyceae: Patterson (1971, 1972); Chlorophyceae: Bean (1973), Dickson *et al.* (1972), Kanazawa *et al.* (1971), Patterson (1971), Patterson *et al.* (1974), Rubinstein and Goad (1974a); higher plants: Bean (1973), Clayton (1965a,b), Patterson (1971); Hyphochytridiomycetes: Bean (1973), Weete (1973); Oomycetes: McCorkindale *et al.* (1969); Chytridiomycetes: Bean (1973); Zygomycetes: Bean (1973), Goulston and Mercer (1969), Goulston *et al.* (1975), McCorkindale *et al.* (1969), Safe (1973); Ascomycetes: Alais *et al.* (1974), Bean (1973), Bu'Lock (1973), Weete (1973); imperfect fungi: Weete (1973, 1974), Weete and Laseter (1974); Basidiomycetes: Bean (1973), Goulston *et al.* (1975), Ragsdale (1975), Weete (1973); Protozoa: Bean (1973), Clayton (1965a,b), Korn *et al.* (1969).

Additional data on algal sterols have recently been published by Chardon-Loriaux *et al.* (1976) and Fatturosso *et al.* (1975) for the Rhodophyceae; Fatturosso *et al.* (1976) for the Phaeophyceae; Paoletti *et al.* (1976) for the Cyanophyceae and Mercer and Carrier (1976) for the Zygomycetes. For the Dinophyceae, Ando and Barbier (1975) have shown the presence of cholesterol, desmosterol, and 24-methylenecholesterol. Clionasterol (or sitosterol), porifasterol (or stigmasterol) and fucosterol (or 28-*iso*-fucosterol) are also present.

[b] C-24 substitution: A, 24α (methyl or ethyl); B, 24β (methyl or ethyl); C, no C-24 alkyl side chain; D, R—CH= (R = H or CH_3).

[c] Although not all Oomycetes accumulate sterols (Weete, 1973).

[d] Present in *Physarum polycephalum* (Lenfant *et al.*, 1970). Rumen ciliates require these sterols (exogenously supplied) for growth (Hino *et al.*, 1973).

[e] The 24 S epimer of brassicasterol is present in *Phaeodactylum tricornutum* (= *Nitzschia closterium* forma *minutissima*) (Rubinstein and Goad, 1974b).

[f] Present in a few species only.

[g] Brassicasterol has been reported in marine animals, but it is probably derived from their (algal) diets (Kanazawa *et al.*, 1971).

[h] Present in the Basidiomycetes except rust fungi (Bean, 1973; Weete, 1973; Weete and Laseter, 1974). Some rust fungal sterols may be derived from the host plant (Goodwin, 1973a).

other Euglenophyceae, this character sets the euglenoids apart from all other protists, including "other" algae.

9.5.3 Sterol Side Chains

It has been suggested (Patterson, 1971) that some sterol side chains are of taxonomic importance in the algae. If only the chemical structures of the side chains are considered, this suggestion is oversimplified for the majority of taxa; but chemically identical side chains may be biosynthesized by dif-

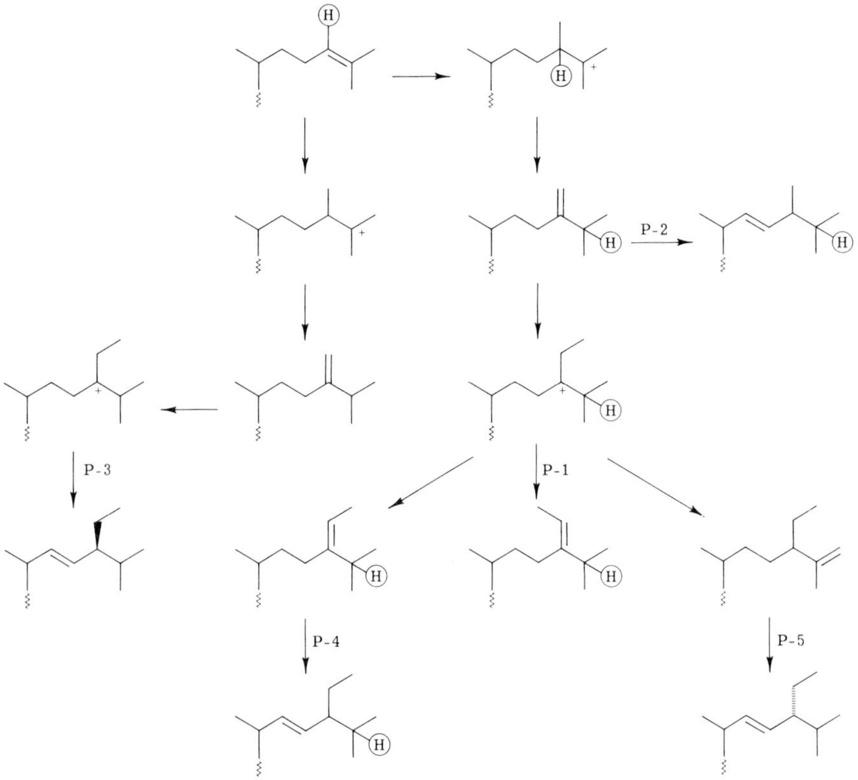

Figure 13. Mechanisms of phytosterol side-chain (C-24) alkylation. Adapted from diagrams in Goad *et al.* (1974). The diagram shows only the empirical carbon skeleton around the C-24 alkyl substitution. The absolute stereochemistry at the C-24 substitution is given in Table 38. The designations of the pathways follow those outlined in more detail in Goad *et al.* (1974), and show the distributions given in the text (Section 9.5.3). P-1, pathway mechanism 1; P-2, pathway mechanism 2; P-3, pathway mechanism 3; P-4, pathway mechanism 4; P-5, pathway mechanism 5.

TABLE 38

C-24 Stereochemistry and Biosynthesis of Sterols

Sterol[a]	C-24(α,β)	C-24(R,S)	Biosynthesis[b]
Campestanol	α	R	
Campesterol	α	R	2, 5, 9
22,23-Dehydrocampesterol	α	S	
Schottenol	α	R	
Sitosterol	α	R	4, 5
α-Spinasterol	α	S	
Stigmastanol (dihydrositosterol)	α	R	
Stigmasterol	α	S	5
Brassicasterol	β	R	5
Cerevisterol	β	R	
Chondrillasterol	β	R	
Δ^7-Chondrillasterol	β	S	
Clionasterol (γ-sitosterol)	β	S	
7-Dehydroclionasterol	β	S	
7-Dehydroporiferasterol	β	R	
22,23-Dihydrobrassicasterol	β	S	
22,23-Dihydroergosterol	β	S	
Ergostanol	β	S	
Ergostenol	β	S	
Δ^7-Ergostenol	β	S	
Ergosterol	β	R	5, 10
Lichesterol	β	R	
Poriferasterol	β	R	2, 5
Cholesterol	—	—	1, 7
Cycloartenol	—	—	3, 10
22-Dehydrocholesterol	—	—	
Desmosterol	—	—	8
Lanosterol	—	—	1
Zymosterol	—	—	
Eburicol	(R=)	(R=)	
Episterol	(R=)	(R=)	6, 10
Fecosterol	(R=)	(R=)	
Fucosterol (E-)	(R=)	(R=)	5, 8
28-Isofucosterol (Z-)	(R=)	(R=)	
Obtusifoliol	(R=)	(R=)	
Sargasterol	(R=)	(R=)	

[a] Our thanks to Dr. L. J. Goad for providing some of these identifications. For formal nomenclature consult Idler and Wiseman (1971).

[b] (1) Anding *et al.*, 1972; (2) Beastall *et al.*, 1972; (3) Ferezou *et al.*, 1974; (4) Goad and Goodwin, 1966; (5) Goad *et al.*, 1974; (6) Lenton *et al.*, 1973; (7) Schroepfer *et al.*, 1972; (8) Smith *et al.*, 1973; (9) Tso and Cheng, 1971; (10) Weete, 1973.

ferent mechanisms, and these different routes of biosynthesis may be of phylogenetic as well as taxonomic value. Although relatively little is known of sterol side chain biosynthetic patterns in most protists, as many as six biosynthetic mechanisms (Figure 13) have been identified, five of which are found in protists (Goad *et al.*, 1974). These characters group Zygomycetes, Ascomycetes, imperfect fungi, Basidiomycetes, and a lichen ("Mechanism 2"); *Physarum polycephalum* and *Dictyostelium discoideum* ("Mechanism 3"); Chlorophyceae and some higher plants ("Mechanism 5"); Oomycetes and Phaeophyceae ("Mechanism 1"); and members of the Chrysophyceae, Bacillariophyceae, Charophyceae, Euglenophyceae, and some other higher plants ("Mechanism 4"). Most Rhodophyceae accumulate sterols lacking side chains.

The phylogenetic isolation of the Dinophyceae appears to be further reinforced by the discovery of a sterol, dinosterol (Shimizu *et al.*, 1976) in *Gonyaulax* that is unique in possessing additional methyl groups on the side chain. Cholesterol is a major dinophycean sterol (Ando and Barbier, 1975).

9.5.4 Stereochemistry

There are two conventions applicable to description of the stereochemistry of the group attached to C-24 of phytosterols. The currently accepted IUPAC/IUB convention describes these sterols as being either 24-*R* or 24-*S*, depending on the relative positions of the four groups attached to C-24; the older convention, still favored by many botanists, describes the C-24 group as being either above (24α) or below (24β) the plane of the rest of the molecule. Groups attached in the 24α configuration are 24-*R* if the bond between C-22 and C-23 is fully saturated; if attached to a Δ^{22} sterol, 24α groups are 24-*S* (Table 38). The stereochemistry of Δ^{22} bond formation is different in investigated algae (*pro-R*) and investigated fungi (*pro-S*) (Weete, 1974).

Phytosterols of the Eumycota, Phaeophyceae, Bacillariophyceae, Chrysophyceae, Xanthophyceae, Charophyceae, and Chlorophyceae tend to have 24β stereochemistry, whereas those of the higher plants usually have 24α side chains. Many exceptions of this generality have been reported, and although reinvestigation will doubtless remove many of these conflicts, others are not due to erroneous identification. Brassicasterol, for instance, has been unambiguously identified in a higher plant (Goad *et al.*, 1974), and 24α-sterols have been reported in various green algae (Table 37). A survey of other members of the Chlorophyceae, particularly of the orders Siphonales, Siphonocladales, and Dichotomosiphonales and of the genus *Klebsormidium*, would be of considerable interest.

10

Metabolites: IV

10.1 THE δ-AMINOLEVULINIC ACID PATHWAY

Many molecules participating in energy-related reactions are biosynthesized by the δ-aminolevulinic acid pathway (Figure 14). Chlorophylls and phycobilins are classes of compounds derived from protoporphyrin IX, and have been examined in some detail for phylogenetically pertinent information. Only a few anaerobic bacteria are believed to be unable to synthesize porphyrin compounds; other protists, including some pathogenic prokaryotes, all investigated trypanosomes, *Physarum polycephalum*, and *Pilobolus* spp., require a dietary source of porphyrins (Lascelles, 1962; Kidder, 1967; De Ley and Kersters, 1975).

δ-Aminolevulinic acid itself is biosynthesized from succinyl coenzyme A and glycine, catalyzed by δ-aminolevulinic acid synthase (EC 2.3.1.37), in investigated bacteria (both photosynthetic and nonphotosynthetic), yeast, and higher animals. Higher plants, however, synthesize δ-aminolevulinate by an incompletely characterized pathway involving glutamic acid and related five-carbon compounds (Beale *et al.*, 1975; Lohr and Friedmann, 1976; Beale, 1976). δ-Aminolevulinic acid synthase has not been detected in extracts from any algae. The biosynthesis of δ-aminolevulinic acid by eukaryotic algae, and especially by Cyanophyceae and Euglenophyceae, may be of considerable phylogenetic interest. Recently Troxler *et al.* (1975) have presented evidence suggesting that algae (e.g., *Porphyridium*) may operate with a mechanism similar to that in higher plants, and in *Cyanidium* (Jurgenson *et al.*, 1976) such a pathway does appear to operate.

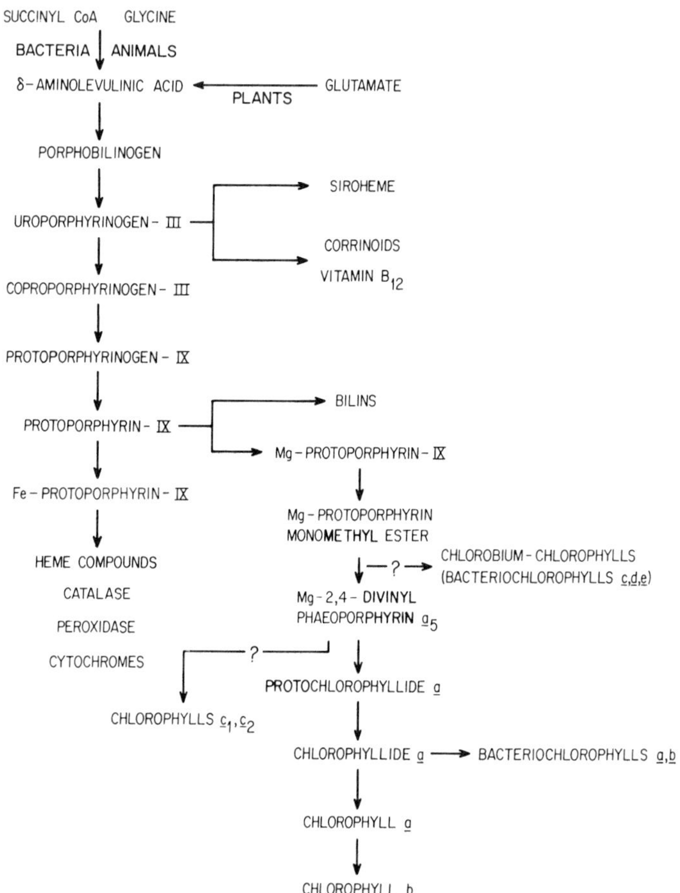

Figure 14. δ-Aminolevulinic acid biosynthetic pathway. The diagram shows the principal tetrapyrrolic intermediates of the biosynthetic pathway. The pathways to the bacteriochlorophyll *c, d, e* and the chlorophylls c_1 and c_2 are marked as speculative since experimental evidence is lacking.

10.2 CHLOROPHYLLS

Chlorophylls (Figure 15) have traditionally been used as taxonomic and phylogenetic markers because they are relatively widely distributed, structurally variable, and relatively easily identified by chromatographic and spectral techniques. The presence of chlorophylls in plants was one of the major arguments in support of the classic distinction between these organisms and animals; moreover, the color of algae, influenced by their chlorophyll content, gave rise to the original classification of the

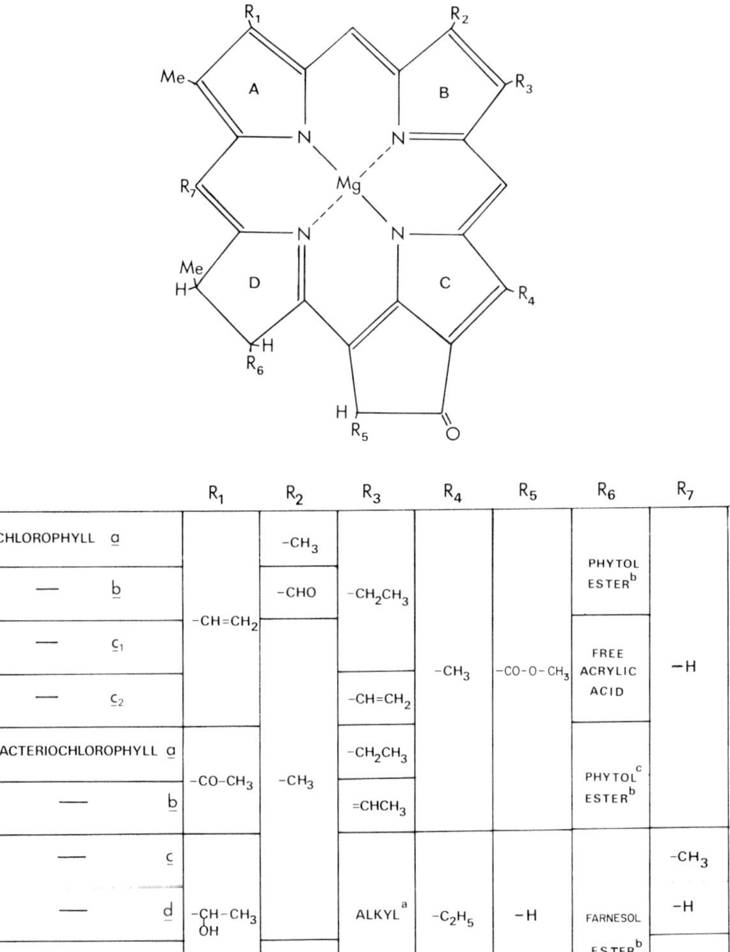

	R_1	R_2	R_3	R_4	R_5	R_6	R_7
CHLOROPHYLL *a*	$-CH=CH_2$	$-CH_3$	$-CH_2CH_3$	$-CH_3$	$-CO-O-CH_3$	PHYTOL ESTER[b]	$-H$
— *b*		$-CHO$					
— *c*$_1$		$-CH_3$				FREE ACRYLIC ACID	
— *c*$_2$			$-CH=CH_2$				
BACTERIOCHLOROPHYLL *a*	$-CO-CH_3$		$-CH_2CH_3$			PHYTOL[c] ESTER[b]	
— *b*			$=CHCH_3$				
— *c*	$-CH(OH)-CH_3$		ALKYL[a]	$-C_2H_5$	$-H$	FARNESOL ESTER[b]	$-CH_3$
— *d*							$-H$
— *e*		$-CHO$					$-CH_3$

Figure 15. Chlorophylls occurring in protists. (a) Alkyl side chain substituents are ethyl, propyl, and isobutyl. (b) Phytol and farnesol esterified to the propionic acid substituent $-CH_2-CH_2-COO-$alcohol. (c) In some members of the Athiorhodaceae (see text) bacteriochlorophyll *a* may be esterified with geranylgeraniol. Bacteriochlorophyll *a* lacks the double bond between R_2 and R_3. Two additional hydrogen atoms present. Bacteriochlorophyll *b* lacks the double bond between R_2 and R_3. Additional hydrogen at carbon substituted with R_2. Adapted from Gloe *et al.* (1975), by permission of Springer-Verlag and the authors.

TABLE 39

Distribution of Chlorophylls in Bacteria[a]

Bacteriochlorophyll[b]	Chlorobacteriaceae	Chlorobacteriaceae[c]	Thiorhodaceae	Athiorhodaceae
a	+	+	+	+
b	−	−	−	+
c	+	−	−	−
d	+	−	−	−
e	−	+	−	

[a] From Brockmann *et al.* (1973); Gloe and Pfennig (1974); Gloe *et al.* (1975); Katz *et al.* (1972); Künzler and Pfennig (1973); Pfennig (1967); Scheer *et al.* (1974).

[b] Bacteriochlorophyll *c*, *Chlorobium* chlorophyll 660 series; bacteriochlorophyll *d*, *Chlorobium* chlorophyll 650 series.

[c] "Brown colored" *Chlorobium phaeobacterioides* and *Chlorobium phaeovibrioides*.

photosynthetic protists during the nineteenth century. The distribution of chlorophylls is given in Tables 39 and 40.

10.2.1 Bacterial Chlorophylls

Bacteriochlorophyll *a* is apparently found in all photosynthetic bacteria (Thiorhodaceae, Athiorhodaceae, Chlorobiaceae, including *Chloroflexus*). In the two former families it is the only chlorophyll, whereas in the latter it is a minor component after bacteriochlorophylls *c, d, e*. The occurrence of this chlorophyll suggests a common phylogenetic thread in the photosynthetic bacteria. With very few exceptions (Katz *et al.*, 1972; Brockmann *et al.*, 1973; Künzler and Pfennig, 1973; Gloe and Pfennig, 1974), the esterifying alcohol is phytol. In some Athiorhodaceae (e.g., *Rhodospirillum rubrum, R. photometricum*) and Thiorhodaceae (*Chromatium okenii, Amoebabacter, Chromatium vinosum* strain de Baer, *Thiopedia, Thiospirillum*) a bacteriochlorophyll *a* esterified with geranylgeraniol has been detected. This second bacteriochlorophyll *a* is usually present in addition to the normal phytol bacteriochlorophyll *a*. Although the distribution is well delineated, no obvious phylogenetic inferences can be drawn. Indeed it is difficult to equate this distribution even with species–genus taxonomy. The esterifying enzyme "chlorophyllase" is probably nonspecific and the similarity of phytol and geranylgeraniol would account for their presence as esterified alcohols. If this is in fact the explanation, the reverse question becomes important: Why isn't the geranylgeraniol bacteriochlorophyll more ubiquitous? It is apparently absent from the Chlorobiaceae. In extension, one must then ask if there are any basic functional differences between the

TABLE 40

Distribution of Chlorophylls in Cyanophyceae and Eukaryotes[a]

Organisms	Chlorophylls			
	a	*b*	c_1	c_2
Cyanophyceae	+			
Cyanellae				
Cyanophora paradoxa	+			
Glaucocystis nostochinearum	+			
Rhodophyceae	+			
Cryptophyceae	+			+
Dinophyceae	+		[b]	+
Bacillariophyceae	+		+	+
Phaeophyceae	+		+	+
Chrysophyceae	+		+	+
Haptophyceae	+		+	+
Xanthophyceae	+		+	+
Eustigmatophyceae	+			
Chloromonadophyceae	+		(+)	
Chlorophyceae	+	+		
Charophyceae	+	+		
Prasinophyceae	+	+		
Euglenophyceae	+	+		
Higher plants	+	+		

[a] Data from the following sources: Cyanophyceae: Parsons (1961), Stransky and Hager (1970b); Cyanellae: Chapman (1966); Rhodophyceae: Allen (1959), Chapman (1966), Stransky and Hager (1970b); Cryptophyceae and Dinophyceae: Hager and Stransky (1970b), Jeffrey (1968a,b), Riley and Wilson (1967); Bacillariophyceae, Phaeophyceae, Chrysophyceae, and Haptophyceae: Guillard and Lorenzen (1972), Hager and Stransky (1970b), Jeffrey (1968a,b), Riley and Wilson (1967), Seely *et al.* (1972), Stransky and Hager (1970c); Xanthophyceae: DeGreef and Caubergs (1970), Guillard and Lorenzen (1972), Norgård *et al.* (1974a), Riley and Wilson (1967), Stransky and Hager (1970a), Whittle and Casselton (1975b); Eustigmatophyceae: Norgård *et al.* (1974a), Whittle and Casselton (1975a); Chloromonadophyceae: Chapman and Haxo (1966), Guillard and Lorenzen (1972); Chlorophyceae: Jeffrey (1968a), Parsons (1961), Riley and Wilson (1967), Hager and Stransky (1970a); Charophyceae: Klein and Cronquist (1967); Prasinophyceae: Ricketts (1967), Riley and Segar (1969); Euglenophyceae: Hager and Stransky (1970b); higher plants: Jeffrey (1968a); general: Jeffrey (1976).

[b] Present only in the dinoflagellate–chrysomonad symbiosis called "*Peridinium foliaceum*" (Withers and Haxo, 1975).

two forms. An alternative explanation can be proposed if one assumes that geranylgeraniol chlorophylls are the precursors to phytol chlorophylls (see Section 9.2), and that the esterification involves chlorophyllides and geranylgeranyl pyrophosphate (Rüdiger *et al.*, 1977). In this regard it should be very profitable to look for geranylgeraniol chlorophyll *a*. Bacteriochlorophyll *b* has a distribution restricted to *Rhodospirillum viridis, R.* sp., and *Thiocapsa*. Like bacteriochlorophyll *a*, it may also exist esterified with geranylgeraniol. It is quite apparent that this particular bacteriochlorophyll represents an isolated anomaly.

Bacteriochlorophylls *c*, *d* ("Chlorobium chlorophylls 660 and 650 series" respectively), and the recently described bacteriochlorophyll *e* series (Gloe *et al.*, 1975) are characteristically the principal chlorophylls of the Chlorbiaceae. Within this family, there is no apparent taxonomic distinction except that the bacteriochlorophyll *e* series has been found only in the "brown" green bacteria *Chlorobium phaeobacterioides* and C. *phaeovibrioides*. The unusual features of these three chlorophylls (farnesyl alcohol, decarboxylated cyclopentanone ring, methyl substituted δ-bridge) indicate a major metabolic deviation from the biosynthetic pathways for the other chlorophylls. This suggests that the Chlorobiaceae represent a phylogenetic offshoot or dead end.

10.2.2 Chlorophyll *a*

The biosynthesis of chlorophyll *a* is one of several evolutionary modifications marking the transition from bacterial photosynthesis to the photosynthesis found in Cyanophyceae and photosynthetic eukaryotes. In the latter organisms, chlorophyll *a* production probably occurs in the chloroplast, but nuclear genes are involved as well (Kirk and Tilney-Basset, 1967; Schiff, 1973). The known facts concerning chlorophyll *a* are consistent with either an endosymbiotic or a gradual evolution of chloroplasts from a common ancestor with the Cyanophyceae.

10.2.3 Chlorophylls *b* and *c**

The biosynthetic transition from chlorophyll *a* to chlorophyll *b* is rather simple, at least by formal chemistry. Only one enzyme may be involved (Shlyk, 1971); however, chlorophyll *c* biosynthesis diverges at an earlier stage from the pathway leading to chlorophylls *a* and *b*, and likely involves several different enzymatic processes. Thus the presence of chlorophylls *c* in

* The term "chlorophylls *c*" will be used instead of the more proper term "chlorophyllides *c*" in order to provide continuity with the bulk of the literature.

an organism may be a more significant biochemical datum than the presence of chlorophyll *b*, given the distribution of chlorophyll *a* biosynthesis. Furthermore, it is much more likely for chlorophyll *b* than for the chlorophylls *c* to have arisen several times during the course of evolution.

The presence of chlorophyll *b* in the higher plants and in most (Allen, 1959) Chlorophyceae is often taken to indicate a common ancestry for these groups. This is indeed consistent with a large body of other evidence, both biochemical and morphological. The Euglenophyceae with chloroplasts also biosynthesize chlorophyll *b*, although the relationship of these organisms to the green algae may be less than straightforward. Moreover, a chlorophyll *b*-like pigment has recently been discovered in an organism reminiscent of the cyanophyte *Synechocystis* (Lewin and Withers, 1975). The phylogenetic significance of the distribution of chlorophyll *b* biosynthetic enzyme(s) is obviously profound. In view of this, the presumptive chlorophyll *b* should be confirmed by more definitive techniques (e.g., mass spectrometry) than electronic absorption spectra and chromatography. Likewise it is important that its prokaryote nature be confirmed with more definitive evidence than has so far been presented (Lewin and Withers, 1975; Lewin, 1975).

In view of the relative biosynthetic complexity of chlorophylls *c*, their presence in the Bacillariophyceae, Phaeophyceae, Chrysophyceae, Haptophyceae (chlorophylls c_1 and c_2), Dinophyceae, and Cryptophyceae (chlorophyll c_2 only) provides reasonably strong evidence for the common ancestry of the chlorophyll *c*-related genes in these organisms. Consistent with the observed distributional data, the chemical structures of chlorophylls *c* suggest the biosynthesis of chlorophyll c_1 from chlorophyll c_2 (Dougherty *et al.*, 1966; Jeffrey, 1968b). Until the biochemical nature of this biosynthetic step is better understood, the absence of chlorophyll c_1 cannot be taken as evidence for the common ancestry of the Dinophyceae and the Cryptophyceae.

The absence of accessory chlorophylls in the Rhodophyceae could be rationalized in so many ways that chlorophyll distributional data serve only to set these organisms apart from other eukaryotes. The Eustigmatophyceae, although also possessing only chlorophyll *a*, are very different from the Rhodophyceae in a number of respects. Chlorophyll *d*, which is related to chlorophyll *a* (perhaps naturally, perhaps as an artifact of extraction) is found only in some Rhodophyceae. Chlorophyll *e* is believed to be an artifact arising from chlorophylls *c* (Holt and Morley, 1959; Meeks, 1974).

Vitamin B_{12} is an example of a biosynthetically related group of compounds known as the corrinoids (Scott *et al.*, 1974). Vitamin B_{12} has been reported in a few bacteria devoid of catalase activity, although most catalase-negative bacteria require B_{12}. The distribution of B_{12} biosynthetic capacity is described in Section 11.4.

10.3 PHOTOSYNTHESIS

Chlorophylls, chlorophyll proteins, chlorophyll aggregates, cytochromes, quinones, various enzymes, accessory pigments, and biomembranes are involved in photosynthesis, and it is difficult to imagine a functional photosynthetic system significantly simpler than that presently seen in some photosynthetic bacteria. There are two major theories of the origin of photosynthesis. One view, propounded by Gaffron (1960a,b) and Uzzell and Spolsky (1974), suggests that photochemically active porphyrin pigments were involved from the earliest stages of the origin of life in trapping energy from light. In this view, photosynthetic processes played an early role in the organization of living systems themselves. If life is considered to be monophyletic, it is necessary that photosynthetic ability must have been lost one or more times during evolution.

Another view, championed by Calvin (1959, 1962), is that the evolution of phosphorylation and pyridine nucleotide reduction preceded the coupling of energy from light to these processes. This coupling would have been selectively advantageous due to the relative efficiency of the photoenergetic process and due to the abundance of the energy source. No losses of photosynthetic abilities would be required by this view.

Modern photosynthesis is a highly complex process, and is one of the most compelling reasons for considering all photosynthetic eukaryotes to be closely related to each other. For this reason, many authorities consider all photosynthetic organisms to be derived from a common ancestor; for instance, Olson (1970) suggests that Cyanophyceae are "obviously" more closely related to photosynthetic bacteria than to any other modern bacteria. For the present purposes, the photosynthetic bacteria and the blue-green algae will be considered to be monophyletic, with the proviso that their early evolutionary divergence be kept in mind.

Assuming their common ancestry, Olson (1970) has suggested a scenario for the evolution of photosynthetic electron transport mechanisms and of novel photoreceptive pigments:

> The common ancestor possessed chlorophyll *a*, and was capable of cyclic phosphorylation coupled to the uptake of organic compounds from the remaining primeval soup. The first step beyond this stage was the appearance of noncyclic electron transport, allowing the reduction of compounds with electron potentials below those of the primitive quinones to be coupled with the oxidation of compounds of higher electron potentials. At this stage, the photosynthetic bacterial line developed bacteriochlorophyll *a*, thereby taking advantage of a radiation "window" in the spectrum of water, but in the process sealed its own fate, because the reduction in electron voltage of the free energy available from the excitation of this new chlorophyll put the oxidation of water forever

> out of reach. Meanwhile, the line to the Cyanophyceae, still with chlorophyll *a*, evolved cytochrome *b* and eventually incorporated the NADP/ferredoxin couple into the photosynthetic electron transport pathway. Selection for the utilization of weaker reducing agents followed, with the emergence of Photosystem II linked to the original photosystem. Hydrazine, hydroxylamine, manganese, nitrous oxide, nitrite, and finally nitrate were utilized as electron donors, until finally water itself was capitalized upon.

This sequence involves many assumptions, many of which are eminently reasonable. Since no chlorophyll *a*-containing organisms are known that cannot utilize water as the proton donor in photosynthesis, apparently the hypothetical intermediate organisms were sufficiently outcompeted as exogenous electron donor supplies diminished, and therefore none of their progeny remains, even in specialized environments. Other assumptions include those concerned with the nonsurvival of other mutations during the process of incorporating weaker reducing agents. Cytochromes *b* have recently been found in a variety of photosynthetic bacteria (Section 5.7). It is also notable that some Cyanophyceae may have an inducible nitrate reductase (Section 6.15), as nitrate is involved in the above scenario; most Cyanophycean enzymes are not inducible.

Despite the complexity of the overall photosynthetic process (i.e., "chloroplast"), it is becoming increasingly apparent that the "heart" of photosynthesis in aerobic photosynthesizers, the reaction center (*P*-700 and immediate primary electron acceptor), has been strongly conserved throughout evolution (Brown *et al.*, 1974). Even though the trapping or antenna pigment system (e.g., accessory chlorophylls, fucoxanthin, peridinin, or biliproteins) has undergone considerable change and evolution, the reaction center has remained basically unchanged.

11

Miscellaneous Simple Molecules

11.1 SULFATE REDUCTION

In simplified form, the sulfur cycle in protists involves the conversion of sulfate and sulfate esters into amino acids or thiols, and the reoxidation of these compounds to sulfur, sulfite, or (more commonly) sulfate. Different portions of this cycle are carried out by different organisms, with the role of bacteria being especially important. Oxidation of reduced sulfur compounds to sulfur or to sulfate is found in heterotrophs of all major taxa, and consequently is of minor phylogenetic interest. This capacity may date from a very early evolutionary stage, when the atmosphere was still replete with highly reduced sulfur compounds.

The reduction of sulfates to thiols for nutritional purposes ("assimilatory sulfate reduction") is believed to occur in most organisms except higher animals, some protozoa, and a few bacteria (*Leuconostoc mesenteroides, Staphylococcus aureus, Streptococcus faecalis*). It has been specifically observed in members of the Cyanophyceae, Rhodophyceae, Cryptophyceae, Dinophyceae, Bacillariophyceae, Phaeophyceae, Prasinophyceae, Chlorophyceae, Euglenophyceae, imperfect fungi, in *Leishmania tarentolae*, and in chloroplasts of higher plants (Schiff, 1962; Schiff and Hodson, 1970, 1973; Tsang and Schiff, 1975). Among the bacteria, it has been found in the Athiorhodaceae, Thiorhodaceae, possibly in the Chlorobacteriaceae, and in many nonphotosynthetic bacteria (Peck, 1962, 1974; Kondrat'eva, 1965; Peck *et al.*, 1974). "Dissimilatory" or "respiratory" sulfate reduction, in which large quantities of sulfate are reduced to sulfide, has been reported only among *Desulfovibrio* spp. and *Desulfomaculum* spp. (Schiff and Hodson, 1973).

The enzymology of sulfate reduction is a field of considerable current interest. Adenosine 3′-phosphate 5′-phosphosulfate (PAPS) and its putative biosynthetic precursor, adenosine 5′-phosphosulfate (APS), may occur in all organisms capable of sulfate reduction. Assays for PAPS sulfotransferase ("PAPS reductase") activity have demonstrated this activity in most nonphotosynthetic sulfate reducers, whereas most photosynthetic sulfate reducers, dissimilatory sulfate reducers, and some sulfate oxidizers have been shown to possess APS sulfotransferase ("APS reductase") activity (Peck, 1962; Tsang and Schiff, 1975). This pattern may hold less rigorously among the bacteria than in the eukaryotes (Schiff and Hodson, 1973; Peck *et al.*, 1974). However, many of the earlier enzyme assays did not take into account the possible occurrence of an enzyme converting PAPS to APS ("enzyme A"). Purification and characterization of the enzymes responsible for sulfate reduction could be of significance for a study of bacterial phylogeny or for an examination of phylogenetic patterns among the protozoa and higher animals.

Choline sulfate is biosynthesized from choline and PAPS by certain Euascomycetidae, Basidiomycetes, imperfect fungi, lichens, and by the red alga *Gelidium cartilagineum* (Spencer and Harada, 1960; Harada and Spencer, 1960). It is not accumulated by *Torulopsis utilis*, nor by investigated bacteria, lower fungi, and water molds. Due to the potential simplicity of choline sulfate biosynthesis, its occurrence in some fungi and in some Rhodophyceae cannot be taken as evidence for phylogenetic proximity of these taxa.

11.2 NITROGEN UTILIZATION

All living organisms require nitrogen for production of amino acids, purines, pyrimidines, and other vital cellular metabolites. Differences exist from organism to organism in the forms in which nitrogen can be absorbed from the environment and incorporated into the necessary metabolites. As might be expected, some of the greatest diversity is seen among the bacteria.

Photosynthetic bacteria are able to use exogenously supplied ammonium as a nitrogen source. Amino acids are good nitrogen sources for purple bacteria, and some amino acids can be utilized by the Chlorobiaceae as well. Urea can be utilized by most photosynthetic bacteria, and all are capable of fixation of atmospheric nitrogen. There are reports of the uptake of nitrite and nitrate by some photosynthetic bacteria (Kondrat'eva, 1965; Malofeeva *et al.*, 1975).

The Cyanophyceae differ from photosynthetic bacteria in their ability to use nitrate as a primary nitrogen source. Ammonium and a wide range of organic nitrogen sources are also used (Wyatt *et al.*, 1971; Kapp *et al.*, 1975), and molecular nitrogen fixation by the blue-green algae has been extensively studied.

Nitrogen utilization capabilities of eukaryotic algae have only rarely been systematically investigated. Antia *et al.* (1975) have shown that unicellular algae representing the Cyanophyceae, Rhodophyceae, Chlorophyceae, Prasinophyceae, Chrysophyceae, Bacillariophyceae, Xanthophyceae, Cryptophyceae, and Dinophyceae can grow on ammonium, nitrate, or nitrite; most investigated organisms, furthermore, were capable of growth on urea, hypoxanthine, or glycine. Differences in uptake and metabolic capability do not appear correlated with taxonomic position. Nitrogen requirements of other algae have been discussed by Birdsey and Lynch (1962), Syrett (1962), Burlakova *et al.* (1971), and Neilson and Lewin (1974).

It is interesting that *Euglena gracilis* var. *bacillaris*, *Phacus pyrum*, and *Trachelomonas* spp. are unable to use exogenous nitrate, nitrite, urea, uric acid, allantoin, or creatine as nitrogen sources; only ammonium and certain amino acids are suitable (Provasoli, 1958; Birdsey and Lynch, 1962; Klein and Cronquist, 1967; Kidder, 1967). Additional organic nitrogen sources available to photosynthetic Chlorophyceae are not utilized by these euglenoids. Many Chytridiomycetes and all examined Oomycetes are also unable to utilize exogenously supplied nitrate (Cantino and Turian, 1959), and many zooflagellates do not hydrolyze urea (Seneca *et al.*, 1962). Most colorless Chlorophyceae cannot utilize nitrate (Hall, 1967; Naylor, 1970; Lewin, 1974). Requirements of protozoa have been discussed by Hall (1967), Kidder (1967), and Sleigh (1973).

It appears unlikely that studies of uptake and utilization of nitrogenous compounds by protists will provide further insight into the phylogeny of higher taxa. In some instances nitrogen utilization patterns may be of interest in the taxonomy of lower taxa of protists; however, a large portion of the apparent differences in nitrogen utilization patterns is probably due to evolutionary adaptations by individual varieties and populations of organisms in particular ecological niches.

11.3 CARBON MONOXIDE PRODUCTION

Carbon monoxide is a rather unusual metabolite produced by a wide range of organisms: bacteria, Cyanophyceae, Phaeophyceae, Chlorophyceae (*Spirogyra* sp.), Ascomycetes, higher plants, and the Portuguese man-of-war (Wilks, 1959; Loewus and Delwiche, 1963). It is a by-product of bile pigment formation, and hence probably occurs in Rhodophyceae, mammals, and other organisms producing bile pigments (Troxler *et al.*, 1970; Troxler and Dokos, 1973). The immediate precursor to carbon monoxide is not often known, but flavonoids and glucose (as well as bilins) are possibilities (Westlake *et al.*, 1961; Radler *et al.*, 1974). It is not known whether carbon monoxide production in these different organisms is evolutionarily homologous; consequently, it is not possible to utilize these data in biochemical phylogenetics. The possibility of several unrelated biogenetic origins of carbon monoxide is heightened by its extremely simple molecular structure, exemplifying the relationship of structural and biosynthetic complexity to potential phylogenetic usefulness.

11.4 MINERAL NUTRITION AND VITAMIN REQUIREMENTS

It is necessary to mention comparative studies on mineral nutrition and vitamin requirements in the protists because such comparisons were extensively examined in the early biochemical phylogenetic literature. As increasing numbers of protists were cultured axenically in defined media, a considerable body of data was built up concerning aspects of comparative nutrition. Now that a moderate number of protists are available in culture, it appears that such comparative nutritional studies are shedding little light on phylogeny. It is possible that trends will emerge as further organisms and further requirements are studied, but it is more likely that the value of this research will lie in increasing the numbers of protists available in conditions conducive to biochemical and physiological investigation.

In any case, there are many experimental and interpretational difficulties in this work. Some nutrients are physiologically active in concentrations of less than 10^{-9} grams per liter of culture medium, necessitating very exacting experimental techniques and sophisticated instrumentation. Perhaps more importantly, there is no *a priori* assurance that a nutritional requirement in two protists is due to the same genetic modification.

Although numerous examples of the earlier research could be cited, the story of vitamin B_{12} is particularly illuminating. It was originally hoped that the strict B_{12} requirement by cultured *Euglena* spp. would be of taxonomic or even of phylogenetic value. Considerable interest was aroused by the discovery of a similar requirement by cultures of the green alga *Chlamydomonas chlamydogama*. Various patterns of B_{12} requirements (satisfaction by cobalamin analogues, by methionine, etc.) were subsequently described for various bacteria, Chrysophyceae, zooflagellates, birds, and mammals. *Acanthamoeba castellani, Hartmannella rhysoides*, and *Tetramitis rostratus* were also shown to require B_{12} (Lilly, 1967). However, bacteria, Actinomycetes, most ciliates, and some insects (gut fauna?) show no vitamin B_{12} requirements. Some lactobacilli, usually considered extremely primitive organisms (Section 5.5), produce their own vitamins, including B_6 and B_{12} (Seaman, 1955).

Requirements for B_{12} have been found in members of the Chlorophyceae, Cryptophyceae, Dinophyceae, Chrysophyceae, Bacillariophyceae, Rhodophyceae, and Cyanophyceae, as well as the Euglenophyceae. Vitamin autotrophs have been found in all of these classes of algae as well (Droop, 1962). Some, but not all, photosynthetic bacteria also require B_{12} (Kondrat'eva, 1965).

Thus the picture for vitamin B_{12} has become quite complex; no phylogenetic trends above perhaps the genus level (in some cases) are seen. Other vitamins show much the same (lack of) pattern. A vitamin B_6 requirement, for instance, is found in some (but not all) *Tetrahymena* spp. (Holz, 1964), and the situation in other protozoa is no less confusing (Lilly, 1967). Lipoate (thioctate) is a requirement for growth of the ciliates *Tetrahymena* spp. (six spp.), *Glaucoma chattoni* A, *Paramecium* spp. (three spp.), and *Colpidium* spp. (two spp.), but is biosynthesized by all other organisms investigated (Holz, 1964). Innumerable other examples could be given, but only rarely would taxonomically informative data be found. Although taxonomically or phylogenetically amenable vitamin and mineral requirement distributions may yet be discovered or established, this field does not appear to be fulfilling the early expectations.

In a recent publication, McClendon (1976) has considered organismal mineral requirements and elemental abundance at the time of prebiological synthesis and later. Although there are no evolutionary considerations per se, it might be possible to obtain input into broad evolutionary trends by utilizing his approach with a judicious choice of organisms.

12

A Biochemical Phylogeny

12.1 OTHER PHYLOGENIES IN THE LITERATURE

Since the first phylogenetic tree was drawn by Haeckel (1866; see Frontispiece), a large number of schemes have been constructed purporting to portray the phylogeny of the protists. Some of the more successful and more noteworthy recent phylogenies have been included here for comparison with our own phylogeny (Figure 23). These include phylogenies by Copeland (1938) (Figure 16), Cronquist (1960) (Figure 17), Christensen (1962, 1964) (Figure 18), Sagan (1967; essentially identical with those of Margulis, 1969, 1970, 1974) (Figure 19), Whittaker (1969) (Figure 20), Erwin (1973) (Figure 21), and Cavalier-Smith (1975) (Figure 22).

Whittaker (1969) has provided a more comprehensive compilation of phylogenetic schemes published up to that date. Still other phylogenies have been proposed by Stransky and Hager (1970c), Dodson (1971), Ehrendorfer (1972), Lee (1972), McLaughlin and Dayhoff (1973), Dodge (1974b), Leedale (1974), Cloud *et al.* (1975), Hirose (1975), Hanson (1976), Hutner and Corliss (1976), Taylor (1976b), and by other authorities mentioned previously in this work (see especially Chapters 3 and 5).

12.2 A BIOCHEMICAL PHYLOGENY

The suggestion that one of these phylogenies might be "better" than the others would be difficult to prove, but is not absurd or meaningless. However, the suggestion that there might be such a thing as an optimal or ideal phylogeny of organisms would appear to be strangely Platonic, as if

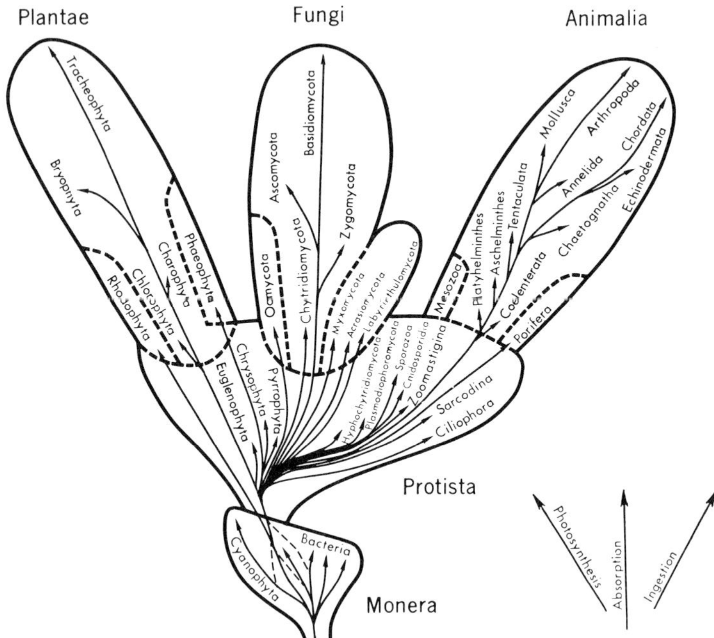

Figure 16. Phylogeny of organisms (Copeland, 1938). Reproduced from Figure 2 from Whittaker (1969) by permission of the American Association for the Advancement of Science and the author. Copyright 1969 by the American Association for the Advancement of Science.

phylogeneticists were gazing heavenward toward the Phylogeny hovering somewhere near the fourth level of the Divided Line. This may prove to be the case. Yet there is the suspicion that the optimal phylogeny is material, although perhaps necessarily unattainable. History may be exceedingly complex and may involve deep interrelationships with the evolution of human consciousness, but it is not entirely an abstraction. Organisms have actually evolved, and phylogeneticists are consequently studying not only philosophical abstractions and academic rationalizations, but also the history of the living world. It may be unrealistic to expect a phylogeny to reflect the exact path of past evolution (Eigen, 1971), but a reasonable goal would be to approach the actual historical process to a high degree. So might the quest for the optimal phylogeny be characterized. The quest, however, will continue until all the appropriate information on all organisms is available.

The potential merits of using biochemical data in phylogenetic reconstruction have been discussed earlier (Sections 1.4 and 2.1). However, biochemistry as a well-developed discipline is hardly three-fourths of a century old, and most modern concepts of evolution have been developed only in the hundred-odd years since the work of Darwin and Wallace (1858). As a result, it might be expected that attempts to reconstruct a satisfactory biochemical phylogeny are premature. The phylogeny we propose (Figure 23) is not presented as the optimal phylogeny, but rather as one step in the dialectical process toward such a goal, subject to the limitations already discussed.

12.3 COMPARISON OF THESE PHYLOGENIES

As stated earlier (Section 2.5), phylogenies based on comparison of morphological data can usefully be employed as a check on phylogenies based on biochemical data. Virtually all phylogenies of the protists, and especially of the algae, have utilized biochemical characters (pigmentation) to some

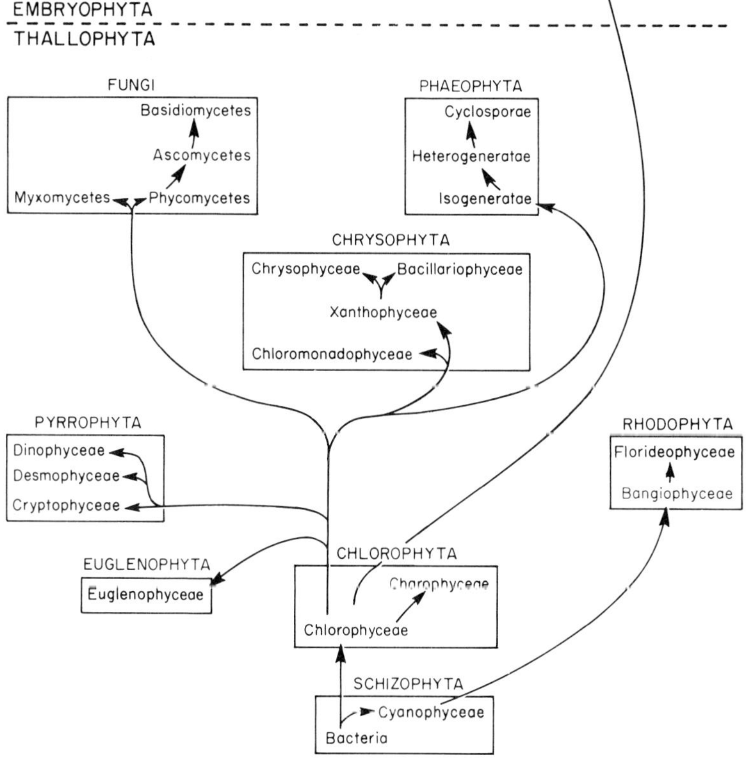

Figure 17. Phylogeny of protists. After figure (p. 434) from Cronquist (1960). Reproduced by permission of the New York Botanical Garden and the author.

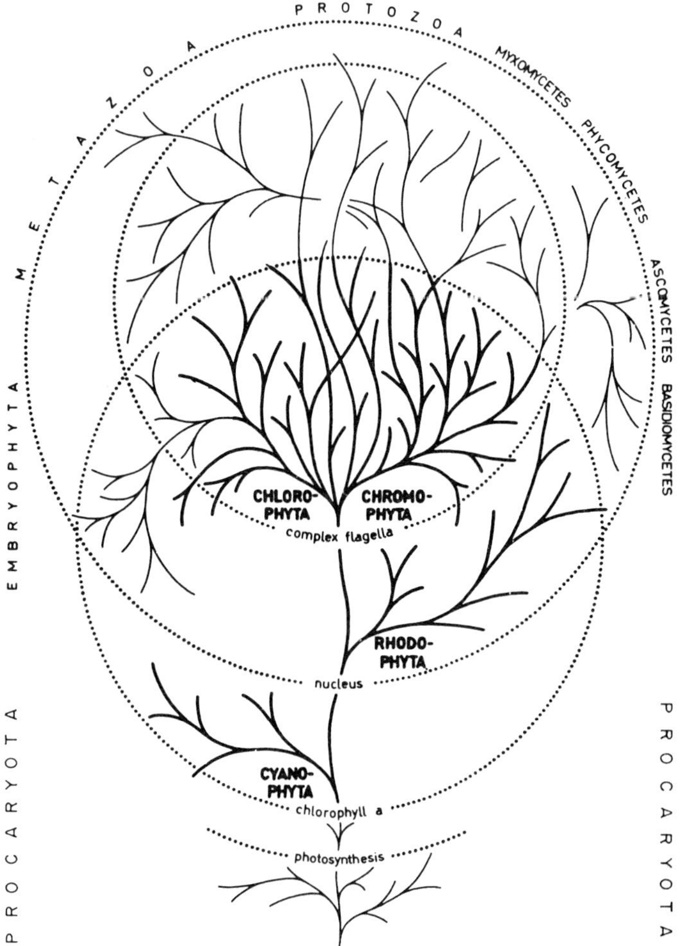

Figure 18. Phylogeny of organisms. Reproduced from Figure 1 from Christensen (1964) by permission of Plenum Press and the author.

degree. Dodge (1974b), however, has considered only morphological and ultrastructural characters in his reassessment of algal classification. Biochemical investigation agrees well with morphology-based classification on numerous points, including by now rarely debated questions such as the derivation of higher plants from Chlorophyceae, the relationship of Chlorophyceae and Charophyceae, and the common ancestry of Ascomycetes and Basidiomycetes. On other questions, e.g., the phylogenetic position of slime molds, more data (both biochemical and morphological) are needed. The following sections in this chapter will examine how various phylogeneticists have treated some of the more interesting problems in protistan phylogeny.

12.3.1 Prokaryotes and Eukaryotes

Virtually all recent phylogenies have emphasized the fundamental importance of the prokaryote–eukaryote dichotomy, although Dillon (1963) has not used this dichotomy as a major part of his classification. Biochemical differences between prokaryotes and eukaryotes include the presence in the former of 70 S ribosomes, characteristic rRNA sedimentation patterns, individual rRNA precursors (active RNase III), fibrillar DNA, separable polyaromatic biosynthetic enzymes, *N*-formylmethionyl-tRNA initiator, mangano- or ferrisuperoxide dismutases, lipopolysaccharides, mucopeptides, teichoic acids, phosphatidylglycerylamino acids, and poly-β-hydroxybutyrate, which are not characteristic of extraorganellar eukaryotic systems.

Whether the Cyanophyceae are labeled "blue-green algae" or "blue-green bacteria" is largely a matter of taste since the algae are a heterogeneous, polyphyletic group that are not, and cannot be, formally defined. Cyanophyceae exhibit numerous evolutionary innovations, including "plant-like" ferredoxins, proline hydroxylase, polyglucoside branching isoenzyme biosynthesis (with BE and Q activities), cytochrome *f*, cytochrome b_6, phycobiliproteins, NADH- and NADPH-linked nitrate reductase activities, plastocyanin, eukaryote-type photosynthesis (chlorophyll *a*, Photosystem II, linkage of two photosystems, utilization of water as electron donor and NADPH production in photosynthesis), digalactosyl diglycerides, characteristic pectin-like substances, 18:1(9) and 18:3 fatty acids and the corresponding enzyme activities, 2-*O*-methylribose cytidine, abundant amino sugars, widespread ability to use exogenous nitrate, one-step reductive amination of pyruvate or 2-ketoglutarate (Neilson and Doudoroff, 1973), loss of NADP-control of citrate synthase activity, loss of fatty acid anaerobic desaturation capability, and loss of the ability to use urea as a nitrogen source. These differences seem significant enough to warrant retention of the term "blue-green algae." However, many features (e.g., lack of membrane-bound organelles, cell wall features) can be used to justify the term "cyanobacteria."

12.3.2 Origin of Chloroplasts

One of the most obvious differences among the phylogenies pictured (Figures 16–23) is that some authors accept an endosymbiotic origin of

VI
(9 + 2) Homologue used both as basal body to flagella and other (9 + 2) homologue permanently differentiated as extranuclear division centers (centrioles).

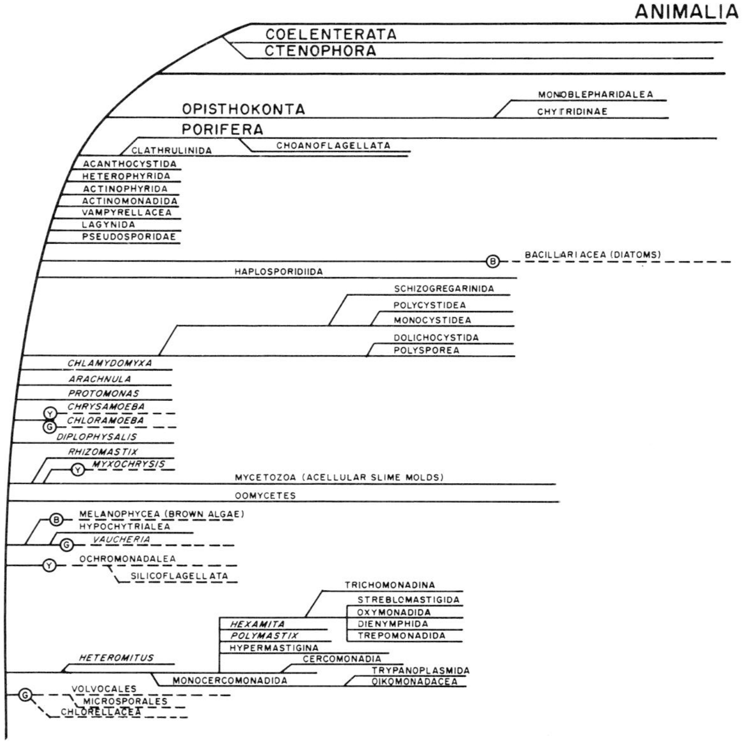

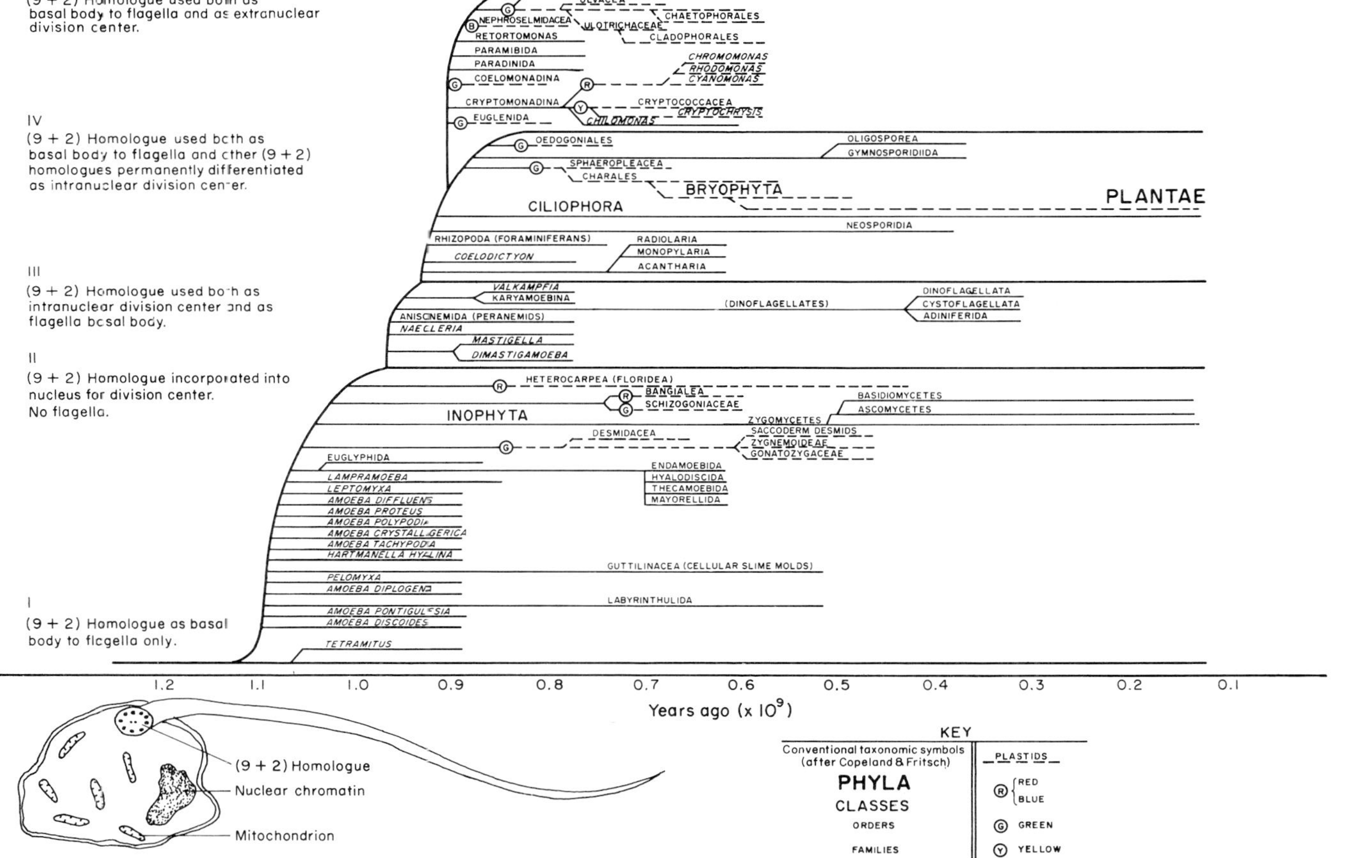

Figure 19. Phylogeny of organisms. After Figure 1 from Margulis (Sagan, 1967) by permission of Academic Press and the author.

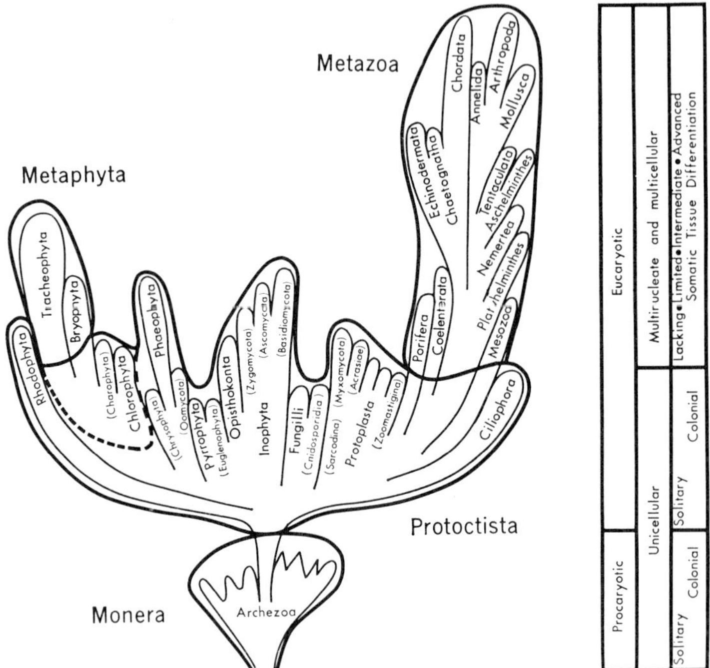

Figure 20. Phylogeny of organisms. Reproduced from Figure 3 from Whittaker (1969) by permission of the American Association for the Advancement of Science and the author. Copyright 1969 by the American Association for the Advancement of Science.

chloroplasts, whether enthusiastically (Sagan, 1967; Margulis, 1969, 1970) or tentatively (Figure 23); others reject this possibility (Copeland, 1938; Cronquist, 1960; Christensen, 1962, 1964; Erwin, 1973; Cavalier-Smith, 1975). Whittaker (1969) has chosen not to deal with the question. Acceptance by some authors of the possibility of an endosymbiotic origin of chloroplasts, and rejection of this possibility by others, makes it difficult to compare details of the resulting phylogenies, because the endosymbiotic theory presents a very different picture of evolutionary history than do the gradualist theories. According to the former view, photosynthetic eukaryotes have arisen from nonphotosynthetic organisms once or possibly several times. Secondary losses of chloroplasts, which have doubtless occurred, need not have contributed greatly to the overall picture. The gradual-development theory as usually stated requires only a single appearance of photosynthetic organisms, but significant losses of photosynthetic capability.

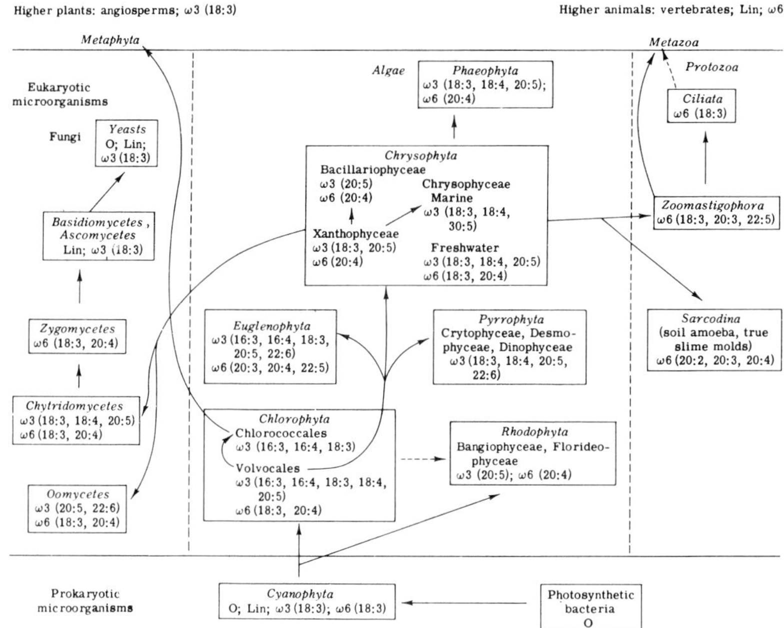

Figure 21. Phylogeny of protists. After Figure 23 from Erwin (1973). Reproduced by permission of Academic Press and the author.

It appears difficult to devise a critical test of either theory of chloroplast origin. Using Popperian logic, it is not obvious what would constitute a falsification of either theory. An alternative approach would involve the examination of subsidiary hypotheses made within the framework of each theory (Lewontin, 1972). Although truth may need more than correct predictions (Van Valen, 1973), the fundamental biochemical similarities observed between chloroplasts and Cyanophyceae—as predicted by the endosymbiotic theory—probably constitute the best demonstration* to date of the validity of this theory. In addition to the biochemical similarities between Cyanophyceae and chloroplasts listed in Section 12.3.6, it is possible to add as further similarities the gene for ribulose-1,5-diphosphate carboxylase (heavy subunit), ribosome subunit hybridization potential, the extent of methylation of rRNA, interconversion of tryptophanyl-tRNA, recognition properties of tRNA synthetases, primary sequence of cytochrome *f* (although its gene location is unknown), phytol biosynthesis, NADP-linked glutamate dehydrogenases, localization of phosphorylase

* "A demonstration is a way to convince a reasonable man, and a proof is a way to convince a stubborn one" (Kac, 1973).

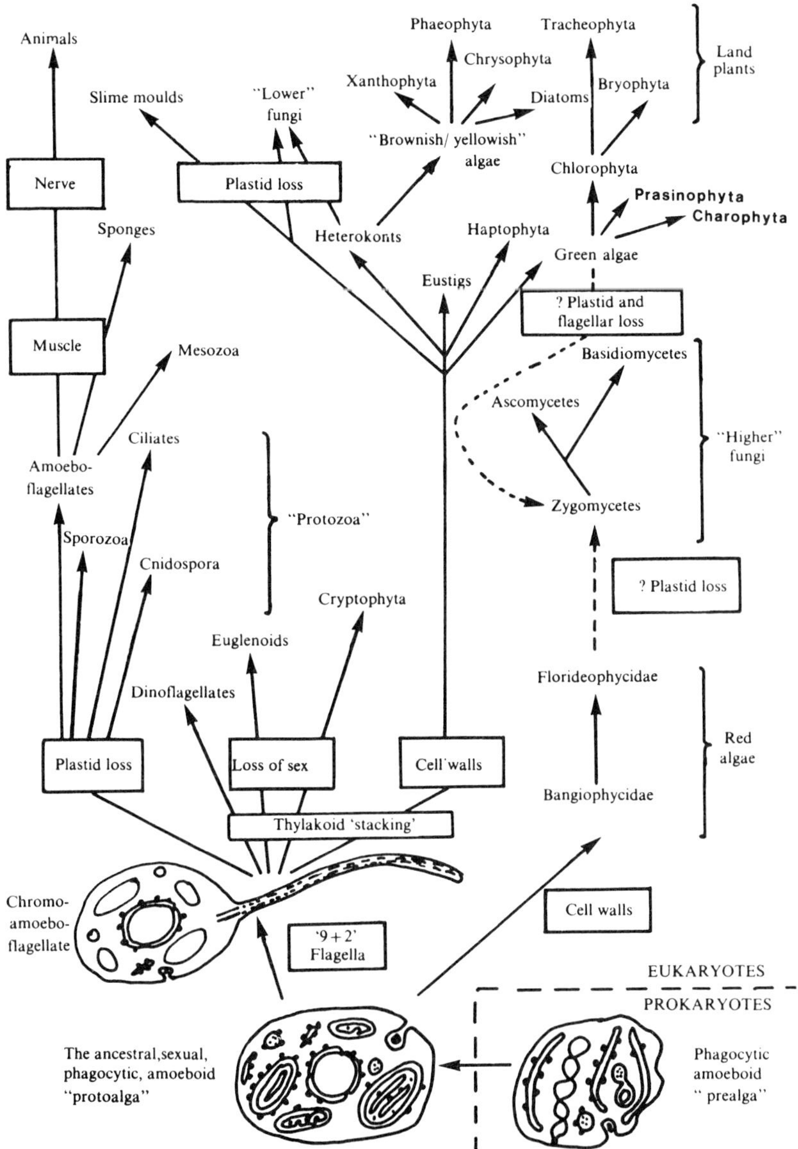

Figure 22. Phylogeny of protists. After Figure 5 from Cavalier-Smith (1975) by permission of Macmillan Journals Ltd. and the author.

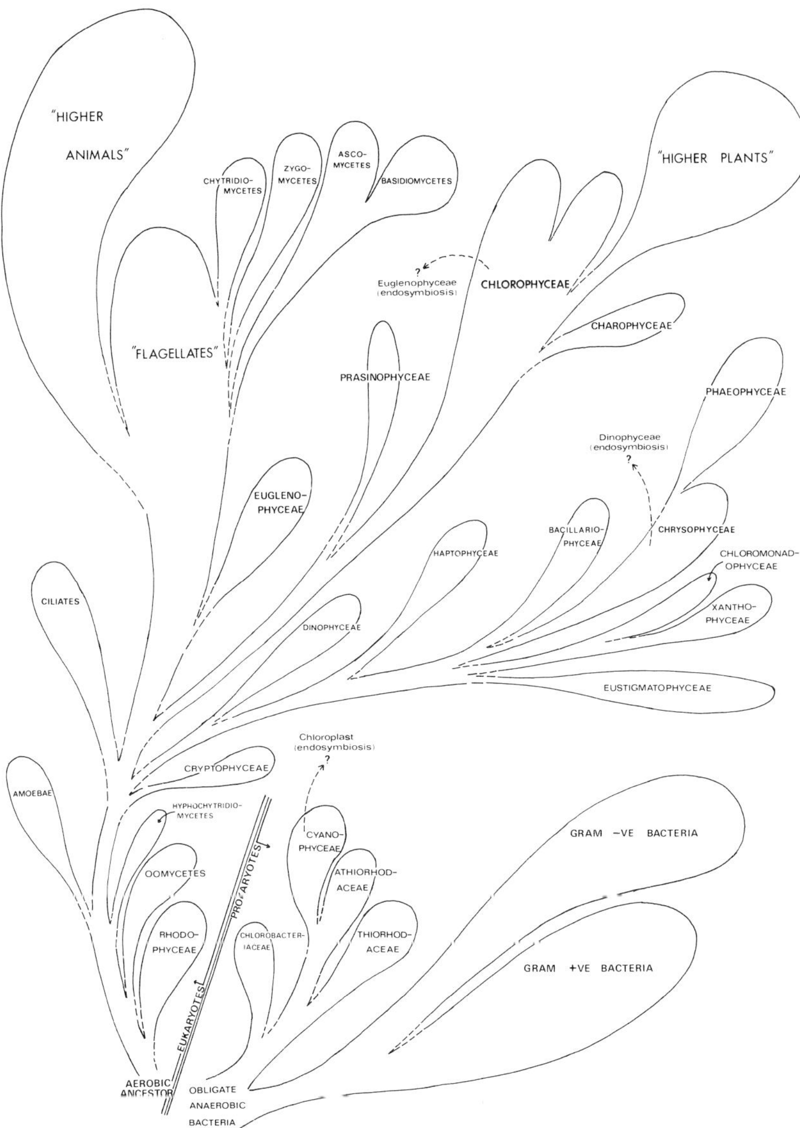

Figure 23. A biochemical phylogeny of protists. In this proposed phylogeny the distance between groups, or the distance from the common ancestor, does not necessarily indicate the degree of divergence or a "time factor" in the evolution over geological time. The size of each group in the scheme is likewise not necessarily indicative of the extent of divergence within the group, nor of the number of extant taxa.

(Yin, 1948), and presence of unconjugated pteridine pigments (Wolf, 1960). Most or all of these characters are explicable in terms of an internal differentiation theory of chloroplast origin, but such an origin would probably not predict that the chloroplast would possess these features in common with Cyanophyceae, or that the extrachlorplastic material would not.

It has been objected that the endosymbiotic theory does not explain why the nuclear genetic system is biochemically different from the genetic systems of the organelles, nor accounts for the origin of the nuclear membrane (Cavalier-Smith, 1975). Indeed, probably all proponents of an endosymbiotic origin of mitochondria and/or chloroplasts readily admit to the development of the nuclear membrane by internal differentiation, perhaps as part of a developing capacity for endocytosis (Stanier, 1970). Cavalier-Smith's suggestion that nuclear genetic systems were modified by "new selective forces" resulting from nucleocytoplasmic separation does not require that these modifications occurred only after the acquisition of organelles (although Cavalier-Smith did propose such a sequence of events). Biochemical modification of the protonuclear genetic material could have occurred equally well (and possibly more rapidly) when the protoeukaryote was still internally undifferentiated, and could have resulted from selection pressures arising from the development of endocytosis and from this organism's exploitation of a substantially different ecological niche than that occupied by photosynthesizers and chemoautotrophs. Moreover, a nearly simultaneous origin of mitochondria and chloroplasts in a single common ancestor would not necessarily rationalize why mitochondrial genetic systems may be less autonomous and less prokaryote-like than are chloroplastic genetic systems (see references in Section 3.3.2).

12.3.3 Rhodophyceae

Most authorities suggest that the Rhodophyceae diverged long ago from lines leading to other organisms. These suggestions are based largely on the unique (among eukaryotes) pigment composition of red algal chloroplasts, and the lack of flagella and 9 + 2 structures (with the concomitant modifications in morphology of the reproductive structures). Margulis (Sagan, 1967), who postulates a diphyletic derivation of the Rhodophyceae, has suggested affinities between Rhodophyceae and Zygomycetes, but does not actually suggest the derivation of one group from the other. Phylogenetic association of Rhodophyceae with Eumycota, especially with Ascomycetes (Bessey, 1950; Denison and Carroll, 1966; Demoulin, 1974; Cavalier-Smith, 1975), is untenable both morphologically (Klein and Cronquist, 1967) and, on the basis of present knowledge, biochemically. Chadefaud (1975) has recently proposed that the Rhodophyceae and Eumycota (Ascomycetes +

Basidiomycetes) evolved in parallel but separate lines, from a common ancestor near the Cyanophyceae–Rhodophyceae connection. The theory, like some before it, relies heavily upon morphology and macrocytology of the reproductive processes. It shows virtually no regard for some very basic fundamental differences. The Rhodophyceae biosynthesize starch and cellulose, different sterols in many cases, numerous polyunsaturated fatty acids (including those of the $\omega 3$ series), histones (where investigated), and, of course, they produce the characters associated with the chloroplast. The cell wall compositions are typically different, as are some fundamental biosynthetic pathways (Chapter 6). The polysaccharide composition and, where investigated, polysaccharide biosynthetic enzymes are different from those in most Eumycota. We favor placing the Rhodophyceae as a specialized group of organisms not closely related phenetically to any other taxa, but in a position that would establish the class as one of the most primitive of eukaryotes. The most compelling evidence for this is the similarity of the biliproteins with those of the Cyanophyceae (Section 5.9) and the absence of secondary chlorophylls, the "simple" structure of the chloroplast (Gibbs, 1970; Bisalputra, 1974), and the simple carotenoid pattern (Section 9.4), among others. The evidence from cytochrome *c* (Meatyard *et al.*, 1975), admittedly sparse, also suggests that the Rhodophyceae are evolutionarily primitive and diverse.

12.3.4 Dinophyceae and Cryptophyceae

Biochemical research has shown that these groups of organisms are quite different from each other and from other photosynthetic eukaryotes. A similar conclusion has been reached by Dodge (1974b) from morphological and ultrastructural considerations. Biochemically, they differ in phycobilin and phycobiliprotein biosynthetic capability, presence of 5-hydroxymethyluracil in DNA, and in carotenoid biosynthetic pathways (Section 9.4). The close phylogenetic relatedness between these two taxa suggested by Pascher (1914, 1927), Cronquist (1960), Klein and Cronquist (1967), Whittaker (1969), and Erwin (1973) is not supported by the majority of biochemical data. The phylogeny of Cavalier-Smith (1975) loses sight of the fact that certain organisms classified as Dinophyceae or as Cryptophyceae (due to typical zoospore morphology) possess perfectly good cell walls and probably have sexual reproduction (Zingmark, 1970; Heywood, 1976). Classification of "Dinophyta" and "Euglenophyta" as the only two major phyla in a single subkingdom "Euglenophytaria" (Dillon, 1963) follows somewhat in the spirit of Haeckel (1866) and Copeland (1938), but on the basis of biochemical investigations is probably unjustified. The demonstration of the DAP pathway of lysine biosynthesis in Dinophyceae, if indeed it exists there, would probably be the final blow to such a classification. Although there is little to suggest a close relationship between the two classes (except chlorophyll *c*), there still remains the placement of these two classes relative to the other algae. Gressel *et al.* (1975) have suggested from a consideration of ribosomal RNA that the Dinophyceae, along with the Rhodophyceae, may be archaic intermediates between prokaryotes and eukaryotes.

The biosynthesis of chlorophyll c_2, but not c_1, by these two classes should not be construed as implying a close phylogenetic relationship.

The presence of biliproteins in the Cryptophyceae would suggest some closeness to the rhodophycean stock, but since the proteins themselves are different, it is perhaps only a distant relationship, and it is perhaps more probable that the biliproteins indicate primitiveness. A primitive positioning for the Cryptophyceae finds support from the apparent simplicity of the chloroplast thylakoids (Gibbs, 1970; Bisalputra, 1974) and the carotenoids. Many of the features of carotenoid biosynthesis of advanced algae (e.g., allenic bonding, epoxidation, polyoxygenation) are absent in the Cryptophyceae. Indeed, the cryptophycean carotenoids are the acetylene-analogs of the rhodophycean carotenoids. The presence of chlorophyll c_2 suggests affinities with the so-called chlorophyll *c* or "Chromophyta" line (Christensen, 1962), whereas starch production suggests an alliance with the chlorophyll *b* or "Chlorophyta" line.

With the admittedly sparse knowledge available, we suggest that the Cryptophyceae represent a primitive group, removed from but derived from an ancestral stock near to the divergence of the two main phylogenetic lines of the algae. This is in keeping with the nonbiochemical evidence; it is perhaps worth noting that the "host" of the cyanelle-containing *Cyanophora paradoxa* is a cryptomonad (Hall and Claus, 1963; Pickett-Heaps, 1972).

The dinoflagellates are anomalous in too many respects, such that any phylogenetic positioning will invite criticism. Some evidence (e.g., acetylenic xanthophylls, chlorophyll c_2) points to affinities with Christensen's (1962) Chromophyta line, but where does the divergence occur?

There are a number of dinoflagellates known to possess fucoxanthin instead of peridinin (see Johansen *et al.*, 1974, Jeffrey *et al.*, 1975). Indeed, the chloroplasts of such species resemble a chrysophycean chloroplast ultrastructurally (Tomas and Cox, 1973; Tomas *et al.*, 1973) and also biochemically. These dinoflagellates are also characterized by the appearance of an additional eukaryote nucleus (Dodge, 1971; Tomas and Cox, 1973; Tomas *et al.*, 1973). Tomas *et al.* (1973) and Tomas and Cox (1973) provided a logical explanation, suggesting that these species were derived from heterotrophic colorless dinoflagellates that ingested a chrysophyte.

Jeffrey *et al.* (1975) have discussed this possibility further. If this did indeed happen, should we consider the possibility that all chloroplast-containing dinoflagellates were derived via such a symbiosis (with loss of all but the chloroplast) and that the present chloroplast is indeed a chrysophyte remnant, whose principal modification has been the "exchange" of a fucoxanthin pathway for a peridinin one? This is not outside the realm of possibility. There are many colorless dinoflagellates. The only restriction would be that the chloroplastic dinoflagellates evolved after the appearance of a chrysophycean type. But then, what did the colorless dinoflagellates evolve from?

12.3.5 Chlorophylls *c*-Containing Organisms Other than Dinophyceae and Cryptophyceae

Haeckel (1866) placed brown algae in the "Plantae," but classifed the diatoms with the "Protista." Copeland (1938) proposed divergent positions for the Bacillariophyceae and Phaeophyceae as well, and Whittaker (1969) did so for Chrysophyceae and Phaeophyceae. Margulis (Sagan, 1967) has suggested a somewhat closer phylogenetic relationship between many Chrysophyceae (but excluding certain amoeboid forms) and the Phaeophyceae; she furthermore suggested that *Vaucheria* is related more closely to the Phaeophyceae than to the Chrysophyceae. Cronquist (1960) has ventured a somewhat closer relationship, and Christensen (1962) lumped all chlorophyll *c*-containing organisms, including Dinophyceae and Cryptophyceae, plus the Raphidophyceae and Craspedophyceae (two minor groups often classified as protozoa), into the "Chromophyta." Dillon (1963) included not only the above-mentioned organisms, but also the ciliates, zooflagellates, Rhodophyceae, sporozoans, and higher animals into the "Chrysophytaria." Finally, Dodge (1974b) has narrowed Christensen's "Chromophyta" slightly (removing the Haptophyceae, Dinophyceae, and Cryptophyceae), in so doing returning more to the grouping proposed by Pascher (1914, 1921).

Biochemically, the Bacillariophyceae, Chrysophyceae, Haptophyceae, Phaeophyceae, Xanthophyceae, and probably the Chloromonadophyceae possess similar β-(1,3)-linked glucans (which are also seen in the Euglenophyceae and occasionally in fungi), are often silicified, and utilize comparable carotenoid biosynthetic routes. These data also accord with those of Erwin (1973), who examined fatty acid distributions. Although there are notable differences in the distributions of individual carotenoids (e.g., vaucheriaxanthin in Xanthophyceae, fucoxanthin absent from Xanthophyceae), the basic biosynthetic pathways (Figure 10, Section 9.4) are present. Most authorities would agree that the Phaeophyceae represent the most advanced class in this line. It is interesting to note that the Phaeophyceae lack the acetylenic pathway and that in the most obvious ancestral stock to the Phaeophyceae, the Chrysophyceae, the presence of the acetylenic pathway is in doubt (Section 9.4). The Bacillariophyceae, with their highly ordered silica metabolism and morphology, are obviously a side branch. Biochemically the Haptophyceae would appear to be very similar to the Chrysophyceae–Phaeophyceae continuum. The establishment of this class in a distinct side branch is based on some very profound morphological differences (e.g., flagellar structure and the haptonema: Dodge, 1974a,b). If an assumption is made that a filamentous Chrysophycean such as *Phaeothamnion* will prove to be biochemically similar to other examined Chrysophyceae, it is a tempting step to propose that the Phaeophyceae, lacking any morphological form more primitive than the filament (e.g., *Ectocarpus*), evolved from such as the filamentous Chrysophyceae. The Xanthophyceae and Chloromonadophyceae are placed at the base of this evolutionary sequence, partly because of the absence of the distinctive fucoxanthin pathway. There is a regrettable lack of biochemical data on the enzymes and nucleic acids of these organisms. The available data point to a relatively ancient, common ancestry of all these organisms, evolving along a distinctive phylogenetic line, that apparently terminated in the Phaeophyceae (with the large kelps and fucoids). This line is often referred to as the "chlorophyll *c*" line or the Chromophyta (Christensen, 1962).

12.3.6 Euglenophyceae

The presence of chlorophyll *b* in the photosynthetic Euglenophyceae has been a major reason for considering these organisms to be phylogenetically close to the Chlorophyceae, Charophyceae, and Prasinophyceae. Haeckel (1866) placed *Euglena* and *Volvox* (and *Peridinium*!) in the "Flagellata." Copeland (1938) collected the Euglenophyceae, Chrysophyceae, Xanthophyceae, Cryptophyceae, Dinophyceae, and Chloromonadophyceae in the "pigmented flagellates," from which he considered diatoms and green algae to have arisen. Cronquist (1960), Margulis (Sagan, 1967), Whittaker (1969), Erwin (1973), and possibly Cavalier-Smith (1975) separated the euglenoids from other taxa (although Cronquist and Erwin suggested a derivation from the Chlorophyceae), whereas Christensen (1962) included the Euglenophyceae within the Chlorophyta.

Biochemically, the Euglenophyceae bear little resemblance to "other" algae, with the exception of characters pertaining to the chloroplast (chlorophylls *a* and *b*, carotenoid biosynthetic pathways, etc.). Lysine biosynthesis, sterol accumulation patterns, Type VII tryptophan biosynthetic enzymes, Type I fatty acid synthase, aggregational properties of shikimate biosynthetic enzymes, ϵ-trimethylation of lysine-94 in cytochrome *c*, and the apparent absence of cellulose differentiate Euglenophyceae from Chlorophyceae and many other organisms. Cytochrome *c* primary structures show a closer relationship between *Euglena gracilis* strain Z (c_{558}), *Crithidia oncopelti* (c_{557}), and *C. fasciculata* (c_{555}) than between the former organism (or Euglenophyceae) and fungi or higher plants, whereas properties of tryptophan biosynthetic enzymes, 28 S rRNA, shikimate pathway enzymes, and lysine biosynthesis suggest a relationship between investigated Euglenophyceae and protozoa and/or simple Eumycota. The sequences of cytochrome *f* (Ambler and Bartsch, 1975), however, suggest that the Euglenophyceae stand apart from the other chloroplast-containing organisms (Andrew *et al.*, 1976).

The similarities between the euglenoid chloroplast and chloroplasts in Chlorophyceae and higher plants, coupled with the striking dissimilarities between euglenoids and these groups in biochemical traits thought not to be associated with the chloroplasts, suggest that euglenoids may be a symbiosis between a chlorophyll-containing chlorophycean and a mastigophorean or chytrid-like colorless eukaryote (see Stewart and Mattox, 1975). Resemblances between euglenoid chloroplasts and Cyanophyceae are numerous: NADP-linked glyceraldehyde-3-phosphate dehydrogenases, DNA–RNA hybridization capacity, rRNA sedimentation patterns, rRNA maturation patterns, 70 S ribosomes with characteristic subunits, similar reactions to antibiotics, *N*-formylmethionyl-tRNA initiator, 5′-monohydroxyphylloquinone, and chlorophyll *a*. Numerous similarities extend to the chlorophycean chloroplast, e.g., synthesis of chlorophyll *b*, synthesis of certain carotenoid types. It may well be that the symbiosis solution offered for the Dinophyceae may have an analogue in the Euglenophyceae. Many colorless Euglenophyceae are known and the chloroplast is surrounded by a triple membrane. With such a solution, however, one would have to assume that symbiosis resulted in the loss of all trace of the invader except the chloroplast and the third membrane (former wall?), and that the chloroplast lost the ability to synthesize chlorophyll *a* in the dark. (Dark chlorophyll synthesis is not a feature of the Euglenophyceae, as it is in the green algae). The ability to synthesize the acetylenic carotenoids is a stumbling block, since this is a feature unknown in chlorophycean chloroplasts. An alternative proposal would involve invasion with a xanthophycean-like organism that had not acquired chlorophyll *c*-synthesizing capabilities. This would resolve the problem of the acetylene carotenoids, but in turn would leave the problem of providing an explanation for the appearance of chlorophyll *b*. One can circumvent these problems by suggesting duplicate evolution, but this is perhaps a little too glib and unparsimonious. These characters do not establish the veracity of the endosymbiotic theory of chloroplast origin or an endosymbiotic origin for Euglenophyceae, but demonstrate basic resemblances that might best be predicted and rationalized by endosymbiosis (see Sections 3.3 and 12.3.2).

We wish to suggest a phylogenetic separation between the Euglenophyceae and the other chlorophyll *b*-containing organisms, and to propose that photosynthetic euglenoids may be the result of an endosymbiotic event as described above (cf. Stewart and Mattox, 1975). We do not pretend to have solved the problem of the origin of the Euglenophyceae. Experiments such as examination of lysine biosynthetic pathways, tryptophan biosynthetic enzymes, shikimate pathway enzymes, cytochrome *c* sequences, and (ideally) nucleotide base sequences in other euglenoids and in Chrysophyceae, Xanthophyceae, Chytridiomycetes, and protozoa would be of great interest.

Finally, in this regard, brief mention should be made of the eyespot. Although it is not a "biochemical" character, it may be considered as quasi-biochemical by virtue of its carotenoid biosynthesis. Numerous morphological forms exist (Dodge, 1974a), but most notably it is in the Euglenophyceae, Dinophyceae, and Eustigmatophyceae that the eyespot exists outside the chloroplast. The two former classes are candidates for the so-called symbiosis theory, from possible invaders whose eyespots would have been within the chloroplast. Did symbiosis result in elimination of the "chloroplast-internal" eyespot? What is the explanation for the "chloroplast-external" eyespot? Although the chemical pigment composition of Euglenophyceae (Batra and Tollin, 1964) and certain Dinophyceae (Withers and Haxo, 1975) is known, nothing is known about the relationship of pigment biosynthesis in chloroplasts and chloroplast-external eyespots. An understanding of this might help elucidate the phylogenetic enigma of the Euglenophyceae and Dinophyceae.

12.3.7 Eustigmatophyceae

The establishment of this class taxonomically has rested principally on morphology at the ultrastructural level (Hibberd and Leedale, 1970, 1972). There appears no question that this class is distinct from all others, but, unfortunately, the reasons for the taxonomic separation provide little in the way of phylogenetic help. Chloroplast pigment distributions are the principal biochemical data. The absence of secondary chlorophylls is unusual. The carotenoids show the biosynthetic patterns typical of the Chlorophyceae. Although there is a slight affinity with the Xanthophyceae (e.g., vaucheriaxanthin), the absence of acetylenic biosynthetic capability is noteworthy. At this stage, in the absence of further data, we have proposed that the Eustigmatophyceae should be placed near the "divergence" of the chlorophyll *c* and chlorophyll *b* lines.

12.3.8 Organisms Containing Chlorophyll *b* Other than Euglenophyceae

The three classes here are the Prasinophyceae, Chlorophyceae, and Charophyceae. The mass of biochemical data on these three classes suggests an obvious continuum running through to the bryophytes and higher plants. The phylogenetic separation of the Prasinophyceae is based principally on the morphological, cytological, and ultrastructural evidence (Dodge, 1974a; Pickett-Heaps and Marchant, 1972; Pickett-Heaps, 1975). The division of the Chlorophyceae into two phylogenetic lines, one of which proceeds through to higher plants, is based solely on the glycolate oxidation pathways (Section 6.5). In itself this would be a shaky ground, but the cytological evidence (Pickett-Heaps, 1975; Stewart and Mattox, 1975) supports this. The cytological evidence is, in fact, the kingpin of this proposal; the biochemical data, which are sparse, lend support to the proposals.

The phylogenetic positioning of the Charophyceae is again based largely on the nonbiochemical characters. One of the noteworthy features of this line is the high degree of biochemical uniformity, which continues through into the higher plants (excepting "acquisitions" such as lignin, flavonoid, and alkaloid biosynthesis). Similarly, the biochemical divergence from the "chlorophyll *c*" line is very sharp. The algae show, at the very least, a distinct diphyletic development, lending strong support to the view that they are not a natural taxonomic grouping.

12.3.9 Eumycota and Oomycota

Biochemical investigation has revealed that organisms once ascribed to the "Phycomycetes" (Scherffel, 1925) exhibit such significant differences among themselves that they can no longer be classified in a single class, and perhaps may require two different divisions or phyla. One group of these organisms, which we term "Oomycota" (water molds), consists of organisms usually possessing cellulosic cell walls (usually without chitin), Type III NAD-linked glutamate dehydrogenases, Class I D-lactate dehydrogenases, NADP-linked isocitrate dehydrogenases catalyzing an essentially irreversible reaction, DAP-lysine biosynthesis, Type IV tryptophan biosynthetic enzyme organization where investigated (Oomycetes), C_{27}, C_{28}, and C_{29} Δ^5-sterols, a heavy 28 S rRNA molecule, proline hydroxylase (where investigated), and histones (where investigated). Investigated Oomycetes and Hyphochytridiomycetes exhibit these characters.

In contrast, Chytridiomycetes and Zygomycetes differ from the Oomycetes by having chitinous cell walls (without cellulose), NAD-linked glutamate dehydrogenases showing no glucose repression of synthesis, Class II D-lactate dehydrogenase, NAD-linked isocitrate dehydrogenase, NADP-linked isocitrate dehydrogenase with easily reversible catalytic action, AAA-lysine biosynthesis, typical "fungal" 28 S rRNA molecular weights, complete aggregation of polyaromatic (shikimate) biosynthetic enzymes, Types I and III tryptophan biosynthetic enzymes, C_{28} Δ^7-sterols based on lanosterol with Mechanism II side chain biosynthesis where investigated, atypical or no histones, and no detectable hydroxyproline. Zygomycetes resemble Ascomycetes and Basidiomycetes in many of the above characteristics. The class Chytridiomycetes has in the past "been used as a dumping ground for poorly understood parasites" (Copeland, 1938); nonetheless, investigated chytrids also resemble Zygomycetes rather closely in many biochemical characters. It seems eminently justifiable to classify Zygomycetes, Chytridiomycetes, Ascomycetes, Basidiomycetes (and imperfect fungi) as "Eumycota" (true fungi).

Other authors, including Alexopoulos (1962), Klein and Cronquist (1967), Margulis (Sagan, 1967), and Whittaker (1969), have stressed to varying degrees that either the Oomycetes, or both the Oomycetes and the Hyphochytridiomycetes, are taxonomically distinct from the Eumycota. This raises the problem of finding separate phylogenetic origins for these two taxa. The situation is not clarified by the differing opinions among authorities as to phylogenetic relationships within each of these two groups. Relationships between Oomycetes and Chrysophyceae or Xanthophyceae (Bessey, 1950; Klein and Cronquist, 1967; Erwin, 1973; and perhaps Cavalier-Smith, 1975) are possibly consistent with some of the morphological data, but unequivocal data of any kind are scarce. Bu'Lock and Osagie (1976) have reemphasized that the Oomycetes should be moved away from the true fungi and toward the algae. In addition to the morphological, cytological, and life history bases, they pointed out that there are strong chemotaxonomic grounds for such a proposal, and the retention of the Oomycetes with the true fungi results "more from tradition and convenience than conviction." Phylogeneticists have often sought to derive fungi (particularly the Ascomycetes) from Rhodophyceae, or vice versa (Section 12.3.3), whereas Klein and Cronquist (1967) and Whittaker (1969) have suggested that the Chytridiomycetes may be derived from a common ancestor with Euglenophyceae and "Pyrrophyta" (Section 12.3.4). Most biochemical data that have been collected from unicellular and filamentous algae have described either carbohydrate biochemistry or chloroplast-associated characters. As a result, the data currently available do not allow a clear choice between homologous and convergent evolution as an explanation for the morphological resemblances among Oomycetes and some nongreen algae, or between certain Eumycota and Rhodophyceae (to be sure, these morphological resemblances are often overstated). The available biochemical data point out fundamental differences among the Eumycota, Oomycetes, and Rhodophyceae (especially red algal chloroplast-related characters). Characterization of lysine biosynthetic pathways, shikimate and tryptophan biosynthetic enzymes, and nucleotide sequences may be more useful in examining possible phylogenetic relationships. However, we suspect that the phylogenetic relationship, if any, between Oomycetes and chlorophylls *c*-containing organisms dates to a very early evolutionary stage.

12.3.10 The Photosynthetic Prokaryotes

This group includes the Thiorhodaceae, Athiorhodaceae, Chlorobacteriaceae, and Cyanophyceae. The common thread here is, of course, the prokaryotic organization. Obviously the Cyanophyceae, with eukaryote-like aerobic (H_2O photolysis) photosynthesis, are the most advanced. The Chlorobacteriaceae are separated as a terminal side line, mainly because of the nature of the chlorophyll and carotenoid biosynthetic patterns. The relationship between the Thiorhodaceae and Athiorhodaceae is complicated by the various views pertaining to the evolution of bioenergetic systems, photosynthetic electron donors, and energy and carbon sources. The placement of the Thiorhodaceae close to the Chlorobacteriaceae with a common ancestor is based admittedly on the common nature of the electron donors and the ability to synthesize aromatic carotenoids. The Athiorhodaceae are proposed as a group divergent and distinct from these two families. The one intriguing organism is the presumptive prokaryote discovered and described by Lewin (1975) and examined and analyzed by Lewin and Withers (1975). Although this undoubtedly represents a significant discovery of phylogenetic importance, we would add a few words of caution.

The absence of biliproteins may not be the manifestation of a basic genetic difference, but rather a result of high incident light intensity. The biosynthesis of biliproteins is strongly influenced by environmental factors.

The evidence presented for chlorophyll *b* is convincing, but the possibility of a chlorophyll *a* oxidation product should not be excluded. In view of the importance of the presence of this presumptive chlorophyll *b*, absolute identification by means other than chromatography and electronic absorption spectra is needed. Lewin (1976) has recently suggested the establishment of a division "Prochlorophyta" to accommodate this *Synechocystis*-like organism. We feel that such a move is premature and unwarranted on the available evidence. The formal descriptions of the division, class

(Prochlorophyceae), order (Prochlorales), family (Prochloraceae), and genus and species (*Prochloron didemni*) have recently been published in *Phycologia* **16,** 217 (1977). The reader is referred to Section 12.3.10 for a further discussion of this paradoxical organism.

12.3.11 Other Protists

Although there are many other protistan divisions and classes (see Appendix), the paucity of information precludes anything but highly speculative proposals. Many or all of these would be severely prejudiced by nonbiochemical or morphological data. In view of this, we make no suggestions concerning a biochemical phylogeny of these protists (De Ley, 1968, 1974).

13

Conclusions and Speculations

13.1 REFLECTIONS ON THE BIOCHEMICAL PHYLOGENY

At this point it is fitting to examine the biochemical phylogeny (Figure 23) as from a distance, momentarily ignoring its less consequential details.

First, it is seen that a reasonable phylogeny can be constructed using only biochemical characters and a biosynthetically based approach. This phylogeny bears some resemblance to some based largely on morphological characters. Not all difficulties, by any means, have been resolved using biochemical data. Virtually all unresolved questions arise from lack of, or inadequacy of data rather than from conflicts among the data in hand. The biosynthetic approach has allowed the identification of characters, such as the distribution of chlorophyll *b*, that could reasonably have arisen by convergent evolution.

Second, the use of biochemical data has not yet definitively resolved the question of the origin of chloroplasts and mitochondria. However, the distinctive biochemistry of the Euglenophyceae *vis-à-vis* the Chlorophyceae, coupled with the fundamental biochemical similarities of chloroplasts in these two groups, may be best rationalized by an endosymbiotic origin of at least the euglenoid chloroplast. If all chloroplasts are considered to have arisen by endosymbiotic events, between two (Euglenophyceae, other photosynthetic eukaryotes) and six (Euglenophyceae; Rhodophyceae; Bacillariophyceae, Chrysophyceae, Haptophyceae, Xanthophyceae, Phaeophyceae, Eustigmatophyceae, and perhaps Chloromonadophyceae;

Dinophyceae; Cryptophyceae; Chlorophyceae, Charophyceae, Prasinophyceae, and higher plants) separate endosymbioses may have occurred. This number is still considerably lower than that hypothesized by Margulis (Sagan, 1967), and preserves Occam's Razor slightly less dulled.

Third, it is by now obvious that very few data bear directly on the possible relationships between eukaryotic algae and protozoa, especially between unicellular algae and zooflagellates. The obvious morphological resemblances between certain algae and certain protozoa could have arisen either by common ancestry or by convergent evolution; if by common ancestry, the ancestor could have been either photosynthetic or nonphotosynthetic. Although we personally doubt that protozoa, especially ciliates and most amoebae, have arisen from photosynthetic ancestors, the data germane to this question are far from unequivocal. It is hoped that this discussion will have pointed out biochemical characters more likely to be of interest in this regard.

Finally, the biochemical phylogeny pictured in Figure 23 is largely consistent with Loeblich's interpretation of the protistan fossil record (Figure 24) (Loeblich, 1974). Particularly interesting is the fact that the earliest known euglenoids are from the Eocene, long after the appearance of several chlorophycean and prasinophycean organisms. Admittedly, this indicates only the minimum age for the Euglenophyceae, but one might expect earlier fossils (they may yet, of course, be merely undiscovered) had euglenoids arisen as a distinct *de novo* group rather than through symbiosis of a green chloroplast with a protozoan (see Section 13.2.1). The Rhodophyceae, Dinophyceae, Phaeophyceae, Ciliatea, and radiolaria (silicified amoebae) have long and distinguishable fossil records. Early Haptophyceae are, perhaps significantly, confined to the family Coccolithaceae of the Prymnesiales, represented today by unicellular marine flagellates.

13.2 TIME COURSE OF EVOLUTION

Very few phylogenies ever incorporate a time scale. If a phylogeny itself is speculative, the time scale is even more so. Organized living matter has apparently been in existence for on the order of 3.5×10^9 years. Within this time span the whole range of biological evolution has occurred. We do not need to consider, or speculate on, the point in time at which every "event" took place. Many events probably did not take place at a given definable time. This would be particularly true for changes that were gradual and involved organisms with no fossil record, and having no extant counterpart.

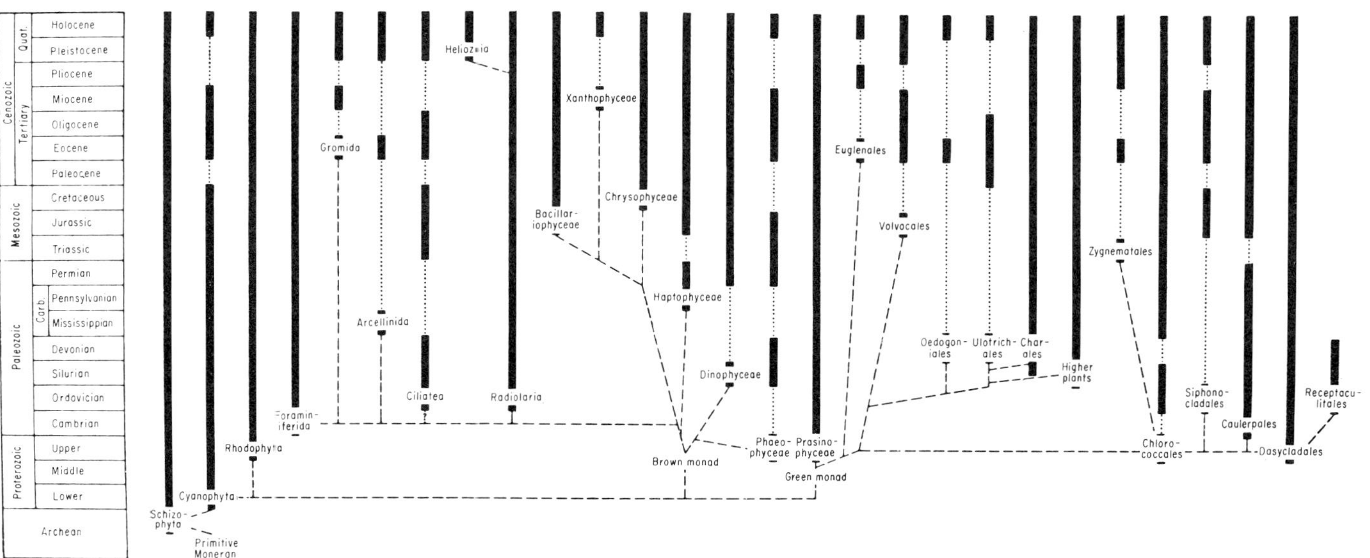

Figure 24. Fossil record of protists. Reproduced from Figure 2 from Loeblich (1974) by permission of the International Bureau for Plant Taxonomy and Nomenclature and the author.

There are, however, a number of key events, essential to protistan phylogeny, that should be considered.

1. Acquisition of the capacity to bring about the photolysis of water. This relates to the origin of blue-green algae and probably the beginning of aerobic biology.
2. The prokaryote–eukaryote divergence.
3. The divergence of the algae into two lines.
4. The appearance of the Eumycota.
5. The appearance of the protozoa.
6. The transition to vascular plants, or the appearance of a "land flora."

There are currently two approaches to the solutions: the fossil record, and the biochemical "record" from such data as the presumed rates of amino acid substitutions in proteins.

13.2.1 The Fossil Record

This is the objective record, in that the results are open to objective analysis and review. Unfortunately, however, it is perhaps prejudiced in that it naturally favors those organisms with mineralization (calcification or silicification) or some form of impervious or resistant covering (e.g., sporopollenin).

Schopf (1974, 1975a,b) has considered the evidence relating to the appearance of blue-green algae and aerobiosis and appearance of the autotrophic mode. The evidence suggests that a capacity for autotrophic CO_2 fixation had been established at least 3.2×10^9 years ago. The oldest bona fide fossils are approximately 3.1×10^9 years old. These appear to be both "alga-" and "bacteria-like." In view of this, Schopf (1974) suggested that photoautotrophic organization appeared, probably, about 3.3×10^9 years ago. This would represent the time of appearance of the chlorophylls or magnesium-porphyrins. This does not imply an oxygenic photoautotrophy. Schopf (1975b) has pointed out that there was a major change in the atmospheric oxygen concentrations about 2×10^9 years ago. This would suggest that oxygen-producing photosynthesis (capacity for biological photolysis of water) appeared about $2.3–2.0 \times 10^9$ years ago (Schopf, 1975b). The coccoid blue-green algae, presumably the most primitive oxygenic photosynthesizers, were established and diversified about 2.5×10^9 years ago. Such early evolvers of oxygen presumably needed a tolerance to anaerobic atmospheres and reducing conditions. The recent discovery of blue-green algae capable of a bacterial-type photosynthesis (Cohen *et al.*, 1975) lends credence to this view. The oldest identifiable eukaryotes appear to date back to about 1.3×10^9 years ago (Schopf, 1970). These presumably were algae, but a fungal origin cannot be totally eliminated. Although this does not represent the limit for eukaryotes, the timing ties in well with the biochemical evidence for the prokaryote–eukaryote divergence (Delauney and Schapira, 1974a, Kimura and Ohta, 1973b). The available biochemical evidence suggests that these primitive eukaryotes were probably red algae (Rhodophyceae). The earliest recorded times for the appearance of the other principal algal classes are illustrated in Figure 24. As we have pointed out, this is probably prejudiced toward the "mineralized" representatives (e.g., Bacillariophyceae, Haptophyceae, Dasycladales, Siphonales). However, this does afford some interesting insights.

The Bacillariophyceae are an apparently recent evolutionary offshoot, having been preceded by the Haptophyceae, Phaeophyceae (and probably Chrysophyceae). The Euglenophyceae presumably have a very recent history. Although we have previously commented on this, apropos of a symbiotic origin, the true span for this class probably extends further back, but almost certainly not as far back as the Prasinophyceae. Loeblich (1974) has suggested that the basic dichotomy of the algae into chromophytan and chlorophytan lines occurred very soon after the emergence of the Rhodophyceae. Schopf (1975a,b) has indicated that mitotic, nucleated cells (i.e., primitive eukaryotes) appeared about 1.5×10^9 years ago, and meiotic sexuality evolved about 1×10^9 years back. If this is correct, as is most likely, we can postulate that the fungi (with their heterotrophic nutrition) probably arose a little less than a billion years ago. This would coordinate well with Schopf's (1975a,b) time scale for the appearance of non-photosynthetic protists, about 800×10^6 years ago. The fossil record, however, is weakest when it comes to dating the fungi. They do not lend themselves to fossilization.

13.2.2 The Biochemical Record

This approach has emerged in the last few years as a powerful ally to the fossil record. In its essence, this approach consists of examining amino acid or nucleotide sequences from extant organisms, and then "back calculating" the age of divergence based upon assumptions about the rate of mutations or the rate of substitution in the sequence. A partial discussion (Section 2.4) has already been presented, and for a more complete discussion the reader is referred to Fitch and Margoliash (1970) and Dayhoff *et al.* (1974).

One feature of this approach is a factor referred to as the "unit evolutionary period" (Nolan and Margoliash, 1968; Dickerson, 1971), which may be defined as the time needed for a 1% change in amino acid sequence to

TABLE 41

Rates of Mutations and Unit Evolutionary Period for Certain Selected Macromolecules[a]

Macromolecule	Mutations per 100 million years	Unit evolutionary period
Cytochrome *c*	3	20×10^6
Hemoglobins	12	5.8×10^6
Fibrinopeptides	90	1.1×10^6

[a] From: Dayhoff (1972b); Dickerson (1971).

show up between divergent lines. The rate of evolution must be assumed to be constant. It is also to be expected that the more rigid the specifications for a given molecule, the slower the accumulation of acceptable substitutions, and hence a longer UEP (Dickerson, 1971). Some values are given in Table 41.

There have been suggestions, however, that changes may be species/time dependent. Jukes and Holmquist (1972) indicated that the divergence of cytochrome *c* of rattlesnakes from that of species of birds is faster than the divergence of cytochrome *c* of rattlesnakes from that of snapping turtles. Penny (1974; see also Jukes and Holmquist, 1972), however, has argued that the rate of change is not species dependent.

Goodman *et al.* (1971) have suggested, using evidence from primates, that steadily increasing generation times may have been an important parameter in slowing molecular evolution. Admittedly, this approach has been developed primarily with "animal" molecular evolution, and the above caveats have been established on the basis of animal systems. However, the system would appear ideally suited to protists, especially such problem groups as the fungi (Sussman, 1974). Two groups (Kimura and Ohta 1973a,b; Delauney and Shapira, 1974a), independently using a biochemical record, have estimated the divergence of the eukaryotes and prokaryotes at 1.9×10^9 years ago. This may be an overestimation, but it can be considered in good agreement with the fossil record. The fossil and biochemical records may well prove to be an ideal check and balance system.

13.3 IMPORTANT UNRESOLVED QUESTIONS

It is a truism that another book could easily be filled with "important unresolved questions." This listing is designed to point out only a few areas of particular interest in the field of biochemical phylogenetics.

1. Is it possible to construct a phylogenetic tree consistent with absolutely all biochemical and morphological data?

2. How many organisms must be examined before any confidence can be placed in the currently postulated phylogeny of organisms?

3. Are all organisms monophyletic, or is "biochemical predestination" a very powerful factor? Is it possible to resolve this question by biochemical methods?

4. What is the phylogeny of the viruses?

5. Is the dichotomy among the bacteria (based on the gram reaction) as fundamental as it appears to be?

6. It is well known that *ecto*symbiotic bacteria play an important role in the physiology of many eukaryotes. How important, in evolutionary terms, was ectosymbiosis?

7. What biochemical characters of the ancestral eukaryote enabled it to become endocytotic? What advantages were found in the endosymbiosis of the protomitochondrion, or alternatively, why did the mitochondrion eventually differentiate from the rest of the cytoplasm?

8. What are the interrelationships among nuclear and cytoplasmic genomes of the eukaryotes? Is it possible to approach this problem by organelle culture techniques (Giles and Sarafis, 1971, 1972; Leech, 1972; Ridley and Leech, 1970)?

9. What is the significance of the reduced degree of autonomy of the modern mitochondrion and its dissimilarities in many respects to modern prokaryotes?

10. Was the 9 + 2 body ever free-living (Sagan, 1967)?

11. Is it possible to distinguish secondarily apochloroplastic protists from primarily (ancestrally) nonphotosynthetic ones by biochemical techniques? How many of the protozoa and fungi, if any, are secondarily nonphotosynthetic?

12. Have the Rhodophyceae evolved from a flagellated ancestor?

13. Is there a reason, other than "convergent evolution," for the morphological similarities seen among *Vaucheria* spp., some Chytridiomycetes, some Rhodophyceae, and some Ascomycetes (Section 12.3.3; also Pickett-Heaps, 1975, p. 569)?

14. Why did the imperfect fungi lose their sexuality? Was this loss selectively advantageous, or have they survived in spite of this loss?

15. What is the relationship among the Cryptophyceae, Dinophyceae, and other chlorophyll *c*-containing organisms?

16. What phylogenetic relationships exist among present-day protozoa, present-day algae, and present-day fungi?

17. Are Euglenophyceae and the fucoxanthin-containing Dinophyceae recently developed symbionts?

18. How did the eukaryotic cell arise? Particular reference here is made to the origin of chloroplasts and mitochondria.

19. Is there a fundamental relationship between the biochemical nature of a particular phylogenetic character and the "weight" best ascribed to it in construction of phylogenies?

20. Is the present representation of phylogenetic history—as a phylogenetic tree—somehow obstructing our approach to actual historical phylogenetic events? Is some other arrangement (perhaps an *n*-dimensional "tree") a better arrangement? Would there be an optimal number of dimensions to this representation?

21. Are other evolutionary phenomena, such as the evolution of morphological form, the evolution of consciousness, and the evolution of social interaction, ultimately reducible to biochemical evolution? Are all these phenomena, including biochemical evolution, merely manifestations of physical or structural properties of matter and energy?

We make no claim that these are the significant "twenty questions" (plus one). In phylogenetics the questions that need answering are probably infinite. It is perhaps impossible to answer the questions, in contrast to providing an explanation, since we are currently trying to solve a three billion-year-old puzzle with ignorance. It is hoped that this little book will prod researchers to pursue phylogenetics with modern approaches and insights. It is, after all, the answer to a very basic question: Where did *we* come from?

Appendix

The following systematic classifications have been followed in the text. They were chosen as convenient reference points, and their selection does not imply that we consider them superior to other classifications.

Algae [modified from Chapman and Chapman (1973)]

Division: -phyta
Class: -phyceae
Order: -ales

Rhodophyta
- Rhodophyceae
  - Porphyridiales, Bangiales, Rhodochaetales, Compsopogonales, Nemalionales, Bonnemaisoniales, Gelidiales, Gigartinales, Cryptonemiales, Rhodymeniales, Ceramiales

Cryptophyta
- Cryptophyceae
  - Cryptococcales, Cryptomonadales

Pyrrophyta
- Dinophyceae
  - Prorocentrales, Gymnodiniales, Peridiniales, Dinophysiales, Dinocapsales, Dinamoebidiales, Dinotrichales, Phytodiniales

Xanthophyta
- Eustigmatophyceae
  - (no orders yet delineated)
- Xanthophyceae
  - Chloramoebales, Rhizochloridales, Heterogloeales, Mischococcales, Tribonematales, Vaucheriales

Chloromonadophyta
  Chloromonadophyceae
    Chloromonadales
Chrysophyta
  Haptophyceae
    Isochrysidales, Prymnesiales
  Chrysophyceae
    Chromulinales, Ochromonadales, Dictyochales, Rhizochrysidales, Chrysosphaerales, Chrysocapsales, Phaeothamniales, Thallochrysidales
Bacillariophyta
  Bacillariophyceae
    Pennales, Centrales
Phaeophyta
  Phaeophyceae
    Ectocarpales, Sphacelariales, Tilopteridales, Dictyotales, Chordariales, Sporochnales, Desmarestiales, Dictyosiphonales, Laminariales, Fucales, Durvilleales, Ascoseirales
Euglenophyta
  Euglenophyceae
    Rhabdomonadales, Sphenomonadales, Heteronematales, Euglenomorphales, Euglenales, Eutreptiales
Charophyta
  Charophyceae
    Charales
Chlorophyta
  Prasinophyceae
    Pyramimonadales, Halosphaerales
  Chlorophyceae
    Chlorococcales, Volvocales, Culotrichales, Chaetophorales, Oedogoniales, Siphonocladales, Dasycladales, Dichotomosiphonales, Siphonales
  Conjugatophyceae
    Desmidiales, Zygnematales

Fungi and Water Molds

Division: -mycota
Class: -mycetes
Subclass: -mycetes
Order: -ales

Myxomycota
  Myxomycetes
    Ceratiomyxales, Liceales, Trichiales, Echinosteliales, Stemonitales, Physarales
  Plasmodiophoromycetes
    Acrasiales, Labyrinthulales
Eumycota (Mastigomycotina)
  Chytridiomycetes
    Chytridiales, Blastocladiales, Monoblepharidales

Eumycota (Zygomycotina)
  Zygomycetes
    Mucorales, Endogonales, Entomophthorales
  Ascomycetes
   Plectomycetes
    Eurotiales, Gymnascales, Microascales, Meliolales, Coronophorales, Caliciales, Laboulbeniales
   Discomycetes
    Pezizales, Helotiales, Phacidiales, Tuberales
   Pyrenomycetes
    Erysiphales, Hypocreales, Sphaeriales
   Loculoascomycetes
    Myrangiales, Dothideales, Pleosporales, Microthyriales
   Hemiascomycetes*
    Endomycetales, Taphrinales
  Basidiomycetes
   Hemibasidiomycetes†
    Tremellales, Uredinales, Ustilaginales
   Hymenomycetes
    Agaricales, Aphyllophorales
   Gasteromycetes
    Mymenogastrales, Lycoperdales, Phallales, Sclerodermatales, Nidulariales, Melanogastrales
Oomycota
  Hyphochytridiomycetes
    Hyphochytriales
  Oomycetes
    Saprolegniales, Peronosporales

Protozoa

After Honigberg *et al.* (1964). (The photosynthetic protozoa, class Phytomastigophorea, have been retained within the algae, and subclass Mycetozoida (Rhizopodea) has been retained within the fungi.) No attempt has been made to standardize the choice of taxa (e.g., superclass; subphylum) with the taxa of the algal and fungal classifications.

  Class: -ea
  Order: -ida

Subphylum: Sarcomastigophora
Superclass: Mastigophora
  Class: Zoomastigophorea
    Choanoflagellida, Bicosoesida, Rhizomastigida, Kinetoplastida, Retortamonadida, Diplomonadida, Oxymonadida, Trichomonadida, Hypermastigida

* Hemiascomycetes are sometimes referred to as the Hemiascomycetidae, while the other four subclasses of the Ascomycetes are grouped together as the Euascomycetidae.

† Sometimes referred to as Heterobasidiomycetidae. The other two subclasses of the Basidiomycetes are sometimes assembled into the Homobasidiomycetidae.

Superclass: Opalinata
  Class: Opalinidea
Superclass: Sarcodina
  Class: Rhizopodea
    Amoebida, Arcellinida, Aconchulinida, Gromiida, Athalamida, Foraminiferida, Xenophyphorida
  Class: Piroplasmea
    Piroplasmida
  Class: Actinopodea
    Porulosida, Occulosida, Acanthometrida, Acanthopractida, Actinophryda, Centrohelida, Desmothoracida, Proteomyxida
Subphylum: Sporozoa
  Class: Telosporea
    Archigregarinida, Eugregarinida, Neogregarinida, Protococcida, Eucoccida
  Class: Toxoplasmea
    Toxoplasmida
Subphylum: Cnidospora
  Class: Myxosporidea
    Myxosporida, Actinomyxida, Helicosporida
  Class: Microsporidea
    Microsporida
Subphylum: Ciliophora
  Class; Ciliatea
    Gymnostomatida, Trichostomatida, Chonotrichida, Apostomatida, Astomatida, Hymenostomatida, Thigmotrichida, Peritrichida, Suctorida, Heterotrichida, Oligotrichida, Tintinnida, Entodiniomorphida, Odontostomatida, Hypotrichida

Bibliography

Aaronson, S. (1970). *Ann. N.Y. Acad. Sci.* **175,** 531–540.

Aaronson, S., and Hutner, S. H. (1966). *Q. Rev. Biol.* **41,** 13–46.

Abdel-Fattah, A. F., Hussein, M. M.-D., and Salem, H. M. (1974). *Carbohydr. Res.* **33,** 209–215.

Ackman, R. G., and Sipos, J. C. (1965). *Comp. Biochem. Physiol.* **15,** 445–456.

Ackman, R. G., Tocher, C. S., and McLachlan, J. (1968). *J. Fish. Res. Board Can.* **25,** 1603–1620.

Ackman, R. G., Addison, R. F., Hooper, S. N., and Prakash, A. (1970). *J. Fish. Res. Board Can.* **27,** 251–255.

Adanson, M. (1763). "Familles des Plantes." Vincent, Paris.

Ahmed, S. I. (1973). *Arch. Mikrobiol.* **92,** 143–152.

Ahmed, S. I., and Campbell, J. J. R. (1973). *J. Bacteriol.* **115,** 205–212.

Ahmed, S. I., and Giles, N. H. (1969). *J. Bacteriol.* **99,** 231–237.

Aitken, A. (1975). *Biochem. J.* **149,** 675–683.

Aitken, A. (1976). *Nature (London)* **263,** 793–798.

Alais, J., Lablache-Combier, A., Lacoste, L., and Vandewalle, B. (1974). *Phytochemistry* **13,** 2833–2837.

Alam, M., Chakravarti, A., and Ikawa, M. (1971). *J. Phycol.* **7,** 267–268.

Alcaide, A., Barbier, M., Potier, P., Magueur, A. M., and Teste, J. (1969). *Phytochemistry* **8,** 2301–2303.

Alexopoulos, C. J. (1962). "Introductory Mycology." Wiley, New York.

Aliev, K. A., and Filippovich, I. I. (1968). *Mol. Biol. (USSR)* **2,** 297–304.

Allen, C. F., Good, P., and Holton, R. W. (1970). *Plant Physiol.* **46,** 748–751.

Allen, D. M., and Northcote, D. H. (1975). *Protoplasma* **83,** 389–412.

Allen, H. J., Ault, C., Winzler, R. J., and Danielli, J. F. (1974). *J. Cell Biol.* **60,** 26–38.

Allen, M. B. (1959). *Arch. Mikrobiol.* **32,** 270–277.

Allsopp, A. (1969). *New Phytol.* **68,** 591–612.

Altman, L. J., Kowerski, R. C., and Rilling, H. C. (1971). *J. Am. Chem. Soc.* **93,** 1782–1784.

Altman, L. J., Ash, L., Kowerski, R. C., Epstein, W. W., Larsen, B. R., Rilling, H. C., Muscio, F., and Gregoris, D. E. (1972). *J. Am. Chem. Soc.* **94,** 3257–3259.

Altmann, R. (1890). "Die Elementarorganismen und ihre Beziehung zu den Zellen." Veit, Leipzig.

Amarasingham, C. R., and Davis, B. D. (1965). *J. Biol. Chem.* **240,** 3664–3668.

Ambler, R. P., and Bartsch, R. G. (1975). *Nature* (*London*) **253,** 285–288.

Ambler, R. P., Meyer, T. E. and Kamen, M. D. (1976). *Proc. Natl. Acad. Sci. U.S.A.* **73,** 472–475.

Amico, V., Oriente, G., Piatelli, M., Tringali, C., Fatturosso, E., Magno, S., Mayol, L., Santacroce, C., and Sica, D. (1976). *Biochem. Syst. Ecol.* **4,** 143–146.

Amy, N. K., and Garrett, R. H. (1974). *Plant Physiol.* **54,** 629–637.

Anderson, L. E. (1971). *Biochim. Biophys. Acta* **235,** 237–244.

Anderson, L. E. (1972). *Photosynth. Two Centuries Its Discovery Joseph Priestley, Proc. Int. Congr. Photosynth. Res., 2nd, 1971* Vol. 3, pp. 1855–1860.

Anderson, R. F. (1963). *In* "Biochemistry of Industrial Microorganisms" (C. Rainbow and A. H. Rose, eds.), pp. 300–319. Academic Press, New York.

Anding, C., Brandt, R. D., Ourisson, G., Pryce, R. J., and Rohmer, M. (1972). *Proc. R. Soc. London, Ser. B* **180,** 115–124.

Ando, T. and Barbier, M. (1975). *Biochem. Syst. Ecol.* **3,** 245.

Andrewes, A. G., Hertzberg, S., Liaaen-Jensen, S., and Starr, M. P. (1973). *Acta Chem. Scand.* **27,** 2383–2395.

Andrews, P. W., Rogers, L. J., Boulter, D., and Haslett, B. G. (1976). *Eur. J. Biochem.* **69,** 243–248.

Anthony, C. (1975). *Sci. Progr.* (*Oxford*) **62,** 167–206.

Antia, N. J. (1967). *J. Phycol.* **3,** 81–85.

Antia, N. J., and Kripps, R. S. (1973). *Arch. Mikrobiol.* **94,** 29–46.

Antia, N. J., Lee, R. F., Nevenzel, J. C., and Cheng, J. Y. (1974). *J. Protozool.* **21,** 768–771.

Antia, N. J., Berland, B. R., Bonin, D. J., and Maestrini, S. Y. (1975). *J. Mar. Biol. Assoc. U.K.* **55,** 519–539.

Apel, K., and Schweiger, H.-G. (1973). *Eur. J. Biochem.* **38,** 373–383.

Arpin, N. (1968). Thesis. No. 527. University of Lyons.

Asada, K., Yoshikawa, K., Takahashi, M. A., Maeda, Y., and Enmanji, K. (1975). *J. Biol. Chem.* **250,** 2801–2807.

Ashmarin, I. P., Loitsyanskaya, M. S., and Kharchenko, E. P. (1970). *Microbiology* (*USSR*) **38,** 659–664.

Aspinall, G. O. (1967). *Pure Appl. Chem.* **14,** 43–55.

Atkinson, A. W., Jr., Gunning, B. E. S., and John, P. C. L. (1972). *Planta* **107,** 1–32.

Augier, J. (1965). *Mem. Soc. Bot. Fr.* **45,** 8–14.

Auriol, P. (1974). *C. R. Hebd. Seances Acad. Sci., Ser. D* **279,** 1867–1869.

Avivi, L., Iaron, O., and Halevy, S. (1967). *Comp. Biochem. Physiol.* **21,** 321–326.
Bachmayer, H. (1970). *Biochim. Biophys. Acta* **209,** 584–586.
Balbinder, E. (1964). *Biochem. Biophys. Res. Commun.* **17,** 770–774.
Baldwin, M. W., and Braven, J. (1968). *J. Mar. Biol. Assoc. U.K.* **48,** 603–608.
Ballio, A., and Barcellona, S. (1968). *Ann. Inst. Pasteur, Paris* **114,** 121–137.
Bandoni, R. J., Moore, K., Subba Rao, P. V., and Towers, G. H. N. (1968). *Phytochemistry* **7,** 205–207.
Barashkov, G. K. (1956). *Dokl. Akad. Nauk SSSR* **111,** 148–150.
Bartnicki-Garcia, S. (1968). *Annu. Rev. Microbiol.* **22,** 87–108.
Bartnicki-Garcia, S. (1970). *In* "Phytochemical Phylogeny" (J. B. Harborne, ed.), pp. 81–103. Academic Press, New York.
Basden, E. H., II, Tourtellotte, M. E., Plastridge, W. N., and Tucker, J. S. (1968). *J. Bacteriol.* **95,** 439–443.
Baseman, J. B., and Cox, C. D. (1969). *J. Bacteriol.* **97,** 992–1000.
Batra, P. P., and Tollin, G. (1964). *Biochim. Biophys. Acta* **79,** 371–378.
Bauer, K. (1971). *Int. J. Protein Res.* **3,** 165–172.
Baur, E. (1909). *Z. Indukt. Abstamm. Vererbungsl.* **1,** 330–351.
Beach, D. H., Harrington, G. W., and Holz, G. G., Jr. (1970). *J. Protozool.* **17,** 501–510.
Beadle, G. W., and Tatum, E. L. (1941). *Proc. Natl. Acad. Sci. U.S.A.* **27,** 499–506.
Beale, S. I. (1976). *Phil. Trans. Roy. Soc. London* **273,** 99–108.
Beale, S. I., and Castelfranco, P. A. (1974). *Plant Physiol.* **53,** 297–303.
Beale, S. I., Gough, S. P., and Granick, S. (1975). *Proc. Natl. Acad. Sci. U.S.A.* **72,** 2719–2723.
Bean, G. A. (1973). *Adv. Lipid Res.* **11,** 193–218.
Bean, G. A., Patterson, G. W., and Motta, J. J. (1972). *Comp. Biochem. Physiol. B* **43,** 935–939.
Bean, R. C., and Hassid, W. Z. (1955). *J. Biol. Chem.* **212,** 411–425.
Bean, R. C., and Hassid, W. Z. (1956). *J. Biol. Chem.* **218,** 425–436.
Beastall, G. H., Rees, H. H., and Goodwin, T. W. (1972). *Biochem. J.* **128,** 179–181.
Beastall, G. H., Tyndall, A. M., Rees, H. H., and Goodwin, T. W. (1974). *Eur. J. Biochem.* **41,** 301–309.
Beauchemin, N., Larue, B., and Cedergren, R. J. (1973). *Arch. Biochem. Biophys.* **156,** 17–25.
Beevers, H. (1961). "Respiratory Metabolism of Plants." Harper, New York.
Beining, P. R., Huff, E., Prescott, B., and Theodore, T. S. (1975). *J. Bacteriol.* **121,** 137–143.
Belozersky, A. N. (1963). *Proc. Int. Congr. Biochem., 5th, 1961* Vol. 3, pp. 198–214.
Belozersky, A. N., and Spirin, A. S. (1958). *Nature (London)* **182,** 111–112.
Bender, V. J. (1975). *Aust. J. Biol. Sci.* **28,** 227–231.
Bendich, A. J., and McCarthy, B. J. (1970). *Proc. Natl. Acad. Sci. U.S.A.* **65,** 349–356.
Bennett, A., and Bogorad, L. (1973). *J. Cell Biol.* **58,** 419–435.

Bennett, A., and Siegelman, H. W. (1977 in preparation). *In* "The Porphyrins" (D. Dolphin, ed.), Vol. 6, Biochemistry, Part A, Ch. 7. Academic Press, New York.

Bennett, T. P., and Frieden, E. (1962). *Comp. Biochem.* **4,** 483–556.

Benson, A. A., and Maruo, B. (1958). *Biochim. Biophys. Acta* **27,** 189–195.

Benson, A. A., and Strickland, E. H. (1960). *Biochim. Biophys. Acta* **41,** 328–333.

Benson, A. A., Wintermans, J. F. G. M., and Wiser, R. (1959). *Plant Physiol.* **34,** 315–317.

Berlyn, M. B., and Giles, N. H. (1969). *J. Bacteriol.* **99,** 222–230.

Berlyn, M. B., and Giles, N. H. (1973). *J. Gen. Microbiol.* **74,** 337–341.

Berlyn, M. B., Ahmed, S. I., and Giles, N. H. (1970). *J. Bacteriol.* **104,** 768–774.

Berns, D. S. (1967). *Plant Physiol.* **42,** 1569–1586.

Berns, D. S., Holohan, P., and Scott, E. (1966). *Science* **152,** 1077–1078.

Bessey, E. A. (1950). "Morphology and Taxonomy of Fungi." McGraw-Hill, New York.

Bezanson, G. S., Desaty, D., Emes, A. V., and Vining, L. C. (1970). *Can. J. Microbiol.* **16,** 147–151.

Birch, A. J. (1973a). *Pure Appl. Chem.* **33,** 17–38.

Birch, A. J. (1973b). *Proc. Nobel Symp.* **25,** 261–270.

Birdsey, E. C., and Lynch, V. H. (1962). *Science* **137,** 763–764.

Birkinshaw, J. H. (1965). *In* "The Fungi: An Advanced Treatise" (G. C. Ainsworth and A. S. Sussman, eds.), Vol. 1, pp. 179–228. Academic Press, New York.

Bisalputra, T. (1974). *In* "Algal Physiology and Biochemistry" (W. D. P. Stewart, ed.), pp. 124–160. Blackwell, Oxford.

Bishop, D. H. L., Pandya, K. P., and King, H. K. (1962). *Biochem. J.* **83,** 606–614.

Bisset, K. A. (1962). *Symp. Soc. Gen. Microbiol.* **12,** 361–373.

Biswas, S. B., and Myers, J. (1960). *Nature* (*London*) **186,** 238–239.

Biswas, S. B., and Sarkar, A. K. (1970). *Phytochemistry* **9,** 2425–2430.

Bjørnland, T., and Aguilar-Martinez, M. (1976). *Phytochemistry* **15,** 291–296.

Blair, G. E., and Ellis, R. J. (1973). *Biochim. Biophys. Acta* **319,** 223–234.

Blake, C. C. F. (1974). *Nature* (*London*) **250,** 284–285.

Blumenthal, H. J. (1965). *In* "The Fungi: An Advanced Treatise" (G. C. Ainsworth and A. S. Sussman, eds.), Vol. 1, pp. 229–268. Academic Press, New York.

Blumer, M., Guillard, R. R. L., and Chase, T. (1971). *Mar. Biol.* **8,** 183–189.

Boardman, N. K., Linnane, A. W., and Smillie, R. M., eds. (1970). "Autonomy and Biogenesis of Mitochondria and Chloroplasts." North-Holland Publ., Amsterdam.

Bock, W. J. (1969). *Ann. N.Y. Acad. Sci.* **167,** 71–73.

Bock, W. J. (1974). *Syst. Zool.* **22,** 375–392.

Bogorad, L. (1975). *Science* **188,** 891–898.

Bohlmann, F., Burkhardt, T., and Zdero, C. (1973). "Naturally Occurring Acetylenes." Academic Press, New York.

Bonen, L., and Doolittle, W. F. (1975). *Proc. Natl. Acad. Sci. U.S.A.* **72,** 2310–2314.

Bonhomme-Florentin, A. (1971). *Protistologica* **7,** 411–419.

Bonner, J., and Ts'o, P., eds. (1964). "The Nucleohistones." Holden-Day, San Francisco, California.

Bonotto, S., Lurquin, P., Baeyens, W., Charles, P., Hoursiangou-Neubrun, D., Mazza, A., Tramontano, G., and Felluga, B. (1975). *Protoplasma* **83,** 172–173.

Borst, P., and Grivell, L. A. (1971). *FEBS Lett.* **13,** 73–78.

Bouck, G. B. (1972). *Adv. Cell Mol. Biol.* **2,** 237–272.

Boulter, D. (1972). *Progr. Phytochem.* **3,** 199–229.

Boulter, D., Ellis, R. J., and Yarwood, A. (1972). *Biol. Rev. Cambridge Philos. Soc.* **47,** 113–175.

Bovarnick, J. G., Schiff, J. A., Freedman, Z., and Egan, J. M., Jr. (1974). *J. Gen. Microbiol.* **83,** 63–71.

Bradbury, E. M., Inglis, R. J., Matthews, H. R., and Sarner, N. (1973). *Eur. J. Biochem.* **33,** 131–139.

Bradley, D. M., Goldin, H. H., and Claybrook, J. R. (1974). *FEBS Lett.* **41,** 219–222.

Bradley, S. G., and Bond, J. S. (1974). *Adv. Appl. Microbiol.* **18,** 131–190.

Brändén, C.-I., Eklund, H., Nordström, B., Boiwe, T., Söderlund, G., Zeppezauer, E., Ohlsson, I., and Åkeson, Å. (1973). *Proc. Natl. Acad. Sci. U.S.A.* **70,** 2439–2442.

Bresinsky, A., and Scheider, G. (1975). *Biochem. Syst. Ecol.* **3,** 129–135.

Brierley, J. A., and Brierley, C. L. (1968). *J. Bacteriol.* **96,** 573.

Britton, G. (1976). *In* "Chemistry and Biochemistry of Plant Pigments" 2nd. Ed. (T. W. Goodwin, Ed.) Vol. 1, 262–327. Academic Press, New York.

Broad, T. E., and Dawson, R. M. C. (1975). *Biochem. J.* **146,** 317–328.

Brockmann, H., Jr. (1971). *Ann. Chem.* **754,** 139–148.

Brockmann, H., Jr., and Kleber, I. (1969). *Angew. Chemie* **81,** 626–627.

Brockmann, H., Jr., Knobloch, G., Schweer, I., and Trowitzsch, W. (1973). *Arch. Mikrobiol.* **90,** 161–164.

Broda, E. (1970). *Progr. Biophys. Mol. Biol.* **21,** 143–208.

Broda, E. (1971a). *In* "Biochemical Evolution and the Origin of Life" (E. Schoffeniels, ed.), Vol. 2, 224–235. Elsevier, Amsterdam.

Broda, E. (1971b). *In* "Chemical Evolution and the Origin of Life" (R. Buvet and C. Ponnamperuma, eds.), pp. 446–452. North-Holland Publ., Amsterdam.

Broda, E. (1975a). "The Evolution of Biosynthetic Processes." Pergamon, Oxford.

Broda, E. (1975b). *J. Mol. Evol.* **7,** 87–100.

Broda, E. (1975c). *Origins Life* **6,** 247–251.

Brooks, A. E., and Nasatir, M. (1966). *J. Phycol.* **2,** 140–144.

Brown, A. E., and Lascelles, J. (1972). *Plant Physiol.* **50,** 747–749.

Brown, A. S., Foster, J. A., Voynow, P. V., Franzblau, G., and Troxler, R. F. (1975). *Biochemistry* **14,** 3581–3588.

Brown, J. S., Alberte, R. S., and Thornber, J. P. (1974). *Proc. Int. Congr. Photosynth. 3rd, 1974,* pp. 1951–1962.

Brown, R. D., and Haselkorn, R. (1971). *J. Mol. Biol.* **59,** 491–503.

Brush, P., and Percival, E. (1972). *Phytochemistry* **11,** 1847–1849.

Bryan-Jones, D. C., and Whittenbury, R. (1969). *J. Gen. Microbiol.* **58,** 247–260.

Bryant, D. A., Glazer, A. N., and Eiserling, F. A. (1976). *Arch. Microbiol.* **110,** 61–75.

Bryce, T. A., Welti, D., Walsby, A. E., and Nichols, B. W. (1972). *Phytochemistry* **11,** 295–302.

Buchanan, B. B. (1972). *In* "The Enzymes" (P. D. Boyer, ed.), 3rd ed., Vol. 6, pp. 193–216. Academic Press, New York.

Buchanan, R. E., and Gibbons, N. E., eds. (1974). "Bergey's Manual of Determinative Bacteriology," 8th ed. Williams & Wilkins, Baltimore, Maryland.

Budzikiewicz, H., and Taraz, K. (1971). *Tetrahedron* **27,** 1447–1460.

Buehner, M., Ford, G. C., Moras, D., Olsen, K. W., and Rossmann, M. G. (1973). *Proc. Natl. Acad. Sci. U.S.A.* **70,** 3052–3054.

Buetow, D. E. (1976). *J. Protozool.* **23,** 41–47.

Buggy, M. J., Britton, G., and Goodwin, T. W. (1974). *Phytochemistry* **13,** 125–129.

Bukowiecki, A. C., and Anderson, L. E. (1974). *Plant Sci. Lett.* **3,** 381–386.

Bu'Lock, J. D. (1973). *Pure Appl. Chem.* **34,** 435–461.

Bu'Lock, J. D. and Osagie, A. U. (1976). *Phytochemistry* **15,** 1249–1251.

Burkard, G., Guillemaut, P., Steinmetz, A., and Weil, J. H. (1973). *Biochem. Soc. Symp.* **38,** 43–56.

Burlakova, Z. O., Kondrat'eva, T. M., and Khailov, K. M. (1971). *In* "Ecological Physiology of Marine Planktonic Algae (in Culture Conditions)" (K. M. Khailov, ed.), pp. 93–142. Naukova Dumka, Kiev (in Russian).

Bush, K. J., and Sweeney, B. M. (1972). *Plant Physiol.* **50,** 446–451.

Bush, P. B., and Grunwald, C. (1974). *Plant Physiol.* **53,** 131–135.

Butterworth, P. H. W., and Bloch, K. (1970). *Eur. J. Biochem.* **12,** 496–501.

Byrne, P. F. S., and Brennan, P. J. (1975). *J. Gen. Microbiol.* **89,** 245–255.

Callely, A. G., Rigopoulos, N., and Fuller, R. C. (1968). *Biochem. J.* **106,** 615–622.

Calvin, M. (1959). *Science* **130,** 1170–1174.

Calvin, M. (1962). *In* "Horizons in Biochemistry" (M. Kasha and B. Pullman, eds.), pp. 23–57. Academic Press, New York.

Cammarano, P., Romeo, A., Gentile, M., Felsani, A., and Gualerzi, C. (1972). *Biochim. Biophys. Acta* **281,** 597–624.

Cantino, E. C., and Turian, G. F. (1959). *Annu. Rev. Microbiol.* **13,** 97–124.

Carr, N. G., and Craig, I. W. (1970). *In* "Phytochemical Phylogeny" (J. B. Harborne, ed.), pp. 119–143. Academic Press, New York.

Casperson, G., and Purz, H.-J. (1974). *Z. Allg. Mikrobiol.* **14,** 645–654.

Cavalier-Smith, T. (1975). *Nature (London)* **256,** 463–468.

Cedergren, R. J., and Cordeau, J. R. (1973). *J. Theor. Biol.* **39,** 477–486.

Cedergren, R. J., Cordeau, J. R., and Robillard, P. (1972). *J. Theor. Biol.* **37,** 209–220.

Chadefaud, M. (1972). *In* "Taxonomy and Biology of Blue-Green Algae" (T. V. Desikachary, ed.), pp. 139–144. Univ. of Madras Press, Madras.

Chadefaud, M. (1974). *C. R. Hebd. Seances Acad. Sci., Ser. D* **278,** 3079–3081.

Chadefaud, M. (1975). *Ann. Sci. Nat. Bot. Biol. Veg. [12]* **16,** 217–247.

Chapman, D. J. (1965). Ph.D. Thesis, University of California, San Diego.

Chapman, D. J. (1966). *Arch. Mikrobiol.* **55,** 17–25.

Chapman, D. J. (1973). *In* "The Biology of Blue-Green Algae" (N. G. Carr and B. A. Whitton, eds.), pp. 162–185. Blackwell, Oxford.

Chapman, D. J. (1975). *Nova Hedwigia* **25,** 673–682.

Chapman, D. J., and Antia, N. J. (1977). *Bot. Mar.* (in press).

Chapman, D. J., and Chapman, V. J. (1973). "The Algae," 2nd ed. Macmillan, London.

Chapman, D. J., and Haxo, F. T. (1963). *Plant Cell Physiol.* **4,** 57–63.

Chapman, D. J., and Haxo, F. T. (1966). *J. Phycol.* **2,** 89–91.

Chapman, D. J., and Ragan, M. A. (1977). *Proc. Int. Symp. Taxon. Algae, 1974* (in press).

Chapman, D. J., Cole, W. J., and Siegelman, H. W. (1968). *Am. J. Bot.* **55,** 314–316.

Chardon-Loriaux, I., Morisaki, M., and Ikekawa, N. (1976). *Phytochemistry* **15,** 723–725.

Charlesworth, M. C., and Parish, R. W. (1975). *Eur. J. Biochem.* **54,** 307–316.

Charlton, J. M., Treharne, K. J., and Goodwin, T. W. (1967). *Biochem. J.* **105,** 205–212.

Charon, N. W., Johnson, R. C., and Peterson, D. (1974). *J. Bacteriol.* **117,** 203–211.

Chattaway, F. W., Holmes, M. R., and Barlow, A. J. E. (1968). *J. Gen. Microbiol.* **51,** 367–376.

Chatton, E. (1937). "Titres et trouvaux scientifiques." Sète, Paris.

Chernyad'ev, I. I., Kondrat'eva, E. N., and Doman, N. G. (1975). *Microbiology* **43,** 806–810 (Engl. Transl.).

Christensen, T. (1962). "Alger" *In* "Systematisk Botanik" (T. W. Böcher, M. Lange, and T. Sørensen, eds.), Vol. II, No. 2, 128–146. Munksgaard, Copenhagen.

Christensen, T. (1964). *In* "Algae and Man" (D. F. Jackson, ed.), pp. 59–64. Plenum, New York.

Chuecas, L., and Riley, J. P. (1966). *J. Mar. Biol. Assoc. U.K.* **46,** 153–159.

Chuecas, L., and Riley, J. P. (1969a). *J. Mar. Biol. Assoc. U.K.* **49,** 97–116.

Chuecas, L., and Riley, J. P. (1969b). *J. Mar. Biol. Assoc. U.K.* **49,** 117–120.

Citharel, J. (1971). Ph.D. Thesis, Université de Rennes, France.

Citharel, J., and Gueune, Y. (1974). *C. R. Hebd. Seances Acad. Sci., Ser. D* **278,** 2437–2440.

Clarke, B. (1970). *Science* **168,** 1009–1011.

Clarke, L., and Carbon, J. (1974). *J. Biol. Chem.* **249,** 6874–6885.

Clarke, P. H. (1974). *Symp. Soc. Gen. Microbiol.* **24,** 183–217.

Clayton, R. B. (1965a). *Q. Rev., Chem. Soc.* **19,** 168–200.

Clayton, R. B. (1965b). *Q. Rev., Chem. Soc.* **19,** 201–230.

Cleveland, L. R., and Grimstone, A. V. (1963). *Proc. R. Soc. London, Ser. B* **159,** 668–686.

Cloud, P., Moorman, M., and Pierce, D. (1975). *Q. Rev. Biol.* **50,** 131–150.

Codd, G. A., and Merrett, M. J. (1971). *Planta* **100,** 124–130.

Codd, G. A., and Schmid, G. H. (1972). *Plant Physiol.* **50,** 769–773.

Codd, G. A., and Stewart, W. D. P. (1973). *Arch. Mikrobiol.* **94,** 11–28.

Codd, G. A., and Stewart, W. D. P. (1974). *Plant Sci. Lett.* **3,** 199–205.

Codd, G. A., and Turnbull, F. (1975). *Arch. Microbiol.* **104,** 155–158.

Codd, G. A., Lord, J. M., and Merrett, M. J. (1969). *FEBS Lett.* **5,** 341–342.

Cohen, P. P., and Brown, G. W., Jr. (1960). *Comp. Biochem.* **2,** 161–244.

Cohen, R. J., and Stein, G. S. (1975). *Exp. Cell Res.* **96,** 247–254.

Cohen, S. S. (1950). *Biol. Bull.* **99,** 369.

Cohen, S. S. (1970). *Am. Sci.* **58,** 281–289.

Cohen, S. S. (1973). *Am. Sci.* **61,** 437–445.

Cohen, Y., Padan, E., and Shilo, M. (1975). *J. Bacteriol.* **123,** 855–861.

Cohn, F. (1867). *Arch. Mikrosk. Anat.* **3,** 1–60.

Cohn, F. (1875). *Beitr. Biol. Pflanz.* **1,** 141 (referenced by Stanier, 1970).

Collins, R. P., and Kalnins, K. (1969). *Comp. Biochem. Physiol.* **30,** 779–782.

Comes, P., and Kleinig, H. (1973). *Biochim. Biophys. Acta* **316,** 13–18.

Conrad, M. (1974). *J. Theor. Biol.* **45,** 585–590.

Constantopoulos, G., and Bloch, K. (1967). *J. Bacteriol.* **93,** 1788–1793.

Conti, S. F., and Benedict, C. R. (1962). *J. Bacteriol.* **83,** 929–930.

Cooksey, K. E. (1974). *J. Phycol.* **10,** 253–257.

Copeland, H. F. (1938). *Q. Rev. Biol.* **13,** 383–420.

Correns, C. (1909). *Z. Indukt. Abstamm. Vererbungsl.* **1,** 291–329.

Corry, M. J., Payne, P. I., and Dyer, T. A. (1974). *FEBS Lett.* **46,** 67–70.

Coudert, M., and Vandecasteele, J.-P. (1975). *C. R. Hebd. Seances Acad. Sci., Ser. D* **280,** 129–131.

Craig, I. W., and Carr, N. G. (1968). *Arch. Mikrobiol.* **62,** 167–177.

Craig, N., and Goldstein, L. (1969). *J. Cell Biol.* **40,** 622–632.

Craigie, J. S. (1974). *In* "Algal Physiology and Biochemistry" (W. D. P. Stewart, ed.), pp. 206–235. Blackwell, Oxford.

Craigie, J. S., and Gruenig, D. E. (1967). *Science* **157,** 1058–1059.

Craigie, J. S., McLachlan, J., and Tocher, R. D. (1968). *Can. J. Bot.* **46,** 605–611.

Craigie, J. S., Leigh, C., Chen, L. C.-M., and McLachlan, J. (1971). *Can. J. Bot.* **49,** 1067–1074.

Craigie, J. S., McInnes, A. G., Ragan, M. A., and Walter, J. A. (1977). *Can. J. Chem.* (in press).

Crato, E. (1892). *Ber. Dsch. Bot. Ges.* **10,** 295–302.

Crato, E. (1896). *Beitr. Biol. Pflanz.* **7,** 407–535.

Crawford, I. P. (1975). *Bacteriol. Rev.* **39,** 87–120.

Crawford, I. P., and Gunsalus, I. C. (1966). *Proc. Natl. Acad. Sci. U.S.A.* **56,** 717–724.

Crawford, I. P., Sikes, S., and Melhorn, D. K. (1967). *Arch. Mikrobiol.* **59,** 72–81.

Crick, F. H. C. (1968). *J. Mol. Biol.* **38,** 367–379.

Cronquist, A. (1960). *Bot. Rev.* **26,** 425–482.

Crook, E. M., and Johnston, I. R. (1962). *Biochem. J.* **83,** 325–331.

Crouse, E. J., Vandrey, J. P., and Stutz, E. (1974). *FEBS Lett.* **42,** 262–266.

Cruden, D. L., and Stanier, R. Y. (1970). *Arch. Mikrobiol.* **72,** 115–134.

Cummings, D. J., Tait, A., and Goddard, J. M. (1974). *Biochim. Biophys. Acta* **374,** 1–11.

Cummins, C. S., and Harris, H. (1956). *J. Gen. Microbiol.* **14,** 583–600.
Curgy, J.-J., Ledoigt, G., Stevens, B. J., and André, J. (1974). *J. Cell Biol.* **60,** 628–640.
Czygan, F.-C. (1966). *Z. Naturforsch. Teil B* **21,** 197–198.
Czygan, F.-C. (1968). *Arch. Mikrobiol.* **61,** 81–102.
Dales, R. P. (1960). *J. Mar. Biol. Assoc. U.K.* **39,** 693–699.
Danforth, W. F. (1967). *Res. Protozool.* **1,** 202–306.
Danielli, J. F., and Davson, H. (1935). *J. Cell. Comp. Physiol.* **5,** 495–508.
Darwin, C. (1872). "The Origin of Species." Murray, London.
Darwin, C., and Wallace, A. R. (1858). *Proc. Linn. Soc. London* **3,** 45–62.
Das, S. K., and Smith, E. D. (1968). *Ann. N.Y. Acad. Sci.* **147,** 413–418.
David, H. L. (1974). *Appl. Microbiol.* **28,** 696–699.
Davies, B. H. (1973). *Pure Appl. Chem.* **35,** 1–28.
Davies, B. H. (1975). *Ber. Dtsch. Bot. Ges.* **88,** 7–25.
Davis, A. R. (1915). *Ann. Mo. Bot. Gard.* **2,** 771–836.
Dayhoff, M. O., ed. (1972a). "Atlas of Protein Sequence and Structure." Natl. Biomed. Res. Found., Washington, D.C.
Dayhoff, M. O. (1972b). *In* "Exobiology" (C. Ponnamperuma, ed.), pp. 266–300. North-Holland Publ., Amsterdam.
Dayhoff, M. O., Barker, W. C., and McLaughlin, P. J. (1974). *Origins Life* **5,** 311–330.
Dayhoff, M. O., McLaughlin, P. J., Barker, W. C., and Hunt, L. T. (1975). *Naturwissenschaften* **62,** 154–161.
De, D. N., and Ghosh, S. N. (1965). *J. Histochem. Cytochem.* **13,** 298.
Dearborn, D. G., and Korn, E. D. (1974). *J. Biol. Chem.* **249,** 3342–3346.
De Beer, G. (1958). "Embryos and Ancestors." Oxford Univ. Press (Clarendon), London and New York.
De Deken, R. H. (1963). *Biochim. Biophys. Acta* **78,** 606–618.
Dedyukhina, E. G., and Bekhtereva, M. N. (1970). *Microbiology* **38,** 653–658 (Engl. Transl.).
De Greef, J. A., and Caubergs, R. (1970). *Naturwissenschaften* **57,** 673–674
de la Guardia, M. D., Aragon, C. M. S., Murillo, F. J., and Cerda-Olmedo, E. (1971). *Proc. Natl. Acad. Sci. U.S.A.* **68,** 2012–2015.
Delaunay, J., and Schapira, G. (1974a). *FEBS Lett.* **40,** 97–100.
Delaunay, J., and Schapira, G. (1974b). *C. R. Hebd. Seances Acad. Sci., Ser. D* **279,** 1777–1780.
Delaunay, J., Creusot, F., and Schapira, G. (1973). *Eur. J. Biochem.* **39,** 305–312.
De Ley, J. (1968). *Evol. Biol.* **2,** 102–156.
De Ley, J. (1974). *Taxon* **23,** 291–300.
De Ley, J., and Kersters, K. (1975). *Compr. Biochem.* **29B,** 1–77.
Demoulin, V. (1974). *Bot. Rev.* **40,** 315–345.
Denis, F. A., D'Oultremont, P. A., Debacq, J. J., Cherel, J. M., and Brisou, J. (1975). *C. R. Seances Soc. Biol. Ses Fil.* **169,** 380–383.
Denison, W. C., and Carroll, G. C. (1966). *Mycologia* **58,** 249–269.

De Rosa, M., Gambacorta, A., Minale, L., and Bu'Lock, J. D. (1972). *Biochem. J.* **128,** 751–754.

De Rosa, M., Gambacorta, A., and Bu'Lock, J. D. (1974). *Phytochemistry* **13,** 1793–1794.

De Souza, N. J., and Nes, W. R. (1968). *Science* **162,** 363.

Detchon, P., and Possingham, J. V. (1973). *Biochem. J.* **136,** 829–836.

Dewey, V. C. (1967). *Chem. Zool.* **1,** 161–274.

Dhawale, M. R., Creaser, E. H., and Loper, J. C. (1972). *J. Gen. Microbiol.* **73,** 353–358.

Dickerson, R. E. (1971). *J. Mol. Evol.* **1,** 26–45.

Dickson, L. G., Patterson, G. W., Cohen, C. F., and Dutky, S. R. (1972). *Phytochemistry* **11,** 3473–3477.

Dillon, L. S. (1963). *Syst. Zool.* **12,** 71–82.

Dillon, L. S. (1973). *Bot. Rev.* **39,** 301–345.

Dodd, J. L., and McCracken, D. A. (1972). *Mycologia* **64,** 1341–1343.

Dodge, J. D. (1965). *Excerpta Found. Int. Congr. Ser.* **91,** 339.

Dodge, J. D. (1971). *Protoplasma* **73,** 145–157.

Dodge, J. D. (1974a). "The Fine Structure of Algal Cells." Academic Press, New York.

Dodge, J. D. (1974b). *Sci. Progr. (Oxford)* **61,** 257–274.

Dodge, J. D., and Crawford, R. M. (1971). *Protistologica* **7,** 295–304.

Dodson, E. O. (1971). *Syst. Zool.* **20,** 265–281.

Döhler, G. (1974a). *Ber. Dtsch. Bot. Ges.* **87,** 229–238.

Döhler, G. (1974b). *Planta* **118,** 259–269.

Doelle, H. W. (1969). "Bacterial Metabolism." Academic Press, New York.

Doolittle, W. F. (1972). *J. Bacteriol.* **111,** 316–324.

Douce, R. (1974). *Science* **183,** 852–853.

Dougherty, E. C. (1955). *Syst. Zool.* **4,** 145–169 and 190.

Dougherty, E. C., and Allen, M. B. (1953). *Experientia* **14,** 78.

Dougherty, R. C., Strain, H. H., Svec, W. A., Uphaus, R. A., and Katz, J. J. (1966). *J. Am. Chem. Soc.* **88,** 5037–5038.

Doyle, W. T. (1970). "Nonseed Plants: Form and Function," 2nd ed. Wadsworth, Belmont, California.

Droop, M. R. (1962). *In* "Physiology and Biochemistry of Algae" (R. A. Lewin, ed.), pp. 141–154. Academic Press, New York.

Drucker, D. B. (1974). *Can. J. Microbiol.* **20,** 1723–1728.

Duffus, J. H. (1971). *Biochim. Biophys. Acta* **228,** 627–635.

Duffus, J. H., Penman, C. S., and Webb, N. W. C. (1973). *Experientia* **29,** 632–633.

Ebringer, L. (1972). *J. Gen. Microbiol.* **72,** 35–52.

Echlin, P. (1966). *Br. Phycol. Bull.* **3,** 150–151.

Echlin, P. (1967). *Brit. Phycol. Bull.* **3,** 225–240.

Echlin, P., and Morris, I. (1965). *Biol. Rev. Cambridge Philos. Soc.* **40,** 143–187.

Edelman, M., Swinton, D., Schiff, J. A., Epstein, H. T., and Zeldin, B. (1967). *Bacteriol. Rev.* **31,** 315–331.

Egami, F. (1973). *Z. Allg. Mikrobiol.* **13,** 177–181.

Egami, F. (1976). *Origins of Life* **7,** 71–72.

Egan, A. F., and Gibson, F. (1967). *Biochim. Biophys. Acta* **136,** 573–576.

Egger, K., Nitsche, H., and Kleinig, H. (1969). *Phytochemistry* **8,** 1583–1585.

Ehrendorfer, F. (1972). *Symp. Biol. Hung.* **12,** 227–231.

Eigen, M. (1971). *Naturwissenschaften* **58,** 465–523.

Eigen, M. (1973). *In* "The Physicist's Conception of Nature" (J. Mehra, ed.), pp. 594–632. Reidel, Boston, Massachusetts.

Elbein, A. D. (1974). *Adv. Carbohydr. Chem. Biochem.* **30,** 227–256.

Ellingson, J. S. (1974). *Biochim. Biophys. Acta* **337,** 60–67.

Ellis, R. J., Blair, G. E., and Hartley, M. R. (1973). *Biochem. Soc. Symp.* **38,** 137–162.

Elton, R. A. (1973). *Nature (London) New Biol.* **243,** 287–288.

Enatsu, T., and Crawford, I. P. (1968). *J. Bacteriol.* **95,** 107–112.

Erdtman, H. G. H. (1968). *Recent Adv. Phytochem.* **1,** 13–56.

Erickson, M., and Miksche, G. E. (1974). *Phytochemistry* **13,** 2295–2299.

Ernst-Fonberg, M. L., Dubinskas, F., and Jonak, Z. L. (1974). *Arch. Biochem. Biophys.* **165,** 646–655.

Erwin, J. A. (1973). *In* "Lipids and Biomembranes of Eukaryotic Microorganisms" (J. A. Erwin, ed.), pp. 41–143. Academic Press, New York.

Erwin, J., and Bloch, K. (1964). *Science* **143,** 1006–1012.

Erwin, J., Hulanicka, D., and Bloch, K. (1964). *Comp. Biochem. Physiol.* **12,** 191–207.

Evans, E. A., Jr. (1960). *In* "Evolution after Darwin" (S. Tax, ed.), Vol. I, pp. 85–93. Univ. of Chicago Press, Chicago, Illinois.

Evans, L. V. (1974). *In* "Algal Physiology and Biochemistry" (W. D. P. Stewart, ed.), pp. 86–123. Blackwell, Oxford.

Eventoff, W., and Rossman, M. G. (1975). *Crit. Rev. Biochem.* **3,** 111–140.

Evreinova, T. N., Davydova, I. M., Sukover, A. P., and Goryunova, S. V. (1961). *Dokl. Biochem.* **137,** 43–46.

Falk, M., Smith, D. G., McLachlan, J., and McInnes, A. G. (1966). *Can. J. Chem.* **44,** 2269–2281.

Famintzin, A. (1907). *Biol. Centralbl.* **27,** 353–364.

Farber, F. E., and Rawls, W. E. (1975). *J. Gen. Virol.* **26,** 21–31.

Farquhar, M. N., and McCarthy, B. J. (1973). *Biochemistry* **12,** 4113–4122.

Fatturosso, E., Magno, S., Santacroce, C., Sica, D., Impellizzeri, G., Mangiafico, S., Oriente, G., Piatelli, M., and Sciuto, S. (1975). *Phytochemistry* **14,** 1579–1582.

Fatturosso, E., Magno, S., Santacroce, C., Sica, D., Impellizzeri, G., Mangiafico, S., Piatelli, M., and Sciuto, S. (1976). *Biochem. Syst. Ecol.* **4,** 135–138.

Feige, G. B. (1975). *Z. Pflanzenphysiol.* **75,** 339–345.

Felden, R. A., Sanders, M. M., and Morris, N. R. (1976). *J. Cell Biol.* **68,** 430–439.

Fellows, F. C. I., and Lewis, M. H. R. (1973). *Biochem. J.* **136,** 329–334.

Fenical, W. (1975). *J. Phycol.* **11,** 245–259.

Ferezou, J. P., Devys, M., Allais, J. P., and Barbier, M. (1974). *Phytochemistry* **13,** 593–598.

Fewson, C. A., Al-Hafidh, M., and Gibbs, M. (1962). *Plant Physiol.* **37,** 402–406.
Fiasson, J.-L. (1968). Thesis No. 352. University of Lyons.
Fincham, J. R. S., and Boylen, J. B. (1955). *Biochem. J.* **61,** xxiii–xxiv.
Fischer, H., and Wenderoth, H. (1939). *Ann. Chem.* **537,** 170–177.
Fitch, W. (1976). *J. Mol. Evol.* **8,** 13–40.
Fitch, W. M., and Margoliash, E. (1967). *Biochem. Genet.* **1,** 65–71.
Fitch, W. M., and Margoliash, E. (1970). *Evol. Biol.* **4,** 67–109.
Fitch, W. M., and Margoliash, E. (1971). *Taxon* **20,** 51–53.
Flavell, R. (1972). *Biochem. Genet.* **6,** 275–291.
Fleming, M., Manners, D. J., and Masson, A. J. (1967). *Biochem. J.* **104,** 32p–33p.
Forget, P. (1974). *Eur. J. Biochem.* **42,** 325–332.
Forin, M.-C., Maume, B., and Baron, C. (1975). *C. R. Hebd. Seances Acad. Sci. Ser. D* **281,** 195–198.
Fowden, L. (1965). *In* "Plant Biochemistry" (J. Bonner and J. E. Varner, eds.), 2nd ed., pp. 361–390. Academic Press, New York.
Fowler, C. F. (1974). *Biochim. Biophys. Acta* **357,** 327–331.
Fox, S. W., ed. (1965). "The Origins of Prebiological Systems and of their Molecular Matrices." Academic Press, New York.
Franco, L., Johns, E. W., and Navlet, J. M. (1974). *Eur. J. Biochem.* **45,** 83–89.
Frank, G., Zuber, H., and Lergier, W. (1975). *Experientia* **31,** 23–26.
Frederick, S. E., Gruber, P. J., and Tolbert, N. E. (1973). *Plant Physiol.* **52,** 318–323.
Fredrick, J. F. (1968a). *Physiol. Plant.* **21,** 176–182.
Fredrick, J. F. (1968b). *Phytochemistry* **7,** 1573–1576.
Fredrick, J. F. (1970). *Ann. N.Y. Acad. Sci.* **175,** 524–530.
Fredrick, J. F. (1972). *In* "Taxonomy and Biology of Blue-Green Algae" (T. V. Desikachary, ed.), pp. 566–570. Univ. of Madras Press, Madras.
Freundt, E. A. (1974). *In* "Bergey's Manual of Determinative Bacteriology" (R. E. Buchanan and N. E. Gibbons, eds.), 8th ed., pp. 949–952. Williams & Wilkins, Baltimore, Maryland.
Fridovich, I. (1974a). *Life Sci.* **14,** 819–826.
Fridovich, I. (1974b). *Adv. Enzymol.* **41,** 35–98.
Fridovich, I. (1974c). *Horiz. Biochem. Biophys.* **1,** 1–37.
Fridovich, I. (1975). *Annu. Rev. Biochem.* **44,** 147–159.
Fujita, Y., and Shimura, S. (1974). *Plant Cell Physiol.* **15,** 939–942.
Fujiwara, T., and Akabori, S. (1954). *J. Chem. Soc. Jpn.* **75,** 900–903; *Chem. Abstr.* **49,** 3325h (1955).
Fulco, A. J. (1970). *J. Biol. Chem.* **245,** 2985–2990.
Fuller, R. C. (1971). *In* "Biochemical Evolution and the Origin of Life" (E. Schoffeniels, ed.), Vol. 2, pp. 259–273. Elsevier, Amsterdam.
Fuller, R. C., and Gibbs, M. (1959). *Plant Physiol.* **34,** 324–329.
Gaffron, H. (1960a). *In* "Evolution after Darwin" (S. Tax, ed.), Vol. I, pp. 39–84. Univ. of Chicago Press, Chicago, Illinois.
Gaffron, H. (1960b). *In* "Plant Physiology: A Treatise" (F. C. Steward, ed.), Vol. 1B, pp. 3–277. Academic Press, New York.

Galling, G. (1974). *Planta* **118,** 283–295.
Galling, G., and Ssymank, V. (1970). *Planta* **94,** 203–212.
Gantt, E., and Conti, S. F. (1966). *J. Cell Biol.* **29,** 423–434.
Gantt, E., and Lipschultz, C. A. (1973). *Biochim. Biophys. Acta* **292,** 858–861.
Gantt, E., Edwards, M. R., and Provasoli, L. (1971). *J. Cell Biol.* **48,** 280–290.
Gapochka, L. D., and Danilova, T. I. (1973). *Dokl. Biochem.* **210,** 217–219.
Garber, E. D., Baird, M. L., and Chapman, D. J. (1975). *Bot. Gaz.* (*Chicago*) **136,** 341–346.
Geissman, T. A., and Crout, D. H. G. (1969). "Organic Chemistry of Secondary Plant Metabolism." Freeman, San Francisco, California.
Geitler, L. (1959). *In* "Handbuch der Pflanzenphysiologie" (W. Ruhland, ed.), Vol. 11, pp. 530–545. Springer-Verlag, Berlin and New York.
Gerasimova, N. M., Aizina, A. F., Zlatoust, M. A., Balabanova, Zh. I., Razumovskii, P. N., and Bekhtereva, M. N. (1974). *Microbiology* **43,** 375–377 (Engl. Trans.).
Gershengorn, M. C., Smith, A. R. H., Goulston, G., Goad, L. J., Goodwin, T. W., and Haines, T. H. (1968). *Biochemistry* **7,** 1698–1706.
Getz, G. S. (1972). *In* "Membrane Molecular Biology" (C. F. Fox and A. D. Keith, eds.), pp. 386–438. Sinauer Assoc., Stamford, Connecticut.
Gibbs, M., Latzko, E., Harvey, M. J., Plaut, Z., and Shain, Y. (1970). *Ann. N.Y. Acad. Sci.* **175,** 541–554.
Gibbs, S. P. (1970). *Ann. N.Y. Acad. Sci.* **175,** 454–473.
Gibor, A., and Granick, S. (1964). *Science* **145,** 890–897.
Gibson, I. (1966). *J. Protozool.* **13,** 650–653.
Giles, K. L., and Sarafis, V. (1971). *Cytobios* **4,** 61–74.
Giles, K. L., and Sarafis, V. (1972). *Nature* (*London*), *New Biol.* **236,** 56–58.
Gill, J. W., and Vogel, H. J. (1963). *J. Protozool.* **10,** 148–152.
Ginsburg, D., and Steitz, J. A. (1975). *J. Biol. Chem.* **250,** 5647–5654.
Givan, A. L., and Criddle, R. S. (1972). *Arch. Biochem. Biophys.* **149,** 153–163.
Givan, C. V., and Leech, R. M. (1971). *Biol. Rev. Cambridge Philos. Soc.* **46,** 409–428.
Glasl, H., and Pohl, P. (1974). *Z. Naturforsch. Teil C* **29,** 399–406.
Glazer, A. N. (1976). *Photochem. Photobiol. Rev.* **1,** 71–115.
Glazer, A. N., and Hixson, C. S. (1975). *J. Biol. Chem.* **250,** 5487–5495.
Glazer, A. N., Cohen-Bazire, G., and Stanier, R. Y. (1971). *Proc. Natl. Acad. Sci. U.S.A.* **68,** 3005–3008.
Glazer, A. N., Aspell, G. S., Hixson, C. S., Bryant, D. A., Rimón, S., and Brown, D. M. (1976). *Proc. Natl. Acad. Sci. U.S.A.* **73,** 428–431.
Gloe, A., and Pfennig, N. (1974). *Arch. Microbiol.* **96,** 93–101.
Gloe, A., Pfennig, N., Brockman, H., Jr., and Trowitzsch, W. (1975). *Arch. Microbiol.* **102,** 103–109.
Glombitza, K.-W., and Rösener, H.-U. (1974). *Phytochemistry* **13,** 1245–1248.
Glombitza, K.-W., and Sattler, E. (1973). *Tetrahedron Lett.* No. 43, pp. 4277–4280.
Glombitza, K.-W., Rösener, H.-U., Vilter, H., and Rauwald, W. (1973). *Planta Med.* **24,** 301–303.

Glombitza, K.-W., Rösener, H.-U., and Müller, D. (1975a). *Phytochemistry* **14,** 1115–1116.

Glombitza, K.-W., Rauwald, H.-W., and Eckhardt, G. (1975b). *Phytochemistry* **14,** 1403–1405.

Goad, L. J., and Goodwin, T. W. (1966). *Biochem. J.* **99,** 735–746.

Goad, L. J., and Goodwin, T. W. (1969). *Eur. J. Biochem.* **7,** 502–508.

Goad, L. J., and Goodwin, T. W. (1972). *Progr. Phytochem.* **3,** 113–198.

Goad, L. J., Lenton, J. R., Knapp, F. F., and Goodwin, T. W. (1974). *Lipids* **9,** 582–595.

Goksøyr, J. (1967) *Nature* (*London*) **214,** 1161.

Goldberg, A. L., and Wittes, R. E. (1966). *Science* **153,** 420–424.

Goldberg, I., and Bloch, K. (1972). *J. Biol. Chem.* **247,** 7349–7357.

Goldfine, H. (1972). *Adv. Microb. Physiol.* **8,** 1–58.

Goldfine, H., and Bloch, K. (1961). *J. Biol. Chem.* **236,** 2596–2601.

Goldman, P., Alberts, A. W., and Vagelos, P. R. (1963). *J. Biol. Chem.* **238,** 1255–1261.

Gol'man, L. P., Mikhaseva, M. F., and Reznikov, V. M. (1973). *Dokl. Akad. Nauk BSSR* **17,** 1031–1033.

Gooday, G. W., Fawcett, P., Green, D., and Shaw, G. (1973). *J. Gen. Microbiol.* **78,** 233–239.

Goodman, M., and Moore, G. W. (1974). *Syst. Zool.* **32,** 508–532.

Goodman, M., Barnabas, J., Matsuda, G., and Moore, G. W. (1971). *Nature* (*London*) **233,** 604–613.

Goodwin, T. W. (1952). *Bot. Rev.* **18,** 291–316.

Goodwin, T. W. (1966). *In* "Comparative Phytochemistry" (T. Swain, ed.), pp. 121–137. Academic Press, New York.

Goodwin, T. W. (1967). *In* "Biochemistry of Chloroplasts" (T. W. Goodwin, ed.), Vol. 2, pp. 721–733. Academic Press, New York.

Goodwin, T. W. (1971a). *In* "Aspects of Terpenoid Chemistry and Biochemistry" (T. W. Goodwin, ed.), pp. 315–356. Academic Press, New York.

Goodwin, T. W. (1971b). *Biochem. J.* **123,** 293–329.

Goodwin, T. W. (1971c). *In* "Carotenoids" (O. Isler, ed.), pp. 577–636. Birkhaeuser, Basel.

Goodwin, T. W. (1973a). *In* "Lipids and Biomembranes of Eukaryotic Microorganisms" (J. A. Erwin, ed.), pp. 1–41. Academic Press, New York.

Goodwin, T. W. (1973b). *Recent Adv. Phytochem.* **6,** 97–115.

Goodwin, T. W. (1974). *In* "Algal Physiology and Biochemistry" (W. D. P. Stewart, ed.), pp. 266–280. Blackwell, Oxford.

Gotelli, I. B., and Cleland, R. (1968). *Am. J. Bot.* **55,** 907–914.

Goulston, G., and Mercer, E. I. (1969). *Phytochemistry* **8,** 1945–1948.

Goulston, G., Mercer, E. I., and Goad, L. J. (1975). *Phytochemistry* **14,** 457–462.

Gram, C. (1884). *Fortschr. Med.* **2,** 185–189.

Grant, V., and Flake, R. H. (1974). *Proc. Natl. Acad. Sci. U.S.A.* **71,** 3863–3865.

Gray, M. W. (1975). *Can. J. Biochem.* **53,** 735–746.

Green, B. R. (1974). *In* "Algal Physiology and Biochemistry" (W. D. P. Stewart, ed.), pp. 281–313. Blackwell, Oxford.

Green, J. C., and Jennings, D. H. (1967). *J. Exp. Bot.* **18,** 359–370.
Greenblatt, G. A., and Sarkissian, I. V. (1973). *Physiol. Plant.* **29,** 361–364.
Gressel, J., Berman, T., and Cohen, N. (1975). *J. Mol. Evol.* **5,** 307–313.
Grierson, D., Rogers, M. E., Sartirana, M. L., and Loening, U. E. (1970). *Cold Spring Harbor Symp. Quant. Biol.* **35,** 589–598.
Grivell, L. A., and Walg, H. L. (1972). *Biochem. Biophys. Res. Commun.* **49,** 1452–1458.
Grodzinski, B., and Colman, B. (1970). *Plant Physiol.* **45,** 735–737.
Groot, G. S. P., Flavell, R. A., and Sanders, J. P. M. (1975). *Biochim. Biophys. Acta* **378,** 186–194.
Gross, J. A., Stroz, R. J., and Britton, G. (1975). *Plant Physiol.* **55,** 175–177.
Gruber, P. J., Frederick, S. E., and Tolbert, N. E. (1974). *Plant Physiol.* **53,** 167–170.
Grunwald, C. (1975). *Annu. Rev. Plant Physiol.* **26,** 209–236.
Guderian, R. H., Pulliam, R. L., and Gordon, M. P. (1972). *Biochim. Biophys. Acta* **262,** 50–65.
Guest, J. R., Drapeau, G. R., Carlton, B. C., and Yanofsky, C. (1967). *J. Biol. Chem.* **242,** 5442–5446.
Guillard, R. R. L., and Lorenzen, C. J. (1972). *J. Phycol.* **8,** 10–14.
Guillemaut, P., Steinmetz, A., Burkard, G., and Weil, J. H. (1973). *C. R. Seances Soc. Biol. Ses Fil.* **167,** 961–966.
Guilliermond, A. (1914). *Ber. Dtsch. Bot. Ges.* **32,** 282–301.
Guilliermond, A. (1921). *Rev. Gen. Bot.* **33,** 401 and 449; summarized in *Bull. Soc. Bot. Fr.* **69,** 407 (1922).
Gunnison, D., and Alexander, M. (1975). *Appl. Microbiol.* **29,** 729–738.
Gupta, K. G., Bhatnagar, L., and Jain, A. K. (1974). *Ann. Microbiol.* (*Inst. Pasteur*) *A* **125,** 419–425.
Guttman, H. N. (1967). *J. Protozool.* **14,** 267–271.
Guttman, H. N., and Eisenman, R. N. (1965). *Nature* (*London*) **206,** 113–114.
Haeckel, E. (1866). "Generelle Morphologie der Organismen," Vol. 2. Reimer, Berlin.
Haeckel, E. (1898). "The Evolution of Man: a Popular Exposition of the Principal Points of Human Ontogeny and Phylogeny," Vol. II. Appleton, New York.
Haeckel, E. (1904). "The Wonders of Life: a Popular Study of Biological Philosophy." Harper, New York.
Hager, A., and Stransky, H. (1970a). *Arch. Mikrobiol.* **72,** 68–83.
Hager, A., and Stransky, H. (1970b). *Arch. Mikrobiol.* **73,** 77–89.
Haines, T. H (1973). *In* "Lipids and Biomembranes of Eukaryotic Microorganisms" (J. A. Erwin, ed.), pp. 197–232. Academic Press, New York.
Haldane, J. B. S. (1957). *J. Genet.* **55,** 511–524.
Hall, D. O., Cammack, R., and Rao, K. K. (1972). *Photosynth., Two Centuries Its Discovery Joseph Priestley, Proc. Int. Congr. Photosynth. Res., 2nd,* 1971 Vol. 3, pp. 1707–1719.
Hall, D. O., Cammack, R., and Rao, K. K. (1973a). *Pure Appl Chem.* **34,** 553–577.
Hall, D. O., Cammack, R., and Rao, K. K. (1973b). *Space Life Sci.* **4,** 455–468.
Hall, D. O., Cammack, R., and Rao, K. K. (1974). *Origins Life* **5,** 363–386.

Hall, D. O., Cammack, R., Rao, K. K., Evans, M. C. W., and Mullinger, R. (1975a). *Biochem. Soc. Trans.* **3,** 361–368.

Hall, D. O., Rao, K. K., and Cammack, R. (1975b). *Sci. Progr.* (*Oxford*) **62,** 285–317.

Hall, D. O., Rao, K. K., and Mullinger, R. (1975c). *Biochem. Soc. Trans.* **3,** 472–479.

Hall, J. B. (1971). *J. Theor. Biol.* **30,** 429–454.

Hall, J. B. (1973a). *J. Theor. Biol.* **38,** 413–418.

Hall, J. B. (1973b). *Space Life Sci.* **4,** 204–213.

Hall, R. P. (1967). *Res. Protozool.* **1,** 338–404.

Hall, W. T., and Claus, G. (1963). *J. Cell Biol.* **19,** 551–563.

Hall, W. T., and Claus, G. (1967). *J. Phycol.* **3,** 37–51.

Halsall, T. G., and Hills, I. R. (1971). *J. Chem. Soc., Chem. Commun.* pp. 448–449.

Halsey, Y. D., and Parson, W. W. (1974). *Biochim. Biophys. Acta* **347,** 404–416.

Hands, A. R., and Bartley, W. (1962). *Biochem. J.* **84,** 238–239.

Hanic, L. A., and Craigie, J. S. (1969). *J. Phycol.* **5,** 89–102.

Hanson, E. D. (1976). *J. Protozool.* **23,** 4–12.

Harada, T., and Spencer, B. (1960). *J. Gen. Microbiol.* **22,** 520–527.

Harrington, G. W., Beach, D. H., Dunham, J. E., and Holz, G. G., Jr. (1970). *J. Protozool.* **17,** 213–219.

Harris, J. U., and Berns, D. S. (1975). *J. Mol. Evol.* **5,** 153–163.

Harrison, J. S. (1963) *In* "Biochemistry of Industrial Micro-organisms" (C. Rainbow and A. H. Rose, eds.), pp. 9–33. Academic Press, New York.

Hartley, B. S. (1974). *Symp. Soc. Gen. Microbiol.* **24,** 151–182.

Hartman, H. (1975). *J. Mol. Evol.* **4,** 359–370.

Harwood, J. L., and James, A. T. (1975). *Eur. J. Biochem.* **50,** 325–334.

Hase, T., Wada, K., and Matsubara, H. (1976). *J. Biochem.* (*Tokyo*) **79,** 329–343.

Haslam, E. (1974). "The Shikimate Pathway" Wiley, New York.

Hatch, M. D., and Slack, C. R. (1966). *Biochem. J.* **101,** 103–111.

Hattori, A., and Myers, J. (1967). *Plant Cell Physiol.* **8,** 327–337.

Hattori, A., and Uesugi, I. (1968). *In* "Comparative Biochemistry and Biophysics of Photosynthesis" (K. Shibata *et al.*, eds.), pp. 201–205. Univ. of Tokyo Press, Tokyo.

Haug, A. (1974). *Phys. Chem., Ser. One* **11,** 51–88.

Haug, A., and Larsen, B. (1974). *In* "Plant Carbohydrate Biochemistry" (J. B. Pridham, ed.), pp. 207–218. Academic Press, New York.

Havir, E. A., and Hanson, K. R. (1975). *Biochemistry* **14,** 1620–1626.

Hawke, J. C., Rumsby, M. G., and Leech, R. M. (1974). *Phytochemistry* **13,** 403–413.

Haxo, F. T. (1955). *Fortschr. Chem. Org. Naturst.* **12,** 169–197.

Hayes, F., Vasseur, M., Nikolaev, N., Schlessinger, D., Sri Widada, J., Krol, A., and Branlant, C. (1975). *FEBS Lett.* **56,** 85–91.

Healey, F. P. (1970a) *Planta* **91,** 220–226.

Healey, F. P. (1970b) *Plant Physiol.* **45,** 153–159.

Hecker, L. I., Egan, J., Reynolds, R. J., Nix, C. E., Schiff, J. A., and Barnett, W. E. (1974). *Proc. Natl. Acad. Sci. U.S.A.* **71,** 1910–1914.

Hegeman, G. D., and Rosenberg, S. L. (1970). *Annu. Rev. Microbiol.* **24,** 429–462.
Heimer, Y. M. (1975). *Arch. Microbiol.* **103,** 181–183.
Heizmann, P. (1974). *Biochem. Biophys. Res. Commun.* **56,** 112–118.
Hellebust, J. A. (1965). *Limnol. Oceanogr.* **10,** 192–206.
Heller, W. (1973). *Naturwissenschaften* **60,** 460–468.
Hellmann, V., and Kessler, E. (1974a). *Arch. Microbiol.* **95,** 311–318.
Hellmann, V., and Kessler, E. (1974b). *Arch. Microbiol.* **100,** 239–242.
Hennig, W. (1965). *Ann. Rev. Entomol.* **10,** 97–116.
Hennig, W. (1966). "Phylogenetic Systematics." Univ. of Illinois Press, Urbana.
Herout, V. (1973). *Proc. Nobel Symp.* **25,** 55–62.
Herrera, J., and Nicholas, D. J. D. (1974). *Biochim. Biophys. Acta* **368,** 54–60.
Hess, J. L., and Tolbert, N. E. (1967). *Plant Physiol.* **42,** 371–379.
Hewett, M. J., Wicken, A. J., Knox, K. W., and Sharpe, M. E. (1976). *J. Gen. Microbiol.* **94,** 126–130.
Hewitt, J., and Morris, J. G. (1975). *FEBS Lett.* **50,** 315–318.
Heywood, P. (1976). *Nature* (*London*) **259,** 425.
Heywood, V. H. (1973). *Proc. Nobel Symp.* **25,** 41–54.
Hibberd, D. J., and Leedale, G. F. (1970). *Nature* (*London*) **225,** 758–760.
Hibberd, D. J., and Leedale, G. F. (1972). *Ann. Bot.* (*London*) [N.S.] **36,** 49–71.
Hill, D. L. (1972). "The Biochemistry and Physiology of *Tetrahymena*." Academic Press, New York.
Hill, L. R. (1966). *J. Gen. Microbiol.* **44,** 419–437.
Hino, T., Kametaka, M., and Kandatsu, M. (1973). *J. Gen. Appl. Microbiol.* **19,** 397–413.
Hirai, K. (1923). *Biochem. Z.* **135,** 299–307.
Hirose, H. (1958). *Bot. Mag.* **71,** 347–352.
Hirose, H. (1975). *In* "Advance of Phycology in Japan" (J. Tokida and H. Hirose, eds.), pp. 52–65. Junk, The Hague.
Hirschfeld, D. R., Fenical, W., Lin, G. H. Y., Wing, R. M., Radlick, P., and Sims, J. J. (1973). *J. Am. Chem. Soc.* **95,** 4049–4050.
His, W. (1874). "Unsere Körperform und das physiologische Problem ihrer Entstehung." Leipzig (referenced in de Beer, 1958).
Hitchcock, C., and Nichols, B. W. (1971). "Plant Lipid Biochemistry" Academic Press, New York.
Hnilica, L. S. (1973). "The Structure and Biological Functions of Histones." CRC Press, Cleveland, Ohio.
Hoffmann, G. W. (1974). *J. Mol. Biol.* **86,** 349–362.
Holdsworth, R. H. (1971). *J. Cell Biol.* **51,** 499–513.
Holm-Hansen, O. and Brown, G. W., Jr. (1963). *Plant Cell Physiol.* **4,** 299–306.
Holm-Hansen, O., Prasad, R., and Lewin, R. A. (1965). *Phycologia* **5,** 1–14.
Holmquist, R., and Jukes, T. H. (1972). *J. Mol. Evol.* **2,** 10–16.
Holmquist, R., and Jukes, T. H. (1975). *J. Mol. Evol.* **4,** 377–381.
Holmquist, R., Jukes, T. H., and Pangburn, S. (1973). *J. Mol. Biol.* **78,** 91–116.
Holt, A. S., and Morley, H. V. (1959). *Can. J. Chem.* **37,** 507–514.
Holt, A. S., Purdie, J. W., and Wasley, J. W. F. (1966). *Can. J. Chem.* **44,** 88–93.
Holz, G. G., Jr. (1964). *Biochem. Physiol. Protozoa* **3,** 199–242.

Honigberg, B. M. (1967). *Chem. Zool.* **1,** 695–814.

Honigberg, B. M., Balamuth, W., Bovee, E. C., Corliss, J. O., Gojdics, M., Hall, R. P., Kudo, R. R., Levine, N. D., Loeblich, A. R., Jr., Weiser, J., and Wenrich, D. H. (1964). *J. Protozool.* **11,** 7–20.

Hoober, J. K., and Blobel, G. (1969). *J. Mol. Biol.* **41,** 121–138.

Hood, W., and Carr, N. G. (1972). *J. Gen. Microbiol.* **73,** 417–426.

Horecker, B. L. (1963). *Proc. Int. Congr. Biochem., 5th, 1961* Vol. III, pp. 86–93.

Horgen, P. A., and O'Day, D. H. (1973). *Cytobios* **8,** 119–123.

Horgen, P. A., Nagao, R. T., Chia, L. S. Y., and Key, J. L. (1973). *Arch. Mikrobiol.* **94,** 249–258.

Hori, H. (1975). *J. Mol. Evol.* **7,** 75–86.

Hori, H. (1976). *Molec. Gen. Genet.* **145,** 119–123.

Horiguchi, M., and Kandatsu, M. (1959). *Nature* (*London*) **184,** 901–902.

Horowitz, N. H. (1945). *Proc. Natl. Acad. Sci. U.S.A.* **31,** 153–157.

Horowitz, N. H. (1965). *In* "Evolving Genes and Proteins" (V. Bryson and H. J. Vogel, eds.), pp. 15–23. Academic Press, New York.

Horvath, R. S. (1974). *J. Theor. Biol.* **47,** 361–371.

Howard, R. J., Wright, S. W., and Grant, B. R. (1976). *Plant Physiol.* **58,** 459–463.

Howland, G. P., and Ramus, J. (1971). *Arch. Mikrobiol.* **76,** 292–298.

Hsiang, M. W., and Cole, R. D. (1973). *J. Biol. Chem.* **248,** 2007–2013.

Hudson, B. J. F., and Karis, I. G. (1974). *J. Sci. Food Agric.* **25,** 759–763.

Hungate, R. E. (1955). *Biochem. Physiol. Protozoa* **2,** 159–199.

Hungate, R. E. (1966). "The Rumen and its Microbes." Academic Press, New York.

Hurlbert, R. E., and Lascelles, J. (1963). *J. Gen. Microbiol.* **33,** 445–458.

Hurlbert, R. E., and Rittenberg, S. C. (1962). *J. Protozool.* **9,** 170–182.

Hutner, S. H., and Corliss, J. (1976). *J. Protozool.* **23,** 48–56.

Hutner, S. H., and Provasoli, L. (1951). *Biochem. Physiol. Protozoa* **1,** 27–128.

Hütter, R., and DeMoss, J. A. (1967). *J. Bacteriol.* **94,** 1896–1907.

Idler, D. R., and Wiseman, P. M. (1971). *In* "Biology Data Book" (P. L. Altman and D. S. Dittmer, eds.), 2nd ed., Vol. I, pp. 355–367. Fed. Am. Soc. Exp. Biol., Bethesda, Maryland.

Iizuka, H., Iida, M., Teshima, S., and Minemura, Y. (1969). *Z. Allg. Mikrobiol.* **9,** 443–448.

Ikan, R., and Seckbach, J. (1972). *Phytochemistry* **11,** 1077–1082.

Ikawa, M. (1967). *Bacteriol. Rev.* **31,** 54–64.

Ikawa, M., Borowski, P. T., and Chakravarti, A. (1968). *Appl. Microbiol.* **16,** 620–623.

Ikawa, T., Asami, S., and Nisizawa, K. (1972). *Proc. Int. Seaweed Symp., 7th, 1971* pp. 526–531.

Ikekawa, N., Morisaki, M., and Hirayama, K. (1972). *Phytochemistry* **11,** 2317–2318.

Il'in, A. V., Vagabova, L. M., and Bur'yanov, Ya. I. (1972). *Microbiology* (*USSR*) **41,** 436–441.

Impellizzeri, G., Mangiafico, S., Oriente, G., Piatelli, M., Sciuto, S., Fatturosso, E.,

Magno, S., Santacroce, C., and Sica, D. (1975). *Phytochemistry* **14,** 1549–1558.

Indik, Z. K., Keller, B. J., and Marks, D. B. (1975). *Arch. Biochem. Biophys.* **170,** 315–325.

Ingold, C. T. (1959). *Vistas Bot.* **1,** 348 (referenced in Bisset, 1962).

Ingram, L. O., Pierson, D., Kane, J. F., VanBaalen, C., and Jensen, R. A. (1972). *J. Bacteriol.* **111,** 112–118.

Ishida, M. R., Kikuchi, T., Matsubara, T., Hayashi, F., and Yokomura, E. (1969). *Annu. Rep. Res. Reactor Inst., Kyoto Univ.* **2,** 73–75.

Iwai, K. (1964). *In* "The Nucleohistones" (J. Bonner and P. Ts'o, eds.), pp. 59–65. Holden-Day, San Francisco, California.

Iwai, K., Shiomi, H., Ando, T., and Mita, T. (1965). *J. Biochem.* (*Tokyo*) **58,** 312–314.

Iwai, K., Hamana, K., and Yabuki, H. (1970). *J. Biochem.* (*Tokyo*) **68,** 597–601.

Iwanij, V., Chua, N.-H., and Siekevitz, P. (1974). *Biochim. Biophys. Acta* **358,** 329–340.

Jack, R. C. M. (1966). *J. Bacteriol.* **91,** 2101–2102.

Jacobi, G. (1962). *In* "Physiology and Biochemistry of Algae" (R. A. Lewin, ed.), pp. 125–140. Academic Press, New York.

Jamieson, G. R., and Reid, E. H. (1976). *Phytochemistry* **15,** 795–796.

Jayanthi Bai, N., Ramachandra Pai, M., Suryanarayana Murthy, P., and Venkitasubramanian, T. A. (1975). *Arch. Biochem. Biophys.* **168,** 230–234.

Jeffrey, C. (1971). *Kew Bull.* **25,** 291–299.

Jeffrey, S. W. (1968a). *Biochim. Biophys. Acta* **162,** 271–285.

Jeffrey, S. W. (1968b). *Nature* (*London*) **220,** 1032–1033.

Jeffrey, S. W. (1972). *Biochim. Biophys. Acta* **279,** 15–33.

Jeffrey, S. W. (1976). *J. Phycol.* **12,** 349–354.

Jeffrey, S. W., Sielicki, M., and Haxo, F. T. (1975). *J. Phycol.* **11,** 374–384.

Jensen, R. A., and Pierson, D. L. (1975). *Nature* (*London*) **254,** 667–671.

Jensen, R. A., and Stenmark, S. L. (1974). *J. Mol. Evol.* **4,** 249–259.

Johansen, J. E., Svec, W. A., Liaaen-Jensen, S., and Haxo, F. T. (1974). *Phytochemistry* **13,** 2261–2271.

Johmann, C. A., and Gorovsky, M. A. (1976). *Arch. Biochem. Biophys.* **175,** 694–699.

Johnson, C. E., Elstner, E., Gibson, J. F., Benfield, G., Evans, M. C. W., and Hall, D. O. (1968). *Nature* (*London*) **220,** 1291–1293.

Johnston, C. S., and Davies, J. M. (1969). *Proc. Int. Seaweed Symp., 6th, 1968* pp. 501–506.

Joklik, W. K. (1974). *Symp. Soc. Gen. Microbiol.* **24,** 293–320.

Jones, A. S., and Thompson, T. W. (1963). *J. Protozool.* **10,** 91–93.

Jones, G. M. T., and Harris, J. I. (1972). *FEBS Lett.* **22,** 185–189.

Jordan, B. R., Galling, G., and Jourdan, R. (1974). *J. Mol. Biol.* **87,** 205–225.

Jukes, T. H., and Holmquist, R. (1972). *Science* **177,** 530–532.

Jukes, T. H., Holmquist, R., and Moise, H. (1975). *Science* **189,** 50–51.

Jurgenson, J. E., Beale, S. I., and Troxler, R. F. (1976). *Biochem. Biophys. Res. Commun.* **69,** 149–157.

Kac, M. (1973). *In* "The Physicist's Conception of Nature" (J. Mehra, ed.), p. 560, Reidel, Boston, Massachusetts.

Kagawa, T., and Hatch, M. D. (1975). *Arch. Biochem. Biophys.* **167,** 687–696.

Kaklij, G. S., and Nadkarni, G. B. (1970). *Arch. Biochem. Biophys.* **140,** 334–340.

Kaklij, G. S., and Nadkarni, G. B. (1974a). *Arch. Biochem. Biophys.* **160,** 47–51.

Kaklij, G. S., and Nadkarni, G. B. (1974b). *Arch. Biochem. Biophys.* **160,** 52–57.

Kamen, M. D. (1973). *Tampakushitsu Kakusan Koso* **18,** 753–773.

Kanazawa, A., and Yoshioka, M. (1971). *Bull. Jpn. Soc. Sci. Fish.* **37,** 397–403.

Kanazawa, A., and Yoshioka, M. (1972). *Proc. Int. Seaweed Symp., 7th, 1971* pp. 502–505.

Kanazawa, A., Yoshioka, M., and Teshima, S. (1971). *Bull. Jpn. Soc. Sci. Fish.* **37,** 899–903.

Kannangara, C. G., Jacobson, B. S., and Stumpf, P. K. (1973). *Plant Physiol.* **52,** 156–161.

Kapp, R., Stevens, S. E., Jr., and Fox, J. L. (1975). *Arch. Microbiol.* **104,** 135–138.

Katayama, T. (1964). *Mem. Fac. Fish. Kagoshima Univ.* **13,** 58–72.

Kates, M., and Volcani, B. E. (1966). *Biochim. Biophys. Acta* **116,** 264–278.

Katz, J. J., Strain, H. H., Harkness, A. L., Studier, M. H., Svec, W. A., Janson, T. R., and Cope, B. T. (1972). *J. Am. Chem. Soc.* **94,** 7938–7939.

Kawashima, N., and Wildman, S. G. (1972). *Biochim. Biophys. Acta* **262,** 42–49.

Keilin, D. (1970). "The History of Cell Respiration and Cytochrome." Cambridge Univ. Press, London and New York.

Keilin, D., and Ryley, J. F. (1953). *Nature* (*London*) **172,** 451.

Keith, A. D., Wisnieski, B. J., Henry, S., and Williams, J. C. (1973). *In* "Lipids and Biomembranes of Eukaryotic Microorganisms" (J. A. Erwin, ed.), pp. 259–321. Academic Press, New York.

Kelly, D. P. (1974). *Arch. Microbiol.* **100,** 163–178.

Kelly, J., and Ambler, R. P. (1974). *Biochem. J.* **143,** 681–690.

Kempner, E. S., and Miller, J. H. (1972). *J. Protozool.* **19,** 678–681.

Kennedy, G. Y., and Collier, R. (1963). *J. Mar. Biol. Assoc. U.K.* **43,** 613–619.

Kennedy, R. A., and Laetsch, W. M. (1974). *Science* **184,** 1087–1089.

Kenyon, C. N. (1972). *J. Bacteriol.* **109,** 827–834.

Kenyon, D. H., and Steinman, G. (1969). "Biochemical Predestination" McGraw-Hill, New York.

Kessler, E. (1974). *In* "Algal Physiology and Biochemistry" (W. D. P. Stewart, ed.), pp. 456–473. Blackwell, Oxford.

Kidder, G. W. (1967). *Chem. Zool.* **1,** 93–159.

Kieras, J. H., and Chapman, D. J. (1977). *Carboh. Res.* **52,** 169–177.

Kieras, J. H., Roden, L. R., and Chapman, D. J. (1977). *Biochem. J.* **165,** 1–9.

Kimura, M. (1968). *Nature* (*London*) **217,** 624–626.

Kimura, M. (1969). *Proc. Natl. Acad. Sci. U.S.A.* **63,** 1181–1188.

Kimura, M., and Ohta, T. (1971). *J. Mol. Evol.* **1,** 1–17.

Kimura, M., and Ohta, T. (1973a). *Genet., Suppl.* **73,** 19–35.

Kimura, M., and Ohta, T. (1973b). *Nature* (*London*), *New Biol.* **243,** 199–200.

Kindel, P., and Gibbs, M. (1963). *Nature* (*London*) **200,** 260–261.

King, J. L., and Jukes, T. H. (1969). *Science* **164,** 788–798.

Kirk, J. T. O. (1972). *Sub-Cell. Biochem.* **1,** 333–361.
Kirk, J. T. O., and Tilney-Basset, R. A. E. (1967). "The Plastids." Freeman, San Francisco, California.
Kirk, P. R., and Leech, R. M. (1972). *Plant Physiol.* **50,** 228–234.
Kish, Z., and Jack, R. C. (1974). *Lipids* **9,** 264–268.
Kittredge, J. S., and Roberts, E. (1969). *Science* **164,** 37–42.
Kjölberg, O., and Manners, D. J. (1963). *Biochem. J.* **86,** 10–12.
Klein, R. M., and Cronquist, A. (1967). *Q. Rev. Biol.* **42,** 105–296.
Klenk, E., Knipprath, W., Eberhagen, D., and Koof, H. P. (1963). *Hoppe-Seyler's Z. Physiol. Chem.* **334,** 44–59.
Kloppstech, K., and Schweiger, H. G. (1973a). *Exp. Cell Res.* **80,** 63–68.
Kloppstech, K., and Schweiger, H. G. (1973b). *Exp. Cell Res.* **80,** 69–78.
Knaff, D. B., and Buchanan, B. B. (1975). *Biochim. Biophys. Acta* **376,** 549–560.
Knights, B. A. (1970). *Phytochemistry* **9,** 903–905.
Knivett, V. A. (1954). *Biochem. J.* **58,** 480–486.
Knoll, A. H., and Barghoorn, E. S. (1975). *Science* **190,** 52–54.
Kochert, G., and Sansing, N. (1971). *Biochim. Biophys. Acta* **238,** 397–405.
Kondrat'eva, E. N. (1965). "Photosynthetic Bacteria." Israel Program Sci. Transl., Jerusalem.
Korn, E. D., vonBrand, T., and Tobie, E. J. (1969). *Comp. Biochem. Physiol.* **30,** 601–610.
Kornberg, H. L. (1959). *Annu. Rev. Microbiol.* **13,** 49–78.
Köst, H.-P., Rudiger, W., and Chapman, D. J. (1975). *Justus Liebigs Ann. Chem.* **1975,** 1582–1593.
Kozarenko, N. M., and Sukhenko, F. T. (1971). *Izv. Sib. Otd. Akad. Nauk SSSR, Ser. Biol. Med. Nauk* **2,** 119–123; *Biol. Abstr.* **54,** 49907 (1972).
Kozitskaya, V. N. (1974). *Plant. Physiol.* (*Moscow*) **21,** 240–243. (Engl. Transl.).
Krauspe, R., and Parthier, B. (1973). *Biochem. Soc. Symp.* **38,** 111–135.
Krauspe, R., and Parthier, B. (1974). *Biochem. Physiol. Pflanz.* **165,** 18–36.
Krebs, H. A., and Lowenstein, J. M. (1960). *Metab. Pathways, 2nd Ed.* **1,** 129–203.
Kremer, B. (1976a). *Phytochemistry* **15,** 1135–1138.
Kremer, B. (1976b). *Biochem. Syst. Ecol.* **4,** 139–142.
Kremer, B., and Vogl, R. (1975). *Phytochemistry* **14,** 1309–1314.
Kumar, A. (1969). *Biochim. Biophys. Acta* **186,** 326–331.
Küntzel, H., and Schäfer, K. P. (1971). *Nature* (*London*) *New Biol.* **231,** 265–269.
Künzler, A., and Pfennig, N. (1973). *Arch. Mikrobiol.* **91,** 83–86.
Kuriyama, Y., and Luck, D. J. L. (1973). *J. Mol. Biol.* **73,** 425–437.
Kurtz, D. I. (1974). *Biochemistry* **13,** 572–577.
Kurtz, M., and Bhattacharjee, J. K. (1975). *J. Gen. Microbiol.* **86,** 103–110.
Kwanyuen, P., and Wildman, S. G. (1975). *Biochim. Biophys. Acta* **405,** 167–174.
Kwapinski, J. B. G., and Kwapinski, E. H. (1975). *Can. J. Microbiol.* **21,** 146–151.
Laccy, J. C., Jr., and Pruitt, K. M. (1969). *Nature* (*London*) **223,** 799–804.
Lach, H.-J., Ruppel, H. G., and Böger, P. (1973). *Z. Pflanzenphysiol.* **70,** 432–451.
Lamport, D. T. A. (1965). *Adv. Bot. Res.* **2,** 151–218.
Landymore, A. F. (1976). Ph.D. Thesis. University of British Columbia, Vancouver.
Langley, C. H., and Fitch, W. M. (1974). *J. Mol. Evol.* **3,** 161–177.

Lara, J. C., and Mills, S. E. (1972). *J. Bacteriol.* **110,** 1100–1106.

Largen, M., and Belser, W. L. (1975). *J. Bacteriol.* **121,** 239–249.

Lascelles, J. (1962). *Bacteria* **3,** 335–372.

Latzko, E., and Gibbs, M. (1969). *Plant Physiol.* **44,** 295–300.

Lava-Sanchez, P. A., Amaldi, F., and LaPosta, A. (1972). *J. Mol. Evol.* **2,** 44–55.

Law, A., Thomas, G., and Threlfall, D. R. (1973). *Phytochemistry* **12,** 1999–2004.

Laycock, M. V. (1972). *Can. J. Biochem.* **50,** 1311–1325.

Laycock, M. V. (1975). *Biochem. J.* **149,** 271–279.

Laycock, M. V., and Craigie, J. S. (1970). *Can. J. Biochem.* **48,** 699–701.

Leach, C. K., and Herdman, M. (1973). *In* "The Biology of Blue-Green Algae" (N. G. Carr and B. A. Whitton, eds.), pp. 186–200. Blackwell, Oxford.

Leaver, J. L., and Ramponi, G. (1971). *Biochem. J.* **125,** 44p.

Lebherz, H. G., and Rutter, W. J. (1969). *Biochemistry* **8,** 109–121.

Lebherz, H. G., and Rutter, W. J. (1973). *J. Biol. Chem.* **248,** 1650–1659.

Lechevalier, M. P., Horan, A. C., and Lechevalier, H. (1971). *J. Bacteriol.* **105,** 313–318.

Lederer, F. (1972). *Eur. J. Biochem.* **31,** 144–147.

Lederer, F., and Simon, A. M. (1974). *Biochem. Biophys. Res. Commun.* **56,** 317–323.

Lederer, F., Simon, A.-M., and Verdière, J. (1972). *Biochem. Biophys. Res. Commun.* **47,** 55–58.

Lee, R. E. (1972). *Nature (London)* **237,** 44–46.

Lee, R. F., and Loeblich, A. R., III. (1971). *Phytochemistry* **10,** 593–602.

Lee, S. G., and Evans, W. R. (1971). *Science* **173,** 241–242.

Leech, R. M. (1972). *Biol. Radiobiol. Anucleate Syst., Proc. Int. Symp., 1971* Vol. 2, pp. 27–49.

Leech, R. M., and Kirk, P. R. (1968). *Biochem. Biophys. Res. Commun.* **32,** 685–690.

Leedale, G. F. (1970). *Ann. N.Y. Acad. Sci.* **175,** 429–453.

Leedale, G. F. (1974). *Taxon* **23,** 261–270.

Lefebvre, G., Schneider, F., Germain, P., and Gay, R. (1974). *Tetrahedron Lett.* pp. 127–128.

Leftley, J. W., and Syrett, P. J. (1973). *J. Gen. Microbiol.* **77,** 109–115.

Leick, V. (1969). *Eur. J. Biochem.* **8,** 221–228.

Leighton, T. J., Dill, B. C., Stock, J. J., and Phillips, C. (1971). *Proc. Nat. Acad. Sci. U.S.A.* **68,** 677–680.

LéJohn, H. B. (1971a). *Nature (London)* **231,** 164–168.

LéJohn, H. B. (1971b). *Biochem. Biophys. Res. Commun.* **42,** 538–544.

LéJohn, H. B. (1974). *Evol. Biol.* **7,** 79–125.

Lemasson, C., Tandeau de Marsac, N., and Cohen-Bazire, G. (1973). *Proc. Natl. Acad. Sci. U.S.A.* **70,** 3130–3133.

Lemberg, R., and Barrett, J. (1972). "Cytochromes." Academic Press, New York.

Lenfant, M., Lecompte, M. F., and Farrugia, G. (1970). *Phytochemistry* **9,** 2529–2535.

Lenton, J. R., Goad, L. J., and Goodwin, T. W. (1973). *Phytochemistry* **12,** 2249–2253.

Lerner, S. A., Wu, T. T., and Lin, E. C. C. (1964). *Science* **146,** 1313–1315.
Lesk, A. M. (1969). *J. Theor. Biol.* **22,** 537–540.
Levenberg, B. (1962). *J. Biol. Chem.* **237,** 2590–2598.
Levin, E., Lennarz, W. J., and Bloch, K. (1964). *Biochim. Biophys. Acta* **84,** 471–474.
Levitina, T. P., and Pinevich, V. V. (1974). *Dokl. Biol. Sci.* **214,** 20–22.
Lewin, J. C. (1955). *J. Gen. Microbiol.* **13,** 162–169.
Lewin, R. A. (1958). *J. Gen. Microbiol.* **19,** 87–90.
Lewin, R. A. (1974). *In* "Algal Physiology and Biochemistry" (W. D. P. Stewart, ed.), pp. 1–39. Blackwell, Oxford.
Lewin, R. A. (1975). *Phycologia* **14,** 149–160.
Lewin, R. A. (1976) *Nature* (*London*) **261,** 697–698.
Lewin, R. A., and Withers, N. W. (1975). *Nature* (*London*) **256,** 735–737.
Lewis, D. H., and Smith, D. C. (1967). *New Phytol.* **66,** 143–184.
Lewis, E. J., and Gonzalves, E. A. (1960). *New Phytol.* **59,** 109–115.
Lewis, E. J., and Gonzalves, E. A. (1962). *Ann. Bot.* (*London*) [N.S.] **26,** 301–316.
Lewontin, R. C. (1972). *Nature* (*London*) **236,** 181–182.
Li, S.-L., and Yanofsky, C. (1973a). *J. Biol. Chem.* **248,** 1830–1836.
Li, S.-L., and Yanofsky, C. (1973b). *J. Biol. Chem.* **248,** 1837–1843.
Liaaen-Jensen, S., and Andrewes, A. G. (1972). *Annu. Rev. Microbiol.* **26,** 225–248.
Lilly, D. M. (1967). *Chem. Zool.* **1,** 275–307.
Lin, C. C., and Aronson, J. M. (1970). *Arch. Mikrobiol.* **72,** 111–114.
Lindmark, D. G., and Müller, M. (1974). *J. Biol. Chem.* **249,** 4634–4637.
Lipmann, F. (1971). *In* "Chemical Evolution and the Origin of Life" (R. Buvet and C. Ponnamperuma, eds.), pp. 381–391. North-Holland Publ., Amsterdam.
Livermore, B. P., and Johnson, R. C. (1974). *J. Bacteriol.* **120,** 1268–1273.
Lloyd, D. (1974). *In* "Algal Physiology and Biochemistry" (W. D. P. Stewart, ed.), pp. 505–529. Blackwell, Oxford.
Loeblich, A. R., Jr. (1974). *Taxon* **23,** 277–290.
Loening, U. E. (1967). *Biochem. J.* **102,** 251–257.
Loening, U. E. (1968). *J. Mol. Biol.* **38,** 355–365.
Loening, U. E. (1973). *Pure Appl. Chem.* **34,** 579–589.
Loening, U. E., and Ingle, J. (1967). *Nature* (*London*) **215,** 363–367.
Loewus, M. W., and Delwiche, C. C. (1963). *Plant Physiol.* **38,** 371–374.
Lohr, J. B., and Friedmann, R. C. (1976). *Biochem. Biophys. Res. Commun.* **69,** 908–913.
London, J. (1974). *J. Biol. Chem.* **249,** 7977–7983.
London, J., and Kline, K. (1973). *Bacteriol. Rev.* **37,** 453–478.
Lord, J. M., and Brown, R. H. (1975). *Plant Physiol.* **55,** 360–364.
Lord, J. M., Codd, G. A., and Stewart, W. D. P. (1975). *Plant Sci. Lett.* **4,** 377–383.
Lovett, J. S. (1963). *J. Bacteriol.* **85,** 1235–1246.
Lovett, J. S., and Haselby, J. A. (1971). *Arch. Mikrobiol.* **80,** 191–204.
Low, E. M. (1955). *J. Mar. Res.* **14,** 199–204.
Lucas, C., and Weitzman, P. D. J. (1975). *Biochem. Soc. Trans.* **3,** 379–381.
Luft, J. H. (1971). *In* "Biological Ultrastructure: The Origin of Cell Organelles" (P. J. Harris, ed.), pp. 1–23. Oregon State Univ. Press, Corvallis.

Lui, N. S. T., and Roels, O. A. (1970). *Arch. Biochem. Biophys.* **139,** 269–277.

Lumsden, J., and Hall, D. O. (1974). *Biochem. Biophys. Res. Commun.* **58,** 35–41.

Lumsden, J., and Hall, D. O. (1975a). *Biochem. Biophys. Res. Commun.* **64,** 595–602.

Lumsden, J., and Hall, D. O. (1975b). *Nature (London)* **257,** 670–672.

Lwoff, A. (1951). *Biochem. Physiol. Protozoa* **1,** 1–26.

Lynch, V. H., and Calvin, M. (1952). *J. Bacteriol.* **63,** 525–531.

Lynen, F. (1961). *Fed. Proc., Fed. Am. Soc. Exp. Biol.* **20,** 941–951.

Lynen, F. (1967). *Pure Appl. Chem.* **14,** 137–167.

McClendon, J. H. (1976). *J. Mol. Evol.* **8,** 175–195.

McCorkindale, N. J., Hutchinson, S. A., Pursey, B. A., Scott, W. T., and Wheeler, R. (1969). *Phytochemistry* **8,** 861–867.

McCully, E. K., and Robinow, C. F. (1972a). *J. Cell Sci.* **10,** 857–881.

McCully, E. K., and Robinow, C. F. (1972b). *J. Cell Sci.* **11,** 1–31.

McDonald, D. W., and Coddington, A. (1974). *Eur. J. Biochem.* **46,** 169–178.

MacElroy, R. D., Johnson, E. J., and Johnson, M. K. (1968). *Arch. Biochem. Biophys.* **127,** 310–316.

McFadden, B. A. (1973). *Bacteriol. Rev.* **37,** 289–319.

McFadden, B. A., and Tabita, F. R. (1974). *BioSystems* **6,** 93–112.

McGowan, R. E., and Gibbs, M. (1974). *Plant Physiol.* **54,** 312–319.

McGregor-Shaw, J. B. (1971). M.Sc. Thesis, Dalhousie University, Halifax.

Mackay, A. L. (1967). *Nature (London)* **216,** 159–160.

Mackie, W., and Preston, R. D. (1974). *In* "Algal Physiology and Biochemistry" (W. D. P. Stewart, ed.), pp. 40–85. Blackwell, Oxford.

McLachlan, J., and Craigie, J. S. (1966). *In* "Some Contemporary Studies in Marine Science" (H. Barnes, ed.), pp. 511–517. Allen & Unwin, London.

McLaughlin, P. J., and Dayhoff, M. O. (1973). *J. Mol. Evol.* **2,** 99–116.

Madgwick, J. C., and Ralph, B. J. (1972). *Bot. Mar.* **15,** 205–209.

Mahler, H. R. (1973). *Crit. Rev. Biochem.* **1,** 381–460.

Mahler, H. R., and Raff, R. A. (1975). *Int. Rev. Cytol.* **43,** 1–124.

Makino, F., and Tsuzuki, J. (1971). *Nature (London)* **231,** 446–447.

Malofeeva, I. V., Kondratieva, E. N., and Rubin, A. B. (1975). *FEBS Lett.* **53,** 188–189.

Mandel, M. (1967). *Chem. Zool.* **1,** 541–572.

Mandel, M. (1968). *In* "Handbook of Biochemistry" (H. A. Sober, ed.), pp. H-27–H-29. CRC Press, Cleveland, Ohio.

Mandel, M., Bergendahl, J. C., and Pfennig, N. (1965). *J. Bacteriol.* **89,** 917–918.

Mandelli, E. F. (1968). *J. Phycol.* **4,** 347–348.

Mangnall, D., and Getz, G. S. (1973). *In* "Lipids and Biomembranes of Eukaryotic Microorganisms" (J. A. Erwin, ed.), pp. 145–195. Academic Press, New York.

Manners, D. J., Pennie, I. R., and Ryley, J. F. (1967). *Biochem. J.* **104,** 32p.

Marchelli, R., and Vining, L. C. (1973). *Can. J. Biochem.* **51,** 1624–1629.

Margoliash, E. (1970). *Science* **168,** 127.

Margulis, L. (1968). *Science* **161,** 1020–1022.

Margulis, L. (1969). *J. Geol.* **77,** 606–617.

Margulis, L. (1970). "Origin of Eukaryotic Cells." Yale Univ. Press, New Haven, Connecticut.

Margulis, L. (1972). *In* "Exobiology"(C. Ponnamperuma, ed.), pp. 342–368. North-Holland Publ., Amsterdam.

Margulis, L. (1974). *Evol. Biol.* **7,** 45–78.

Marsh, H. V., Galmiche, J. M., and Gibbs, M. (1965). *Plant Physiol.* **40,** 1013–1022.

Marzzoco, A., and Colli, W. (1974). *Biochim. Biophys. Acta* **374,** 292–303.

Massarini, E., and Cazzulo, J. J. (1975). *FEBS Lett.* **57,** 134–138.

Masters, C. J., and Holmes, R. S. (1972). *Biol. Rev. Cambridge Philos. Soc.* **47,** 309–361.

Matchett, W. H., and DeMoss, J. A. (1975). *J. Biol. Chem.* **250,** 2941–2946.

Mautner, H. G. (1954). *Econ. Bot.* **8,** 174–192.

Maynard Smith, J. (1968). *Nature* (*London*) **219,** 1114–1116.

Mayr, E. (1969). "Principles of Systematic Zoology." McGraw-Hill, New York.

Mazelis, M., Miflin, B. J., and Pratt, H. M. (1976). *FEBS Lett.* **64,** 197–200.

Mazin, A. L., and Sulimova, G. E. (1974). *Dokl. Biochem.* **211,** 358–361.

Meatyard, B. T., Scawen, M. D., Ramshaw, J. A. H., and Boulter, D. (1975). *Phytochemistry* **14,** 1493–1498.

Meeks, J. C. (1974). *In* "Algal Physiology and Biochemistry" (W. D. P. Stewart, ed.), pp. 161–175. Blackwell, Oxford.

Meeuse, B. J. D., and Hall, D. M. (1973). *Ann. N.Y. Acad. Sci.* **210,** 39–45.

Mehler, A. H. (1950). *Biol. Bull.* **99,** 371.

Mercer, E. I., and Bartlett, K. (1974). *Phytochemistry* **13,** 1099–1106.

Mercer, E. I., and Carrier, D. J. R. (1976). *Phytochem.* **15,** 283–286.

Mercer, E. I., and Harries, W. B. (1975). *Phytochemistry* **14,** 439–443.

Mercer, E. I., London, R. A., Kent, I. S. A., and Taylor, A. J. (1974). *Phytochemistry* **13,** 845–852.

Mereschkowsky, C. (1905). *Biol. Centralbl.* **25,** 593–604 and 689–691.

Merrett, M. J., and Lord, J. M. (1973). *New Phytol.* **72,** 751–767.

Metzner, H. (1973). *Naturwissenschaften* **60,** 507–515.

Meyer, R. E. (1973). *J. Theor. Biol.* **38,** 647–663.

Meyer, T. E., Ambler, R. P., Bartsch, R. G., and Kamen, M. D. (1975). *J. Biol. Chem.* **250,** 8416–8421.

Midgley, J. E. M. (1962). *Biochim. Biophys. Acta* **61,** 513–525.

Mignot, J-P., Joyon, L., and Pringsheim, E. G. (1969). *J. Protozool.* **16,** 138–145.

Miller, D. H., Lamport, D. T. A., and Miller, M. (1972). *Science* **176,** 918–920.

Miller, J. S., and Allen, M. M. (1972). *Arch. Mikrobiol.* **86,** 1–12.

Miller, M. J., and McMahon, D. (1974). *Biochim. Biophys. Acta* **366,** 35–44.

Miller, S. L. (1953). *Science* **117,** 528–529.

Miller, S. L. (1955). *J. Am. Chem. Soc.* **77,** 2351–2361.

Mills, G. L., Myers, R. B., and Cantino, E. C. (1974). *Phytochemistry* **13,** 2653–2657.

Milne, P. R., Wells, J. R. E., and Ambler, R. P. (1974). *Biochem. J.* **143,** 691–701.

Misra, H. P., and Keele, B. B., Jr. (1975). *Biochim. Biophys. Acta* **379,** 418–425.

Mitchell, P. (1969). *Theor. Exp. Biophys.* **2,** 159–216.

Mittenzwei, H. (1942). *Hoppe-Seyler's Z. Physiol. Chem.* **275,** 93–121.

Miyazawa, K. (1971). *Bull. Jpn. Soc. Sci. Fish.* **37,** 788–794.

Mo, Y., Harris, B. G., and Gracy, R. W. (1973). *Arch. Biochem. Biophys.* **157,** 580–587.

Møller, B. L. (1974). *Plant Physiol.* **54,** 638–643.

Mohberg, J., and Rusch, H. P. (1969). *Arch. Biochem. Biophys.* **134,** 577–589.

Mooney, C. L., Mahoney, E. M., Pousada, M., and Haines, T. H. (1972). *Biochemistry* **11,** 4839–4844.

Moore, P. D. (1974). *Nature* (*London*) **251,** 380.

Moore, R. E., and Yost, G. (1973). *J. Chem. Soc., Chem. Commun.* pp. 937–938.

Morris, I. (1974). *In* "Algal Physiology and Biochemistry" (W. D. P. Stewart, ed.), pp. 583–609. Blackwell, Oxford.

Moshier, S. E., and Chapman, D. J. (1973). *Biochem. J.* **136,** 395–404.

Müller, F. (1864). "Für Darwin." Leipzig (translated into English, 1869, as "Facts and Arguments for Darwin"). Murray, London (referenced in de Beer, 1958).

Müller, M., Hogg, J. F., and de Duve, C. (1968). *J. Biol. Chem.* **243,** 5385–5395.

Mullins, D. W., Jr., Lacey, J. C., Jr., and Hearn, R. A. (1973). *Nature* (*London*) *New Biol.* **242,** 80–82.

Munsche, D., and Wollgiehn, R. (1973). *Biochim. Biophys. Acta* **294,** 106–117.

Murphy, M. J., and Siegel, L. M. (1973). *J. Biol. Chem.* **248,** 6911–6919.

Murphy, M. N., and Lovett, J. S. (1966). *Dev. Biol.* **14,** 68–95.

Murphy, T. M., and Mills, S. E. (1969). *J. Bacteriol.* **97,** 1310–1320.

Nägeli, C. (1846). *Z. Wiss. Bot.* **3,** 94 (referenced by Schenk and Hofer, 1972).

Nakabayashi, T. (1955). *J. Agric. Chem. Soc. Jpn.* **29,** 897–899.

Nakajima, T., and Volcani, B. E. (1969). *Science* **164,** 1400–1401.

Nakase, T., and Komagata, K. (1971a). *J. Gen. Appl. Microbiol.* **17,** 43–50.

Nakase, T., and Komagata, K. (1971b). *J. Gen. Appl. Microbiol.* **17,** 227–238.

Nambudiri, A. M. D., Vance, C. P., and Towers, G. H. N. (1973). *Biochem. J.* **134,** 891–897.

Nasatir, M., and Brooks, A. E. (1966). *J. Phycol.* **2,** 144–147.

Nason, A. (1962). *Bacteriol. Rev.* **26,** 16–41.

Nass, S. (1969). *Int. Rev. Cytol.* **25,** 55–129.

Naylor, A. W. (1959). *Symp. Soc. Exp. Biol.* **13,** 193–209.

Naylor, A. W. (1970). *Ann. N.Y. Acad. Sci.* **175,** 511–523.

Neilson, A. H., and Doudoroff, M. (1973). *Arch. Mikrobiol.* **89,** 15–22.

Neilson, A. H., and Lewin, R. A. (1974). *Phycologia* **13,** 227–264.

Neish, A. C. (1964). *In* "Biochemistry of Phenolic Compounds" (J. B. Harborne, ed.), pp. 295–359. Academic Press, New York.

Netrawali, M. S. (1970). *Exp. Cell Res.* **63,** 422–426.

Nichols, B. W. (1970). *In* "Phytochemical Phylogeny" (J. B. Harborne, ed.), pp. 105–118. Academic Press, New York.

Nichols, B. W. (1973). *In* "The Biology of Blue-Green Algae" (N. G. Carr and B. A. Whitton, eds.), pp. 144–161. Blackwell, Oxford.

Nichols, B. W., and Appleby, R. S. (1969). *Phytochemistry* **8,** 1907–1915.

Nichols, B. W., Harris, R. V., and James, A. T. (1965). *Biochem. Biophys. Res. Commun.* **20,** 256–262.

Niederpruem, D. J. (1965). *In* "The Fungi: An Advanced Treatise" (G. C. Ainsworth and A. S. Sussman, eds.), Vol. I, pp. 269–300. Academic Press, New York.

Nikolaev, N., Birenbaum, M., and Schlessinger, D. (1975). *Biochim. Biophys. Acta* **395,** 478–489.

Nitsche, H. (1974). *Biochim. Biophys. Acta* **338,** 572–576.

Njus, D., Sulzman, F. M., and Hastings, J. W. (1974). *Nature (London)* **248,** 116–120.

Nolan, C., and Margoliash, E. (1968). *Annu. Rev. Biochem.* **37,** 727–790.

Norgård, S., Svec, W. A., Liaaen-Jensen, S., Jensen, A., and Guillard, R. R. L. (1974a). *Biochim. Syst. Ecol.* **2,** 3–6.

Norgård, S., Svec, W. A., Liaaen-Jensen, S., Jensen, A., and Guillard, R. R. L. (1974b). *Biochem. Syst. Ecol.* **2,** 7–9.

Norris, L., Norris, R. E., and Calvin, M. (1955). *J. Exp. Bot.* **6,** 64–74.

Novaes-Ledieu, M., Jiménez-Martínez, A., and Villanueva, J. R. (1967). *J. Gen. Microbiol.* **47,** 237–245.

O'Carra, P., and O'hEocha, C. (1976). *In* "Chemistry and Biochemistry of Plant Pigments," 2nd. ed. (T. W. Goodwin, ed.), Vol. 1, pp. 328–376. Academic Press, New York.

Ogata, K., Uchiyama, K., and Yamada, H. (1967). *Agric. Biol. Chem.* **31,** 200–206.

Ogawa, T., Vernon, L. P., and Yamamoto, H. Y. (1970). *Biochim. Biophys. Acta* **197,** 302–307.

O'hEocha, C. (1965). *Annu. Rev. Plant Physiol.* **16,** 415–434.

Ohta, T., and Kimura, M. (1971). *Nature (London)* **233,** 118–119.

Okunuki, K., Kamen, M. D., and Sekuzu, I., eds. (1968). "Symposium on Structural and Chemical Aspects of Cytochromes." Univ. of Tokyo Press, Tokyo.

Olson, J. M. (1970). *Science* **168,** 438–446.

Onodera, R., and Kandatsu, M. (1973). *Nature (London) New Biol.* **244,** 31–32.

Onodera, R., and Kandatsu, M. (1974). *Agric. Biol. Chem.* **38,** 913–920.

Onodera, R., Shinjo, T., and Kandatsu, M. (1974). *Agric. Biol. Chem.* **38,** 921–926.

Opute, F. I. (1974a). *J. Exp. Bot.* **25,** 810–822.

Opute, F. I. (1974b). *J. Exp. Bot.* **25,** 823–835.

Orgel, L. E. (1968). *J. Mol. Biol.* **38,** 381–393.

Pace, B., and Campbell, L. L. (1967). *Proc. Natl. Acad. Sci. U.S.A.* **57,** 1110–1116.

Pace, N. R. (1973). *Bacteriol. Rev.* **37,** 562–603.

Packter, N. M. (1972). "Biosynthesis of Acetate-Derived Compounds." Wiley, New York.

Pakhomova, M. V. (1974). *Dokl. Biochem.* **214,** 71–73.

Pakhomova, M. V., Zaitseva, G. N., and Belozerskii, A. N. (1968). *Dokl. Biochem.* **182,** 227–229.

Paleontology Correspondent. (1974). *Nature (London)* **248,** 635.

Palmer, F. B. St.C. (1973). *Biochim. Biophys. Acta* **326,** 194–200.

Palmer, F. B. St.C. (1974). *J. Protozool.* **21,** 160–163.

Pannbacker, R. G., and Wright, B. E. (1967). *Chem. Zool.* **1,** 573–616.

Paoletti, C., Pushparaj, B., Florenzano, G., Capella, P., and Lercker, G. (1976). *Lipids* **11,** 266–271.

Papentin, F. (1973a). *J. Theor. Biol.* **39,** 397–415.

Papentin, F. (1973b). *J. Theor. Biol.* **39,** 417–430.

Papentin, F. (1973c). *J. Theor. Biol.* **39,** 431–445.

Parker, P. L., Van Baalen, C., and Maurer, L. (1967). *Science* **155,** 707–708.

Parsons, T. R. (1961). *J. Fish. Res. Board Can.* **18,** 1017–1025.

Parthier, B., and Krauspe, R. (1974). *Biochem. Physiol. Pflanz.* **165,** 1–17.

Parthier, B., Krauspe, R., and Samtleben, S. (1972). *Biochim. Biophys. Acta* **277,** 335–341.

Pascher, A. (1914). *Ber. Dtsch. Bot. Ges.* **32,** 136–160.

Pascher, A. (1921). *Ber. Dtsch. Bot. Ges.* **39,** 236–248.

Pascher, A. (1927). *Arch. Protistenkd.* **58,** 1–54.

Patel, R., Hoare, S. L., Hoare, D. S., and Taylor, B. F. (1975). *J. Bacteriol.* **123,** 382–384.

Patterson, G. W. (1968). *Comp. Biochem. Physiol.* **24,** 501–505.

Patterson, G. W. (1971). *Lipids* **6,** 120–127.

Patterson, G. W. (1972). *Phytochemistry* **11,** 3481–3483.

Patterson, G. W., Thompson, M. J., and Dutky, S. R. (1974). *Phytochemistry* **13,** 191–194.

Paul, J. S., and Volcani, B. E. (1974). *Arch. Microbiol.* **101,** 115–120.

Paul, J. S., and Sullivan, C. W., and Volcani, B. E. (1975). *Arch. Biochem. Biophys.* **169,** 152–159.

Peacock, D., and Boulter, D. (1975). *J. Mol. Biol.* **95,** 513–527.

Pearce, J., and Carr, N. G. (1969). *J. Gen. Microbiol.* **54,** 451–462.

Peck, H. D., Jr. (1962). *Bacteriol. Rev.* **26,** 67–94.

Peck, H. D., Jr. (1974). *Symp. Soc. Gen. Microbiol.* **24,** 241–262.

Peck, H. D., Jr., Tedro, S., and Kamen, M. D. (1974). *Proc. Natl. Acad. Sci. U.S.A.* **71,** 2404–2406.

Pedersén, M., and DaSilva, E. J. (1973). *Planta* **115,** 83–86.

Pedersén, M., and Fries, L. (1975). *Z. Pflanzenphysiol.* **74,** 272–274.

Penny, D. (1974). *J. Mol. Evol.* **3,** 179–188.

Percival, E., and McDowell, R. H. (1967). "Chemistry and Enzymology of Marine Algal Polysaccharides." Academic Press, New York.

Perry, R. P., Cheng, T.-Y., Freed, J. J., Greenberg, J. R., Kelley, D. E., and Tartof, K. D. (1970a). *Proc. Natl. Acad. Sci. U.S.A.* **65,** 609–616.

Perry, R. P., Greenberg, J. R., and Tartof, K. D. (1970b). *Cold Spring Harbor Symp. Quant. Biol.* **35,** 577–587.

Peterkofsky, A., and Racker, E. (1961). *Plant Physiol.* **36,** 409–414.

Peterson, D. H. (1963). *In* "Biochemistry of Industrial Microorganisms" (C. Rainbow and A. H. Rose, eds.), pp. 538–606. Academic Press, New York.

Pettigrew, G. W. (1974). *Biochem. J.* **139,** 449–459.

Pettigrew, G. W., Leaver, J. L., Meyer, T. E., and Ryle, A. P. (1975). *Biochem. J.* **147,** 291–302.

Pfennig, N. (1967). *Annu. Rev. Microbiol.* **21,** 285–324.

Pham Quang, L., and Laur, M. H. (1974). *Biochimie* **56,** 925–935.
Pham Quang, L., and Laur, M. H. (1975). *Bot. Mar.* **18,** 105–113.
Philippe, M., Fournet, B., and Schrével, J. (1975). *C. R. Hebd. Seances Acad. Sci., Ser. D* **280,** 1801–1804.
Pichinoty, F. (1966). *Bull. Soc. Fr. Physiol. Veg.* **12,** 97–104.
Pichinoty, F., and Méténier, G. (1967). *Ann. Inst. Pasteur, Paris* **112,** 701–711.
Pickett-Heaps, J. D. (1972). *New Phytol.* **71,** 561–567.
Pickett-Heaps, J. D. (1975). "Green Algae: Structure, Reproduction and Evolution in Selected Genera." Sinauer, Sunderland, Massachusetts.
Pickett-Heaps, J. D., and Marchant, H. J. (1972). *Cytobios* **6,** 255–264.
Pierson, B. K., and Castenholz, R. W. (1974a). *Arch. Microbiol.* **100,** 5–24.
Pierson, B. K., and Castenholz, R. W. (1974b). *Arch. Microbiol.* **100,** 283–305.
Pigott, G. H., and Carr, N. G. (1972a). *Photosynth., Two Centuries Its Discovery Joseph Priestley, Proc. Int. Congr. Photosynth. Res., 2nd, 1971* Vol. 3, pp. 2685–2689.
Pigott, G. H., and Carr, N. G. (1972b). *Science* **175,** 1259–1261.
Plouvier, V. (1963). *In* "Chemical Plant Taxonomy" (T. Swain, ed.), pp. 313–336. Academic Press, New York.
Pohl, P., Wagner, H., and Passig, T. (1968). *Phytochemistry* **7,** 1565–1572.
Pommier, M. T., and Michel, G. (1973). *Biochem. Syst.* **1,** 3–12.
Ponsinet, G., Ourisson, G., and Oehlschlager, A. C. (1968). *Recent. Adv. Phytochem.* **1,** 271–302.
Posner, H. B., Gressel, J., and Rosner, A. (1974). *Plant Cell Physiol.* **15,** 807–811.
Power, D. M., Towers, G. H. N., and Neish, A. C. (1965). *Can. J. Biochem.* **43,** 1397–1407.
Powls, R., and Redfearn, E. R. (1967). *Biochem. J.* **104,** 24c–26c.
Preddie, D. L., Preddie, E. C., Guerrini, A. M., and Cremona, T. (1973). *Can. J. Biochem.* **51,** 951–954.
Prieur, P., and Tavlitzki, J. (1974). *Ann. Microbiol.* (*Inst. Pasteur*) *A* **125,** 3–16.
Prigogine, I. (1947). "Etude thermodynamique des phénomènes irréversibles." Desoer, Liège.
Prigogine, I. (1969). *In* "Theoretical Physics and Biology" (M. Marois, ed.), pp. 23–52. North-Holland Publ., Amsterdam.
Pringsheim, E. G. (1958). *Stud. Plant Physiol.* (*Prague*) pp. 165–184.
Provasoli, L. (1958). *Annu. Rev. Microbiol.* **12,** 279–308.
Provasoli, L., and Carlucci, A. F. (1974). *In* "Algal Physiology and Biochemistry" (W. D. P. Stewart, ed.), pp. 788–813. Blackwell, Oxford.
Pryce, R. J. (1971). *Phytochemistry* **10,** 2679–2685.
Pryce, R. J. (1972). *Phytochemistry* **11,** 1759–1761.
Punnett, T., and Derrenbacker, E. C. (1966). *J. Gen. Microbiol.* **44,** 105–114.
Purdie, J. W., and Holt, A. S. (1965). *Can. J. Chem.* **43,** 3347–3353.
Rabache, M., Neumann, J., and Lavollay, J. (1974). *Phytochemistry* **13,** 637–642.
Rabinowitch, H. D., Budowski, P., and Kedar, N. (1975). *Planta* **122,** 91–97.
Radler, F., Greese, K. D., Bock, R., and Seiler, W. (1974). *Arch. Microbiol.* **100,** 243–252.
Radunz, A. (1968). *Hoppe-Seyler's Z. Physiol. Chem.* **349,** 1091–1094.

Radunz, A. (1969). *Hoppe-Seyler's Z. Physiol. Chem.* **350,** 411–417.
Rae, P. M. M. (1970). *J. Cell Biol.* **46,** 106–113.
Rae, P. M. M. (1973). *Proc. Natl. Acad. Sci. U.S.A.* **70,** 1141–1145.
Rae, P. M. M. (1976). *Science* **194,** 1062–1064.
Raff, R. A., and Mahler, H. R. (1972). *Science* **177,** 575–582.
Raff, R. A., and Mahler, H. R. (1973). *Science* **180,** 516–517.
Ragan, M. A., and Craigie, J. S. (1976). *Can. J. Biochem.* **54,** 66–73.
Ragsdale, N. N. (1975). *Biochim. Biophys. Acta* **380,** 81–96.
Ramachandran, S., and Gottlieb, D. (1963). *Biochim. Biophys. Acta* **69,** 74–84.
Ramanathan, J. D., Craigie, J. S., McLachlan, J., Smith, D. G., and McInnes, A. G. (1966). *Tetrahedron Lett.* No. 14, pp. 1527–1531.
Ramshaw, J. A. M., Peacock, D., Meatyard, B. T., and Boulter, D. (1974). *Phytochemistry* **13,** 2783–2789.
Rapoport, G., Davis, L., and Horecker, B. L. (1969). *Arch. Biochem. Biophys.* **132,** 286–293.
Raszka, M., and Mandel, M. (1972). *J. Mol. Evol.* **2,** 38–43.
Ratner, S., and Petrack, B. (1953). *J. Biol. Chem.* **200,** 161–174.
Raudaskoski, M. (1971). *Luonnon Tutkija* **75,** 165–171.
Raven, P. H. (1970). *Science* **169,** 641–646.
Ravindrath, S. D., and Fridovich, I. (1975). *J. Biol. Chem.* **250,** 6107–6112.
Rawson, J. R. Y., and Haselkorn, R. (1973). *J. Mol. Biol.* **77,** 125–132.
Reazin, G. H., Jr. (1956). *Plant Physiol.* **31,** 299–303.
Reddi, T. K., and Borovkov, A. V. (1969). *Chem. Natur. Compounds* **5,** 116. (Engl. Transl.).
Reger, B. J., Fairfield, S. A., Epler, J. L., and Barnett, W. E. (1970). *Proc. Natl. Acad. Sci. U.S.A.* **67,** 1207–1213.
Reid, P. D., Havir, E. A., and Marsh, H. V., Jr. (1972). *Plant Physiol.* **50,** 480–484.
Reijnder, L. (1975). *J. Mol. Evol.* **5,** 167–176.
Reisner, A. H., Rowe, J., and Macindoe, H. M. (1968). *J. Mol. Biol.* **32,** 587–610.
Reithel, F. J. (1971). *In* "The Enzymes" (P. D. Boyer, ed.), 3rd ed., Vol. 4, pp. 1–21. Academic Press, New York.
Reitz, R. C., and Hamilton, J. G. (1968). *Comp. Biochem. Physiol.* **25,** 401–416.
Reuter, G. (1961). *Nature (London)* **190,** 447.
Rho, J. H. (1957). Ph.D. Thesis, Duke University, Durham, North Carolina.
Richards, O. C., and Ryan, R. S. (1974). *J. Mol. Biol.* **82,** 57–75.
Richards, R. L., and Quackenbush, F. W. (1974). *Arch. Biochem. Biophys.* **165,** 780–786.
Ricketts, T. R. (1967). *Phytochemistry* **6,** 1375–1386.
Ricketts, T. R. (1971a). *Phytochemistry* **10,** 155–160.
Ricketts, T. R. (1971b). *Phytochemistry* **10,** 161–164.
Ridley, S. M., and Leech, R. M. (1970). *Nature (London)* **227,** 463–465.
Rigano, C. (1971). *Arch. Mikrobiol.* **76,** 265–276.
Rijven, A. H. G. C., and Zwar, J. A. (1973). *Biochim. Biophys. Acta* **299,** 564–567.
Riley, J. P., and Segar, D. A. (1969). *J. Mar. Biol. Assoc. U.K.* **49,** 1047–1056.
Riley, J. P., and Wilson, T. R. S. (1967). *J. Mar. Biol. Assoc. U.K.* **47,** 351–362.
Ripa, R., and Borquez, L. M. (1967). *Scientia* **133,** 36–38.

Rippka, R., Waterbury, J., and Cohen-Bazire, G. (1974). *Arch. Microbiol.* **100,** 419–436.
Ris, H. (1961). *Can. J. Genet. Cytol.* **3,** 95–120.
Ris, H., and Plaut, W. (1962). *J. Cell Biol.* **13,** 383–391.
Rizzo, P. J. (1976), *J. Mol. Evol.* **8,** 79–94.
Rizzo, P. J., and Noodén, L. D. (1972). *Science* **176,** 796–797.
Rizzo, P. J., and Noodén, L. D. (1974a). *Biochim. Biophys. Acta* **349,** 402–414.
Rizzo, P. J., and Noodén, L. D. (1974b). *Biochim. Biophys. Acta* **349,** 415–427.
Robinson, T. (1968). "Biochemistry of Alkaloids." Springer-Verlag, Berlin and New York.
Rocha, V., Crawford, I. P., and Mills, S. E. (1972). *J. Bacteriol.* **111,** 163–168.
Rock, R. C. (1971). *Comp. Biochem. Physiol. B* **40,** 657–669.
Rodríguez-López, M., and Muñoz Calvo, M. L. (1971). *Actas Simp. Int. Zoofilogenia, 1st, 1971* pp. 463–469.
Rodríguez-López, M., and Vasquez, D. (1968). *Life Sci.* **7,** 327–336.
Rogers, H. J., and Perkins, H. R. (1968). "Cell Walls and Membranes." Spon, London.
Rönnerstrand, S. (1968). *Bot. Mar.* **11,** 106–114.
Roon, R. J., and Levenberg, B. (1968). *J. Biol. Chem.* **243,** 5213–5215.
Roon, R. J., and Levenberg, B. (1970). *In "Methods in Enzymology"* (H. Tabor and C. W. Tabor, eds.), Vol. 17A, pp. 317–324. Academic Press, New York.
Rosell, K.-G., and Birkhed, D. (1974). *Acta Chem. Scand. Ser. B* **28,** 589.
Rosen, R. (1973). *In* "Biogenesis Evolution Homeostasis" (A. Locker, ed.), pp. 113–118. Springer-Verlag, Berlin and New York.
Rosenberg, A. (1973). *In* "Lipids and Biomembranes of Eukaryotic Microorganisms" (J. A. Erwin, ed.), pp. 233–257. Academic Press, New York.
Rossmann, M. (1974). *New Sci.* **61,** 266–268.
Roth-Bejerano, N., and Lips, S. H. (1973). *Isr. J. Bot.* **22,** 1–7.
Rubinstein, I., and Goad, L. J. (1974a). *Phytochemistry* **13,** 481–484.
Rubinstein, I., and Goad, L. J. (1974b). *Phytochemistry* **13,** 485–487.
Rüdiger, W. (1971). *Fortshr. Chem. Org. Naturst.* **29,** 59–139.
Rüdiger, W. (1975). *Ber. Dtsch. Bot. Ges.* **88,** 125–139.
Rüdiger, W., Hedden, P., Köst, H.-P., and Chapman, D. J. (1977). *Biochim. Biophys. Res. Comm.* **74,** 1268–1272.
Russell, G. K., and Gibbs, M. (1967). *Biochim. Biophys. Acta* **132,** 145–154.
Russell, G. K., and Gibbs, M. (1968). *Plant Physiol.* **43,** 649–652.
Rutter, W. J. (1964). *Fed. Proc., Fed. Am. Soc. Exp. Biol.* **23,** 1248–1257.
Rutter, W. J. (1965). *In* "Evolving Genes and Proteins" (V. Bryson and H. J. Vogel, eds.), pp. 279–291. Academic Press, New York.
Ryan, F. J., and Tolbert, N. E. (1975). *J. Biol. Chem.* **250,** 4229–4233.
Ryley, J. F. (1967). *Chem. Zool.* **1,** 55–92.
Safe, S. (1973). *Biochim. Biophys. Acta* **326,** 471–475.
Safe, S., and Brewer, D. (1973). *Lipids* **8,** 311–314.
Sagan, L. (1967). *J. Theor. Biol.* **14,** 225–274.
Sagan, L., and Scher, S. (1961). *J. Protozool., Suppl.* **8,** 8.
Sakaguchi, K. (1970). *Biochim. Biophys. Acta* **220,** 580–593.

Sanchez, J. J., Palleroni, N. J., and Doudoroff, M. (1975). *Arch. Microbiol.* **104,** 57–65.

Sanders, J. P. M., Heyting, C., and Borst, P. (1975). *Biochem. Biophys. Res. Commun.* **65,** 699–707.

Sanger, F., Donelson, J. E., Coulson, A. R., Kössel, H., and Fischer, D. (1973). *Proc. Natl. Acad. Sci. U.S.A.* **70,** 1209–1213.

Sapozhnikov, D. I. (1973). *Pure Appl. Chem.* **35,** 47–61.

Sapozhnikov, D. I., Krasovskaya, T. A., and Mayevskaya, A. N. (1957). *Dokl. Bot. Sci.* **113,** 74–76.

Scagel, R. F., Bandoni, R. J., Rouse, G. E., Schofield, W. B., Stein, J. R., and Taylor, T. M. C. (1966). "An Evolutionary Survey of the Plant Kingdom." Wadsworth, Belmont, California.

Scawen, M. D., Ramshaw, J. A. M., and Boulter, D. (1975). *Biochem. J.* **147,** 343–349.

Scheer, H., Svec, W. A., Cope, B. T., Studier, M. H., Scott, R. G., and Katz, J. J. (1974). *J. Am. Chem. Soc.* **96,** 3714–3716.

Schenk, H. E. A., and Hofer, I. (1972). *Photosynth., Two Centuries Its Discovery Joseph Priestley, Proc. Int. Congr. Photosynth. Res., 2nd, 1971* Vol. 3, pp. 2095–2100.

Scher, W. I., Jr., and Vogel, H. J. (1957). *Proc. Natl. Acad. Sci. U.S.A.* **43,** 796–803.

Scherffel, A. (1925). *Arch. Protistenkd.* **52,** 1–141.

Schiff, J. A. (1962). *In* "Physiology and Biochemistry of Algae" (R. A. Lewin, ed.), pp. 239–246. Academic Press, New York.

Schiff, J. A. (1973). *Adv. Morphog.* **10,** 265–312.

Schiff, J. A., and Epstein, H. T. (1965). *In* "Reproduction: Molecular, Subcellular and Cellular" (M. Locke, ed.), pp. 131–189. Academic Press, New York.

Schiff, J. A., and Hodson, R. C. (1970). *Ann. N.Y. Acad. Sci.* **175,** 555–576.

Schiff, J. A., and Hodson, R. C. (1973). *Annu. Rev. Plant Physiol.* **24,** 381–414.

Schildkraut, C. L., Mandel, M., Levisohn, S., Smith-Sonneborn, J. E., and Marmur, J. (1962). *Nature (London)* **196,** 795–796.

Schimper, A. F. W. (1883). *Bot. Z.* **41,** 105–114.

Schimper, A. F. W. (1885). *Jahrb. Wiss. Bot.* **16,** 1–247.

Schmauder, H. P., Seidler, M., and Gröger, D. (1974). *Biochem. Physiol. Pflanz.* **166,** 263–273.

Schmitz, Fr. (1883). *Verh. Naturhist. Ver. Preuss. Rheinl.* **40,** 1–180.

Schnarrenberger, C., and Oeser, A. (1974). *Eur. J. Biochem.* **45,** 77–82.

Schnepf, E. (1965). *Planta* **67,** 213–224.

Schnepf, E., and Brown, R. M., Jr. (1971). *In* "Origin and Continuity of Cell Organelles" (J. Reinert and H. Ursprung, eds.), pp. 299–322. Springer-Verlag, Berlin and New York.

Schnepf, E., and Deichgraber, G. (1966). *Arch. Mikrobiol.* **55,** 149–175.

Schnepf, E., and Koch, W. (1966). *Z. Pflenzenphysiol.* **55,** 97–109.

Schopf, J. W. (1970). *Biol. Rev. Cambridge Philos. Soc.* **45,** 319–352.

Schopf, J. W. (1974). *Evol. Biol.* **7,** 1–43.

Schopf, J. W. (1975a). *Endeavour* **34,** 51–58.

Schopf, J. W. (1975b). *Annu. Rev. Earth Planet. Sci.* **3,** 213–249.

Schroepfer, G. J., Jr., Lutsky, B. N., Martin, J. A., Huntoon, S., Fourcans, B., Lee, W.-H., and Vermilion, J. (1972), *Proc. R. Soc. London, Ser. B* **180,** 425–446.

Schubert, K., Rose, G., Tümmler, R., and Ikekawa, N. (1964). *Hoppe-Seyler's Z. Physiol. Chem.* **339,** 293–296.

Schubert, K., Rose, G., and Hörhold, C. (1967). *Biochim. Biophys. Acta* **137,** 168–171.

Schubert, K., Rose, G., Wachtel, H., Hörhold, C., and Ikekawa, N. (1968). *Eur. J. Biochem.* **5,** 246–251.

Schürhoff, P. N. (1924). *In* "Handbuch der Pflanzenanatomie" (K. Linsbauer, ed.), Vol. I, pp. 1–219. Borntraeger, Berlin.

Schwartz, J. H., Meyer, R., Eisenstadt, J. M., and Brawerman, G. (1967). *J. Mol. Biol.* **25,** 571–574.

Schwertner, H. A., and Biale, J. B. (1973). *J. Lipid Res.* **14,** 235–242.

Scott, A. I., Lee, E., and Townsend, C. A. (1974). *Bioorg. Chem.* **3,** 229–237.

Scott, N. S. (1973). *J. Mol. Biol.* **81,** 327–336.

Scott, N. S. (1974). *Proc. R. Soc. N.Z.* **12,** 345–349.

Scott, N. S. (1976). *Phytochemistry* **15,** 1207–1213.

Seaman, G. R. (1955). *Biochem. Physiol. Protozoa* **2,** 91–158.

Searcy, D. G. (1975). *Biochim. Biophys. Acta* **395,** 535–547.

Sebald, M., and Véron, M. (1963). *Ann. Inst. Pasteur, Paris* **105,** 897–910.

Seely, G. R., Duncan, M. J., and Vidaver, W. E. (1972). *Mar. Biol.* **12,** 184–188.

Seitz, U., and Seitz, U. (1973). *Arch. Mikrobiol.* **90,** 213–222.

Seneca, H., Peer, P., and Nally, R. (1962). *Nature (London)* **193,** 1106–1107.

Sepsenwol, S. (1973). *Exp. Cell Res.* **76,** 395–409.

Serres, M. (1824). *Ann. Sci. Natl., Zool. Biot. Anim.* [*2*] **2,** 248 (referenced in de Beer, 1958).

Shapiro, H. S. (1968). *In* "Handbook of Biochemistry" (H. A. Sober, ed.), 1st ed., p. H-41. CRC Press, Cleveland, Ohio.

Shatilov, V. R., Ambartsumyan, V. G., Kasparova, M. A., and Kretovich, V. L. (1974). *Dokl. Biochem* **216,** 206–208.

Shaw, G. (1970). *In* "Phytochemical Phylogeny" (J. B. Harborne, ed.), pp. 31–58. Academic Press, New York.

Shaw, R. (1965). *Biochim. Biophys. Acta* **98,** 230–237.

Shaw, R. (1966). *Adv. Lipid Res.* **4,** 107–174.

Shibuya, I. (1960). Ph.D. Thesis, Pennsylvania State University, University Park.

Shimizu, Y., Alam, M., and Kobayashi, A. (1976), *J. Amer. Chem. Soc.* **98,** 1059–1060.

Shlyk, A. A. (1971). *Annu. Rev. Plant Physiol.* **22,** 169–184.

Shorb, M. S. (1964). *Biochem. Physiol. Protozoa* **3,** 383–457.

Short, S. A., and White, D. C. (1970). *J. Bacteriol.* **104,** 126–132.

Siegel, B. Z., and Siegel, S. M. (1973). *Crit. Rev. Microbiol.* **3,** 1–26.

Siegel, S. M., and Guimarro, C. (1966). *Proc. Natl Acad. Sci. U.S.A.* **55,** 349–353.

Siegel, S. M., Roberts, K., Nathan, H., and Daly, O. (1967). *Science* **156,** 1231–1234.

Siegelman, H. W., Chapman, D. J. and Cole W. J. (1968). *Biochem. Soc. Symp.* **28,** 107–120.

Sih, C. J., Laval, J., and Rahim, M. A. (1963). *J. Biol. Chem.* **238,** 566–571.

Silverberg, B. A., and Sawa, T. (1973). *Can. J. Bot.* **51,** 2025–2032.

Simon, R. D. (1971). *Proc. Natl. Acad. Sci. U.S.A.* **68,** 265–267.

Simon, R. D. (1973). *Arch. Mikrobiol.* **92,** 115–122.

Singer, S. J., and Nicolson, G. L. (1972). *Science* **175,** 720–731.

Singh, S., and Wildman, S. G. (1973). *Mol. Gen. Genet.* **124,** 187–196.

Sirevåg, R. (1974). *Arch. Microbiol.* **98,** 3–18.

Sirevåg, R., and Levine, R. P. (1972). *J. Biol. Chem.* **247,** 2586–2591.

Sirevåg, R., and Ormerod, J. G. (1970a). *Science* **169,** 186–188.

Sirevåg, R., and Ormerod, J. G. (1970b). *Biochem. J.* **120,** 399–408.

Sleigh, M. A. (1973). "The Biology of Protozoa." Arnold, London.

Smillie, R. M. (1968). *In* "The Biology of *Euglena*" (D. E. Buetow, ed.), Vol. 2, pp. 1–54. Academic Press, New York.

Smillie, R. M., and Evans, W. R. (1963). *In* "Bacterial Photosynthesis" (H. Gest, A. San Pietro, and L. P. Vernon, eds.), pp. 151–159. Antioch Press, Yellow Springs, Ohio.

Smillie, R. M., Rigopoulos, N., and Kelly, H. (1962). *Biochim. Biophys. Acta* **56,** 612–614.

Smith, A. J. (1973). *In* "The Biology and Blue-Green Algae" (N. G. Carr and B. A. Whitton, eds.), pp. 1–38. Blackwell, Oxford.

Smith, A. J., London, J., and Stanier, R. Y. (1967). *J. Bacteriol.* **94,** 972–983.

Smith, J. D., and Dunn, D. B. (1959). *Biochim. Biophys. Acta* **31,** 573–575.

Smith, J. D., Snyder, W. R., and Law, J. H. (1970). *Biochem. Biophys. Res. Commun.* **39,** 1163–1169.

Smith, L. L., Dhar, A. K., Gilchrist, J. L., and Lin, Y. Y. (1973). *Phytochemistry* **12,** 2727–2732.

Smith, P. F. (1957). *J. Bacteriol.* **74,** 801–806.

Smith, S. W., and Lester, R. L. (1974). *J. Biol. Chem.* **249,** 3395–3405.

Sneath, P. H. A., and Sokal, R. R. (1973). "Numerical Taxonomy: The Principles and Practice of Numerical Classification." Freeman, San Francisco, California.

Sogin, M. L., Woese, C. R., Pace, B., and Pace, N. R. (1973). *J. Mol. Evol.* **2,** 167–174.

Sokal, R. R. (1974a). *Science* **185,** 1115–1123.

Sokal, R. R. (1974b). *Syst. Zool.* **22,** 360–374.

Sokatch, J. R. (1968). "Bacterial Physiology and Metabolism." Academic Press, New York.

Soma, S. (1960). *J. Fac. Sci., Univ. Tokyo, Sect. 3* **7,** 535–542.

Sorby, H. C. (1873). *Proc. R. Soc. London* **21,** 442–483.

Sørensen, N. A. (1968). *Recent. Adv. Phytochem.* **1,** 187–227.

Spencer, B., and Harada, T. (1960). *Biochem. J.* **77,** 305–315.

Spencer, R., and Cross, G. A. M. (1975). *Biochim. Biophys. Acta* **390,** 141–154.

Spilsbury, J. F., and Wilkinson, S. (1961). *J. Chem. Soc.* pp. 2085–2091.

Stanier, R. Y. (1970). *Symp. Soc. Gen. Microbiol.* **20,** 1–38.

Stanier, R. Y. (1974). *Symp. Soc. Gen. Microbiol.* **24,** 219–239.
Stanier, R. Y., and van Niel, C. B. (1941). *J. Bacteriol.* **42,** 437–466.
Stanier, R. Y., and van Niel, C. B. (1962). *Arch. Mikrobiol.* **42,** 17–35.
Stanier, R. Y., Doudoroff, M., and Adelberg, E. A. (1970). "The Microbial World," 3rd ed. Prentice-Hall, Englewood Cliffs, New Jersey.
Steiner, S., Conti, S. F., and Lester, R. L. (1969). *J. Bacteriol.* **98,** 10–15.
Steinman, H. M., and Hill, R. L. (1973). *Proc. Natl. Acad. Sci. U.S.A.* **70,** 3725–3729.
Stenmark, S. L., Pierson, D. L., Jensen, R. A., and Glover, G. I. (1974). *Nature (London)* **247,** 290–292.
Stevenson, I. (1972). *J. Gen. Microbiol.* **71,** 69–76.
Steward, F. C., Mott, R. L., Israel, H. W., and Ludford, P. M. (1970). *Nature (London)* **225,** 760–763.
Stewart, J. M., and Beck, J. S. (1967). *J. Protozool.* **14,** 225–231.
Stewart, K. D., and Mattox, K. R. (1975). *Bot. Rev.* **41,** 104–135.
Stoffel, W., Dittmar, K., and Wilmes, R. (1975). *Hoppe-Seyler's Z. Physiol. Chem.* **356,** 715–725.
Stoloff, L., and Silva, P. (1957). *Econ. Bot.* **11,** 327–330.
Storck, R. (1965). *J. Bacteriol.* **90,** 1260–1264.
Storck, R. (1966). *J. Bacteriol.* **91,** 227–230.
Storck, R., and Alexopoulos, C. J. (1970). *Bacteriol. Rev.* **34,** 126–154.
Storck, R., Alexopoulos, C. J., and Phaff, H. J. (1969). *J. Bacteriol.* **98,** 1069–1072.
Storck, R., Nobles, M. K., and Alexopoulos, C. J. (1971). *Mycologia* **63,** 38–49.
Strain, H. H., Benton, F. L., Grandolfo, M. C., Aitzetmüller, K., Svec, W. A., and Katz, J. J. (1970). *Phytochemistry* **9,** 2561–2565.
Stransky, H., and Hager, A. (1970a). *Arch. Microbiol.* **71,** 164–190.
Stransky, H., and Hager, A. (1970b). *Arch. Microbiol.* **72,** 84–96.
Stransky, H., and Hager, A. (1970c). *Arch. Microbiol.* **73,** 315–323.
Straub, O. (1971). *In* "Carotenoids" (O. Isler, ed.), pp. 771–850. Birkhaeuser, Basel.
Stribling, D., and Perham, R. N. (1973). *Biochem. J.* **131,** 833–841.
Strøm, T., Ferenci, T., and Quayle, J. R. (1974). *Biochem. J.* **144,** 465–476.
Stubbe, W. (1971). *In* "Origin and Continuity of Cell Organelles" (J. Reinert and H. Ursprung, eds.), pp. 65–81. Springer-Verlag, Berlin and New York.
Stumm, C., and van Went, J. L. (1968). *Experientia* **24,** 1112–1113.
Stumpf, P. K. (1948). *J. Biol. Chem.* **176,** 233–241.
Sturgeon, R. J. (1974). *In* "Plant Carbohydrate Biochemistry" (J. B. Pridham, ed.), pp. 219–233. Academic Press, New York.
Subak-Sharpe, J. H., Elton, R. A., and Russell, G. J. (1974). *Symp. Soc. Gen. Microbiol.* **24,** 131–150.
Subden, R. E., and Turian, G. (1970). *Mol. Gen. Genet.* **108,** 358–364.
Sueoka, N. (1961). *J. Mol. Biol.* **3,** 31–40.
Sullivan, J. D., Jr., and Ikawa, M. (1973). *Biochim. Biophys. Acta* **309,** 11–22.
Sussman, A. S. (1974). *Taxon* **23,** 301–323.
Sutherland, I. W., and Smith, M. L. (1973). *J. Gen. Microbiol.* **74,** 259–266.
Suzuki, H. (1974). *Phytochemistry* **13,** 1159–1160.
Swain, T. (1974). *Compr. Biochem.* **29A,** 125–302.

Syrett, P. J. (1962). *In* "Physiology and Biochemistry of Algae" (R. A. Lewin, ed.), pp. 171–188. Academic Press, New York.

Szalay, A., Munsche, D., Wollgiehn, R., and Parthier, B. (1972). *Biochem. J.* **129,** 135–140.

Taber, R. L., and Vincent, W. S. (1969). *Biochim. Biophys. Acta* **186,** 317–325.

Tabita, F. R., McFadden, B. A., and Pfennig, N. (1974a). *Biochim. Biophys. Acta* **341,** 187–194.

Tabita, F. R., Stevens, S. E., Jr., and Quijano, R. (1974b). *Biochem. Biophys. Res. Commun.* **61,** 45–52.

Taketomi, T. (1961). *Z. Allg. Mikrobiol.* **1,** 331–340.

Tamura, A. (1967). *J. Bacteriol.* **93,** 2009–2016.

Tanaka, M., Haniu, M., Yasunobu, K. T., Evans, M. C. W., and Rao, K. K. (1974). *Biochemistry* **13,** 2953–2959.

Tanaka, M., Haniu, M., Yasunobu, K. T., Evans, M. C. W., and Rao, K. K. (1975a). *Biochemistry* **14,** 1938–1943.

Tanaka, M., Haniu, M., Yasunobu, K. T., Rao, K. K., and Hall, D. O. (1975b). *Biochemistry* **14,** 5535–5540.

Tanaka, M., Haniu, M., Zeitlin, S., Yasunobu, K. T., Evans, M. C. W., Rao, K. K., and Hall, D. O. (1975c). *Biochem. Biophys. Res. Commun.* **64,** 399–407.

Tanaka, M., Haniu, M., Yasunobu, K. T., Rao, K. K., and Hall, D. O. (1976). *Biochem. Biophys. Res. Commun.* **69,** 759–765.

Taylor, D. L. (1970). *Int. Rev. Cytol.* **27,** 29–64.

Taylor, F. J. R. (1974). *Taxon* **23,** 229–258.

Taylor, F. J. R. (1976a). *Taxon* **25,** 377–390.

Taylor, F. J. R. (1976b). *J. Protozool.* **23,** 28–40.

Taylor, M. M., and Storck, R. (1964). *Proc. Natl. Acad. Sci. U.S.A.* **52,** 958–965.

Tel-Or, E., Cammack, R., and Hall, D. O. (1975). *FEBS Lett.* **53,** 135–138.

Teshima, S., and Kanazawa, A. (1972). *Bull. Jpn. Soc. Sci. Fish.* **38,** 1197–1202.

Tewari, K. K. (1971). *Annu. Rev. Plant Physiol.* **22,** 141–168.

Thiemann, J. E., Zucco, G., and Pelizza, G. (1969). *Arch. Mikrobiol.* **67,** 147–155.

Thomas, D. M., and Goodwin, T. W. (1965). *J. Phycol.* **1,** 118–121.

Thomas, J. R., and Tewari, K. K. (1974). *Biochim. Biophys. Acta* **361,** 73–83.

Thompson, H. A., Baca, O. G., and Paretsky, D. (1971). *Biochem. J.* **125,** 365–366.

Thompson, J. F., and Muenster, A.-M.E. (1971). *Biochem. Biophys. Res. Commun.* **43,** 1049–1055.

Thomson, R. H. (1971). "Naturally Occurring Quinones." Academic Press, New York.

Tiedemann, F. (1810's). Referenced in de Beer (1958).

Timkovich, R., and Dickerson, R. E. (1973). *J. Mol. Biol.* **79,** 39–56.

Tipton, C. L., and Swords, M. D. (1966). *J. Protozool.* **13,** 469–472.

Tolbert, N. E. (1976). *Aust. J. Plant Physiol.* **3,** 129–132.

Tomas, R. N., and Cox, E. R. (1973). *J. Phycol.* **9,** 304–323.

Tomas, R. N., Cox, E. R., and Steidinger, K. A. (1973). *J. Physiol.* **9,** 91–98.

Tonino, G. J. M., and Rozijn, T. H. (1966). *Biochim. Biophys. Acta* **124,** 427–429.

Tornabene, T. G., Kates, M., and Volcani, B. E. (1974). *Lipids* **9,** 279–284.

Tremoliéres, A., and Mazliak, P. (1974). *Plant Sci. Lett.* **2,** 193–201.

Troxler, R. F., and Dokos, J. M. (1973). *Plant Physiol.* **51,** 72–75.

Troxler, R. F., Brown, A., Lester, R., and White, P. (1970). *Science* **167,** 192–193.

Troxler, R. F., Foster, J. A., Brown, A. S., and Franzblau, C. (1975). *Biochemistry* **14,** 268–274.

Trudinger, P. A. (1956). *Biochem. J.* **64,** 274–286.

Tsang, M. L.-S., and Schiff, J. A. (1975). *Plant Sci. Lett.* **4,** 301–307.

Tso, T. C., and Cheng, A. L. S. (1971). *Phytochemistry* **10,** 2133–2137.

Tuan, R. S., and Chang, K. P. (1975). *J. Cell Biol.* **65,** 309–323.

Tulloch, A. P., Heinz, E., and Fischer, W. (1973). *Hoppe-Seyler's Z. Physiol. Chem.* **354,** 879–889.

Turner, B. L. (1967). *Pure Appl. Chem.* **14,** 189–213.

Turner, B. L. (1969). *Taxon* **18,** 134–151.

Turner, W. B. (1971). "Fungal Metabolites." Academic Press, New York.

Twarog, R., and Liggins, G. L. (1970). *J. Bacteriol.* **104,** 254–263.

Työrinoja, K., Nurminen, T., and Suomalainen, H. (1974). *Biochem. J.* **141,** 133–139.

Uchida, K. (1974). *Biochim. Biophys. Acta* **369,** 146–155.

Udaka, S., and Kinoshita, S. (1958). *J. Gen. Appl. Microbiol.* **4,** 283–288.

Uzzell, T., and Spolsky, C. (1973). *Science* **180,** 516–517.

Uzzell, T., and Spolsky, C. (1974). *Am. Sci.* **62,** 334–343.

Valadon, L. R. G., and Mummery, R. S. (1976). *Trans. Brit. Mycol. Soc.* **65,** 485–487.

Vance, C. P., Nambudiri, A. M. D., and Towers, G. H. N. (1973). *Can. J. Biochem.* **51,** 731–734.

Vance, C. P., Bandoni, R. J., and Towers, G. H. N. (1975). *Phytochemistry* **14,** 1513–1514.

van den Berg, A., and Beintema, J. J. (1975). *Nature* (*London*) **253,** 207–210.

Van Pel, B., and Cocito, C. (1973). *Exp. Cell Res.* **78,** 111–117.

Van't Riet, J., and Planta, R. J. (1975). *Biochim. Biophys. Acta* **379,** 81–94.

Van Valen, L. (1973). *Evol. Theory* **1,** 1–30.

Van Valen, L. (1974). *J. Mol. Evol.* **3,** 89–101.

van Wagtendonk, W. J. (1963). *Proc. Int. Congr. Biochem., 5th, 1961* Vol. III, pp. 343–346.

Veronese, F. M., Boccu, E., and Coventi, L. (1975). *Biochim. Biophys. Acta* **377,** 217–228.

Villa, V. D., and Storck, R. (1968). *J. Bacteriol.* **96,** 184–190.

Vining, L. C. (1973). *In* "Genetics of Industrial Microorganisms" (Z. Vaněk, Z. Hoštálek, and J. Cudlín, eds.), pp. 405–419. Akadémiai Kiadó, Budapest.

Vining, L. C., and Taber, W. A. (1964). *Can. J. Microbiol.* **10,** 647–657.

Vogel, H. J. (1953). *Proc. Natl. Acad. Sci. U.S.A.* **39,** 578–583.

Vogel, H. J. (1955). *Symp. Amino Acid Metab.* [*Proc.*] *1954* pp. 335–346.

Vogel, H. J. (1965). *In* "Evolving Genes and Proteins" (V. Bryson and H. J. Vogel, eds.), pp. 25–40. Academic Press, New York.

Vogel, H. J., Thompson, J. S., and Shockman, G. D. (1970). *Symp. Soc. Gen. Microbiol.* **20,** 107–119.

Vogel, R. H., and Vogel, H. J. (1963). *Biochim. Biophys. Acta* **69,** 174–176.

Vollbrecht, D. (1974). *Biochim. Biophys. Acta* **362,** 382–389.
Volpe, J. J., and Vagelos, P. R. (1973). *Annu. Rev. Biochem.* **42,** 21–60.
von Baer, K. E. (1828). "'Über Enwicklungsgeschichte der Thiere', Beobachtung und Reflexion." Königsberg (referenced in de Beer, 1958).
Wada, K., Kagamiyama, H., Shin, M., and Matsubara, H. (1974). *J. Biochem.* (*Tokyo*) **76,** 1217–1225.
Wada, K., Hase, T., Tokunaga, H., and Matsubara, H. (1975a). *FEBS Lett.* **55,** 102–104.
Wada, K., Hase, T., and Matsubara, H. (1975b). *J. Biochem.* (*Tokyo*) **78,** 637–639.
Walker, D. A. (1974). *In* "Plant Carbohydrate Biochemistry" (J. B. Pridham, ed.), pp. 7–26. Academic Press, New York.
Walker, J. B., and Myers, J. (1953). *J. Biol. Chem.* **203,** 143–152.
Wallin, J. E. (1927). "Symbionticism and the Origin of Species" Williams & Wilkins, Baltimore.
Wang, H. S., and LéJohn, H. B. (1974a). *Can. J. Microbiol.* **20,** 567–574.
Wang, H. S., and LéJohn, H. B. (1974b). *Can. J. Microbiol.* **20,** 575–580.
Wang, M. C., and Bartnicki-Garcia, S. (1974). *Carbohydr. Res.* **37,** 331–338.
Watanabe, A., and Yamamoto, Y. (1972). *In* "Taxonomy and Biology of Blue-Green Algae" (T. V. Desikachary, ed.), pp. 556–565. Univ. of Madras Press, Madras.
Watson, J. D. (1970). "Molecular Biology of the Gene," 2nd ed. Benjamin, New York.
Watson, M. J., and Baddiley, J. (1974). *Biochem. J.* **137,** 399–404.
Watts, R. L., and Watts, D. C. (1968a). *Nature* (*London*) **217,** 1125–1130.
Watts, R. L., and Watts, D. C. (1968b). *J. Theor. Biol.* **20,** 227–244.
Weedon, B. C. L. (1971). *In* "Carotenoids" (O. Isler, ed.), pp. 29–50. Birkhaeuser, Basel.
Weete, J. D. (1973). *Phytochemistry* **12,** 1843–1864.
Weete, J. D. (1974). "Fungal Lipid Biochemistry." Plenum, New York.
Weete, J. D., and Laseter, J. L. (1974). *Lipids* **9,** 575–581.
Weidner, M., and Küppers, U. (1973). *Planta* **114,** 365–372.
Weise, G., Drews, G., Jann, B., and Jann, K. (1970). *Arch. Mikrobiol.* **71,** 89–98.
Weisiger, R. A., and Fridovich, I. (1973a). *J. Biol. Chem.* **248,** 3582–3592.
Weisiger, R. A., and Fridovich, I. (1973b). *J. Biol. Chem.* **248,** 4793–4796.
Weiss, B., Stiller, R. L., and Jack, R. C. M. (1973). *Lipids* **8,** 25–30.
Westlake, D. W. S., Roxburgh, J. M., and Talbot, G. (1961). *Nature* (*London*) **189,** 510–511.
Whitfeld, P. R. (1953). *Aust. J. Biol. Sci.* **6,** 234–243.
Whittaker, R. H. (1969). *Science* **163,** 150–160.
Whittle, S. J., and Casselton, P. J. (1975a). *Br. Phycol. J.* **10,** 179–191.
Whittle, S. J., and Casselton, P. J. (1975b). *Br. Phycol. J.* **10,** 192–204.
Whitton, B. A., Carr, N. G., and Craig, I. W. (1971). *Protoplasma* **72,** 325–357.
Wicken, A. J., and Knox, K. W. (1975). *Science* **187,** 1161–1167.
Wilkins, M. H. F., and Zubay, G. (1959). *J. Biophys. Biochem. Cytol.* **5,** 55–58.
Wilks, S. S. (1959). *Science* **129,** 964–966.

Willard, J. M., and Gibbs, M. (1968a). *Biochim. Biophys. Acta* **151,** 438–448.
Willard, J. M., and Gibbs, M. (1968b). *Plant Physiol.* **43,** 793–798.
Willenbrink, J., and Kremer, B. P. (1973). *Planta* **113,** 173–178.
Williams, J. (1974a). *Biochem. Soc. Trans.* **2,** 840–844.
Williams, J. (1974b). *Phys. Chem., Ser. One* **1,** 1–56.
Williams, M. B. (1970). *J. Theor. Biol.* **29,** 343–385.
Williams, V. P., Freidenreich, P., and Glazer, A. N. (1974). *Biochem. Biophys. Res. Commun.* **59,** 462–466.
Willmer, E. N. (1974). *Biol. Rev. Cambridge Philos. Soc.* **49,** 321–369.
Wills, C. (1973). *Am. Nat.* **107,** 23–34.
Wintersberger, U., Smith, P., and Letnansky, K. (1973). *Eur. J. Biochem.* **33,** 123–130.
Withers, N., and Haxo, F. T. (1975). *Plant Sci. Lett.* **5,** 7–15.
Wittmann, H. G. (1970). *Symp. Soc. Gen. Microbiol.* **20,** 55–76.
Woese, C. R. (1961). *Nature (London)* **189,** 920–921.
Woese, C. R. (1967). "The Genetic Code." Harper, New York.
Woese, C. R. (1969). *J. Mol. Biol.* **43,** 235–240.
Woese, C. R. (1973). *Naturwissenschaften* **60,** 447–459.
Wolf, F. T. (1960). *In* "Comparative Biochemistry of Photoreactive Systems" (M. B. Allen, ed.), pp. 53–67. Academic Press, New York.
Wolfersberger, M. G., and Pieringer, R. A. (1974). *J. Lipid Res.* **15,** 1–10.
Wolk, C. P. (1973). *Bacteriol. Rev.* **37,** 32–101.
Wong, J. T.-F. (1975). *Proc. Natl. Acad. Sci. U.S.A.* **72,** 1909–1912.
Wood, B. J. B. (1974). *In* "Algal Physiology and Biochemistry" (W. D. P. Stewart, ed.), pp. 236–265. Blackwell, Oxford.
Wood, B. J. B., Nichols, B. W., and James, A. T. (1965). *Biochim. Biophys. Acta* **106,** 261–273.
Woodin, T. S., and Nishioka, L. (1973). *Biochim. Biophys. Acta* **309,** 224–231.
Wootton, J. C. (1974). *Nature (London)* **252,** 542–546.
Wright, B. E. (1964). *Biochem. Physiol. Protozoa* **3,** 341–381.
Wu, T. T., Lin, E. C. C., and Tanaka, S. (1968). *J. Bacteriol.* **96,** 447–456.
Wurst, M., Vančura, V., and Kalachová, L. (1974). *J. Chromatogr.* **91,** 469–474.
Wyatt, J. T., Lawley, G. G., and Barnes, R. D. (1971). *Naturwissenschaften* **58,** 570–571.
Yamada, Y., and Kondô, K. (1973). *J. Gen. Appl. Microbiol.* **19,** 58–77.
Yarrow, D., and Nakase, T. (1975). *Antonie van Leeuwenhoek* **41,** 81–88.
Yasunobu, K. T., and Tanaka, M. (1974). *Syst. Zool.* **22,** 570–589.
Yin, H. C. (1948). *Nature (London)* **162,** 928–929.
Young, M. R., Towers, G. H. N., and Neish, A. C. (1966). *Can. J. Bot.* **44,** 341–349.
Youngblood, W. W., and Blumer, M. (1973). *Mar. Biol.* **21,** 163–172.
Yourno, J., Kohno, T., and Roth, J. R. (1970). *Nature (London)* **228,** 820–824.
Yurina, N. P., and Odintsova, M. S. (1974). *Plant Sci. Lett.* **3,** 229–234.
Yurina, N. P., and Odintsova, M. S. (1975). *Dokl. Biochem.* **218,** 512–515.
Zablen, L., and Woese, C. R. (1975). *J. Biol. Evol.* **5,** 25–34.

Zablen, L. B., Kissil, M. S., Woese, C. R., and Buetow, D. E. (1975). *Proc. Natl. Acad. Sci. U.S.A.* **72,** 2418–2422.
Zalashko, M. V., and Pidoplichko, G. A. (1974). *Microbiology (USSR)* **43,** 202–205.
Zawada, J. W., and Sutcliffe, J. F. (1974). *Ann. Bot. (London)* [N.S.] **38,** 1093–
Ziegler, H,, and Ziegler, I. (1967). *Planta* **72,** 162–169.
Zingmark, R. G. (1970). *J. Phycol.* **6,** 122–126.
Zubay, G., and Watson, M. R. (1959). *J. Biophys. Biochem. Cytol.* **5,** 51–54.
Zucker, M. (1972). *Annu. Rev. Plant Physiol.* **23,** 133–156.
Zuckerkandl, E., and Pauling, L. (1965a). *J. Theor. Biol.* **8,** 357–366.
Zuckerkandl, E., and Pauling, L. (1965b). *In* "Evolving Genes and Proteins" (V. Bryson and H. J. Vogel, eds.), pp. 97–181. Academic Press, New York.
Zurawski, G., and Brown, K. D. (1975). *Biochim. Biophys. Acta* **377,** 473–481.

Taxonomic Index

I

K

L

M

T

Subject Index

B

F

G

H

I

Q

R

S